W9-BJN-680

MOLECULAR MOVEMENTS AND CHEMICAL REACTIVITY AS CONDITIONED BY MEMBRANES, ENZYMES AND OTHER MACROMOLECULES

XVIth SOLVAY CONFERENCE ON CHEMISTRY

ADVANCES IN CHEMICAL PHYSICS

VOLUME XXXIX

EDITORIAL BOARD

Advances in chemical physics
"

MOLECULAR MOVEMENTS AND CHEMICAL REACTIVITY

AS CONDITIONED BY MEMBRANES ENZYMES AND OTHER MACROMOLECULES

XVIth SOLVAY CONFERENCE ON CHEMISTRY
BRUSSELS, NOVEMBER 22 – NOVEMBER 26, 1976

Edited by

R. LEFEVER and A. GOLDBETER
University of Brussels · Brussels, Belgium

ADVANCES IN CHEMICAL PHYSICS
VOLUME XXXIX

Series editors

Ilya Prigogine
University of Brussels
Brussels, Belgium
and
University of Texas
Austin, Texas

Stuart A. Rice
Department of Chemistry
and
The James Frank Institute
University of Chicago
Chicago, Illinois

AN INTERSCIENCE® PUBLICATION
JOHN WILEY AND SONS
NEW YORK · CHICHESTER · BRISBANE · TORONTO

An Interscience® Publication

Library of Congress Catalog Number: 58-9935

ISBN 0-471-03541-6

Printed in the United States of America

10 9 8 7 6 5 4 3 2 1

INTRODUCTION

Few of us can any longer keep up with the flood of scientific literature, even in specialized subfields. Any attempt to do more, and be broadly educated with respect to a large domain of science, has the appearance of tilting at windmills. Yet the synthesis of ideas drawn from different subjects into new, powerful, general concepts is as valuable as ever, thus the desire to remain educated persists in all scientists. This series, *Advances in Chemical Physics*, is devoted to helping the reader obtain general information about a wide variety of topics in chemical physics, which field we interpret very broadly. Our intent is to have experts present comprehensive analyses of subjects of interest, and to encourage the expression of individual points of view. We hope that this approach to the presentation of an overview of a subject will both stimulate new research and serve as a personalized learning text for beginners in a field.

ILYA PRIGOGINE
STUART A. RICE

PREFACE

The papers and discussions in this volume were presented at the XVIth Solvay Conference on Chemistry held at the Free University of Brussels from November 22 to November 26, 1976.

The main objective of the Scientific Committee was to stress in this meeting the structure-environment relationships that condition the chemical reactivity in molecular and supramolecular assemblies such as polymers, enzymes, and membranes. Remarkable achievements have been made over the last few years. They demonstrate the subtleties of the molecular events that take place in such systems and account for highly specific and coordinated processes of catalysis, transport, and recognition. In usual chemical systems, reactivity and reaction rate are simple functions of the molecular species involved and of their relative abundance. In processes involving macromolecules or supramolecular organizations the dynamic behavior is controlled by a much more complex interplay of short and long range interactions, be it at the intramolecular or supramolecular level. Furthermore, these interactions are at the heart of the fascinating feedback mechanisms by which it becomes possible, under nonequilibrium conditions, for some chemical systems to come to "life," that is, to self-organize and adapt their functional, spatial, and temporal order to their environment.

To explore these mechanisms is of fundamental importance and interest. This field is at present giving rise to great expectations also from the technological point of view. A better understanding of the problems involved is indeed needed for the succesful realization of many progresses envisioned in medical and industrial research.

These problems therefore are great challenges to physicists, chemists, and biologists. We hope that this volume will help in finding the road to their solution.

Great thanks are due to Professors A. R. Ubbelohde, S. Claesson, and V. Prelog for their decisive role in the organization and running of the conference.

Brussels, Belgium
April 1978

I. Prigogine
R. Lefever
A. Goldbeter

Administrative Council
of the
Instituts Internationaux de Physique et de Chimie
founded by E. Solvay

Scientific Committee in Chemistry
of the
Instituts Internationaux de Physique et de Chimie
founded by E. Solvay

Participants

A. Babloyantz, Université Libre de Bruxelles, Belgium

A. D. Bangham, Agricultural Research Council, Babraham, England

B. Baranowski, Polska Akademia Nauk, Warszawa, Poland

H. Benoit, C.N.R.S., Centre de Recherches sur les Macromolécules, Strasbourg, France

G. Bergson, University of Uppsala, Uppsala, Sweden
C. Berliner, Université Libre de Bruxelles, Belgium
P. Bisch, Université Libre de Bruxelles, Belgium
S. R. Caplan, The Weizmann Institute of Science, Rehovot, Israel
J. Caspers, Université Libre de Bruxelles, Belgium
E. Carafoli, Ecole Polytechnique Fédérale, Zürich, Switzerland
R. Cerf, Université Louis Pasteur, Strasbourg, France
J. Chanu, Université Paris VII, France
J. B. Chappell, University of Bristol, Great Britain
P. Chatelain, Université Libre de Bruxelles, Belgium
E. Clementi, Montedison, Novara, Italy
G. Coppens, Solvay & Cie., Bruxelles, Belgium
S. Damjanovich, Medical University School of Debrecen, Hungary
L. de Brouckére, Université Libre de Bruxelles, Belgium
M. Delmotte, Université Paris VII, France
G. Eisenman, University of California, Los Angeles, California, U.S.A.
A. Englert–Schwoles, Université Libre de Bruxelles, Belgium
T. Erneux, Université Libre de Bruxelles, Belgium
S. FörSen, Chemical Center, Lund, Sweden
J. J. Fripiat, Centre National de la Recherche Scientifique, Orléans, France
R. Gabellieri, Solvay & Cie., Bruxelles, Belgium
G. Geuskens, Université Libre de Bruxelles, Belgium
C. Geuskens–David, Université Libre de Bruxelles, Belgium
A. Goldbeter, Université Libre de Bruxelles, Belgium
V. I. Goldanskii, Institute of Chemical Physics, Moscow, U.S.S.R.
G. G. Hammes, Cornell University, Ithaca, New York, U.S.A.
H. Hauser, Eidgenössische Technische Hochschule Zürich, Switzerland
M. Hennenberg, Université Libre de Bruxelles, Belgium
M. Herschkowitz-Kaufman, Université Libre de Bruxelles. Belgium
B. Hess, Max-Planck Institüt für Ernährungsphysiologie, Dortmund, D.B.R.
J. Hiernaux, Université Libre de Bruxelles, Belgium
R. Kinne, Max-Planck Institüt für Biophysik, Frankfurt, D.B.R.
K. Kirschner, Biozentrum der Universität Basel, Switzerland
I. M. Klotz, Northwestern University, Evanston, Illinois, U.S.A.
R. Korenstein, Max-Planck Institüt für Ernährungsphysiologie, Dortmund, D.B.R.

D. Kuschmitz, Max-Planck Institüt für Ernährungsphysiologie, Dortmund, D.B.R.

R. Lefever, Université Libre de Bruxelles, Belgium

J. M. Lehn, Université Louis Pasteur, Strasbourg, France

J. Leonis, Université Libre de Bruxelles, Belgium

M. Magat, Faculté des Sciences, Orsay, France

M. Mandel, Rijksuniversiteit te Leiden, Nederland

V. Mathot, Université Libre de Bruxelles, Belgium

H. McConnell, Stanford University, California, U.S.A.

J. Mehra, Université de Genève, Suisse

K. Mosbach, University of Lund, Sweden

K. Mühlethaler, Eidgenössische Technische Hochschule, Zürich, Switzerland

I. Prigogine, Université Libre de Bruxelles, Belgium

A. Pullman, Institut de Biologie Physico-Chimique, Paris, France

B. Pullman, Institut de Biologie Physico-Chimique, Paris, France

J. Ross, Massachusetts Institute of Technology, Cambridge, Massachusetts, U.S.A.

J. M. Ruysschaert, Université Libre de Bruxelles, Belgium

A. Sanfeld, Université Libre de Bruxelles, Belgium

A. Sanfeld–Steinchen, Université Libre de Bruxelles, Belgium

E. Schoffeniels, Université de Liège, Belgium

W. Simon, Eidgenössische Technische Hochschule, Zürich, Switzerland

R. Somorjai, Conseil National de Recherches, Ottawa, Canada

G. Szasz, General Electric Company, Zürich, Switzerland

L. Ter–Minassian–Saraga, U.E.R. Biomédicale, Paris, France

D. Thomas, Université de Technologie de Compiègne, France

D. Van Lamsweerde-Gallez, Université Libre de Bruxelles, Belgium

R. Welch, Université Libre de Bruxelles, Belgium

R. J. P. Williams, University of Oxford, England

CONTENTS

Reports and Discussions

MOLECULAR MOVEMENTS AND CHEMICAL REACTIVITY AS CONDITIONED BY MEMBRANES, ENZYMES AND OTHER MACROMOLECULES

ADVANCES IN CHEMICAL PHYSICS

VOLUME XXXIX

COUPLING BETWEEN DIFFUSION AND CHEMICAL REACTIONS

I. PRIGOGINE and R. LEFEVER

Service de Chimie-Physique II Brussels, Belgium

I. INTRODUCTION

One of the main themes of this volume is the influence of the environment on chemical reactivity. Such a question is of special interest for chemical systems in far from equilibrium conditions. It is today well known that, far from equilibrium, chemical systems involving catalytic mechanisms may lead to *dissipative structures.*[1,2] It has also been shown—and this is one of the main themes of this discussion—that dissipative structures are very sensitive to global features characterizing the environment of chemical systems, such as their size and form, the boundary conditions imposed on their surface, and so on. All these features influence in a decisive way the type of instabilities, called *bifurcations,* that lead to dissipative structures.

Far from equilibrium, there appears an unexpected relation between chemical kinetics and the *space-time structure* of reacting systems. It is true that the interactions that determine the values of the relevant kinetic constants and transport coefficients result from short-range interactions (valency forces, hydrogen bounds, Van der Waals forces). However, the solutions of the kinetic equations depend *in addition* on global characteristics. This dependence, which near equilibrium on the thermodynamic branch is trivial, becomes decisive in chemical systems working under far from equilibrium conditions. For example, the occurrence of dissipative structures generally requires that the system's size exceed some critical value, the latter being a complex function of the parameters describing the reaction-diffusion processes. Therefore chemical bifurcations imply *long-range order* involving the system as a whole.

We review in Section II the basic results of nonequilibrium thermodynamic stability theory and recall the thermodynamic and kinetic conditions necessary to the occurrence of cooperative coherent behaviors in chemical systems. We briefly indicate some known experimental systems that meet these requirements and in which dissipative structures may be observed. Throughout these notes we focus on qualitative rather than quantitative aspects and draw our conclusions from the study of idealized models involving a minimal number of ingredients. The prototype of these models is the so-called Brusselator.[3] It corresponds to a reaction chain involving two chemical intermediates and a trimolecular step. This is the simplest stoichiometric reaction (in two variables systems) having an instability of the thermodynamic branch.[4,5,6] The Brusselator reads

$$A \to X \tag{1a}$$

$$2X + Y \to 3X \tag{1b}$$

$$B + X \to Y + D \tag{1c}$$

$$X \to E \tag{1d}$$

where X and Y are the intermediate composition variables and A, B, D, E the initial and final products whose concentration remains constant in time. Many remarkable properties of this model are shared qualitatively by numerous experimental (both biological and nonbiological) systems in which dissipative structures have been described.[2,7,8,9,10,11,12]

In Sections III to V we analyze the properties of dissipative structures in one-dimension reaction mediums. As shown originally by Auchmuty and Nicolis,[13] Herschkowitz-Kaufman,[14] Keener,[15] and Mahar and Matkowsky,[16] the methods of bifurcation theory permit us to evaluate the new regimes that bifurcate from a homogeneous equilibrium type of steady state as well as regimes that branch off an already organized state and correspond to secondary bifurcations. We see that a great variety of solutions may be observed by slight modifications in boundary conditions or size parameters. Furthermore, as is also shown in Section V, this multiplicity increases and qualitatively new behaviors appear when chemical and diffusion processes take place in two dimensions.

Bifurcation theory can only provide information on the solutions that appear in the neighborhood of a point of bifurcation and whose amplitude goes to zero at that point. Other approximate analytical methods, like singular perturbation theory, have permitted the calculation of solutions of large amplitude, but for conditions that postulate large differences in the values of the diffusion coefficients of the reacting species.[17,18,19] Only under special conditions can one go beyond the limitations of such descriptions and find an *exact solution* to the steady-state problem in general. These conditions are often met in one-variable systems. Examples of both first- and second-order phase transitions in such systems have been discussed by Schlögl.[20]* In Section VI we

* More recently a remarkably simple experimental example of spatial order in a one-variable system that presents an electrothermal instability has been analyzed by Bedaux et al.[21]

consider a simplified version of the Brusselator, which belongs to a class of systems described by equations of the form:

$$\frac{\partial x}{\partial t}=f(x, y)+D_1\frac{\partial^2 x}{\partial r^2} \tag{2a}$$

$$\frac{\partial y}{\partial t}=cf(x, y)+D_2\frac{\partial^2 y}{\partial r^2} \tag{2b}$$

Although these systems involve two variables, their steady-state solutions can be calculated in general and a more complete mathematical analysis of dissipative structures is possible. From a practical point of view it is interesting to note that systems obeying equations of the form (2) may be found in artificial membrane reactors.[22] Examples are presented by D. Thomas in this volume.

II. THERMODYNAMIC AND KINETIC ASPECTS OF SELF-ORGANIZATION

In open systems the total entropy variation

$$dS=d_iS+d_eS \tag{3}$$

decomposes into an entropy flow d_eS exchanged with the environment and an entropy production d_iS due to the irreversible processes. The explicit evaluation of (3) yields the balance equation[1]

$$\frac{d_iS}{dt}\equiv\int\sum_\rho J_\rho X_\rho\,dV=\frac{\partial S}{\partial t}+\phi[S]\geqslant 0 \tag{4}$$

where the sign of d_iS/dt is imposed by the second law of thermodynamics. The J_ρ's and X_ρ's are the conjugate thermodynamic fluxes and forces of the irreversible processes. The flow term $\phi[S]$ corresponds to the entropy flux across the boundaries. Equation (4) has two immediate consequences:

1. Any decrease of the entropy in the system $[(\partial S/\partial t)<0]$ requires a negative entropy flux with the environment i.e., the system has to be open and $d_eS/dt\equiv-\phi[S]<0$.
2. Once a state of low entropy has been achieved it can only be sustained in an open system, the condition $\partial S/\partial t=0$ requiring necessarily

$$\int\sum_\rho J_\rho X_\rho\,dV=\phi[S]>0 \tag{5}$$

We thus see that nonequilibrium may be a source of order. Two kinds of situations and two kinds of order must then be distinguished. In the first place, we have situations that correspond to slight deviations from an equilibrium state between the system and its environment. In that case the thermodynamic forces X_ρ and fluxes J_ρ obey linear relations

$$J_\rho = \sum_\alpha L_{\alpha\rho} X_\alpha \tag{6}$$

in which the phenomenological coefficients $L_{\alpha\rho}$ are constants satisfying Onsager's reciprocity relations $L_{\alpha\rho} = L_{\rho\alpha}$. This guarantees the existence of a state function, the entropy production, that has the properties of a potential. Accordingly,

$$\frac{\partial X_i}{\partial t} = -\frac{\delta P(\{X_j\})}{\delta X_i}; \qquad P \equiv \frac{d_i S}{dt} \tag{7}$$

and

$$\frac{\partial P}{\partial t} = \sum_i \frac{\delta P}{\delta X_i}\frac{\partial X_i}{\partial t} = -\sum_i \left(\frac{\delta P}{\delta X_i}\right)^2 \leqslant 0 \tag{8}$$

Under those conditions P behaves as a Lagrangian in mechanics. Furthermore, as P is a nonnegative function for any positive value of the concentrations $\{X_j\}$, by a theorem due to Lyapounov, the *asymptotic stability* of nonequilibrium steady states is ensured (theorem of minimum entropy production.[1,23] These steady states are thus characterized by a minimum level of the dissipation; in the linear domain of nonequilibrium thermodynamics the systems tend to states approaching equilibrium as much as their constraints permit. Although entropy may be lower than at equilibrium, the equilibrium type of order still prevails. The steady states belong to what has been called the *thermodynamic branch*, as it contains the equilibrium state as a particular case.

Beyond the domain of validity of the minimum entropy production theorem (i.e., far from equilibrium), a new type of order may arise. The stability of the thermodynamic branch is no longer automatically ensured by the relations (8). Nevertheless it can be shown that even then, with fixed boundary conditions, nonequilibrium systems always obey to the inequality[1]

$$\frac{d_x P}{dt} = \int \sum_\rho J_\rho \frac{dX_\rho}{dt} \leqslant 0 \tag{9}$$

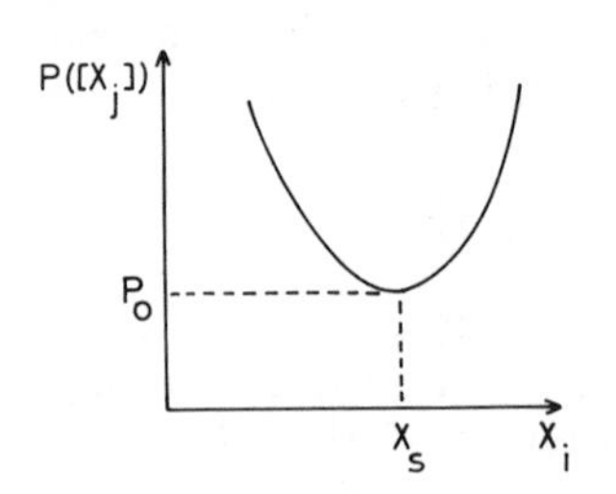

Fig. 1. Entropy production as a function of some concentration X_i (in the linear range). Here P_0 and X_s are the steady-state values of P and X_i.

This is the so-called *evolution criterion.* The equality sign applies only to steady states. Two cases may then arise:

1. Here $d_xP/dt = d\phi/dt \leq 0$ is reducible to the total differential of some scalar potential ϕ (kinetic potential). We have a generalization of the behavior found near equilibrium with the important difference that the steady-state solutions are no longer necessarily unique for a set of values of the constraints.
2. In this case d_xP/dt cannot be reduced to an exact differential and the criterion simply expresses that in the course of time the variation of the thermodynamic forces tends to diminish the entropy production.

A stability criterion for nonequilibrium steady states can readily be deduced from (9). Suppose that for all small departures from the steady state considered we have

$$\delta_x P \geq 0 \qquad \text{or} \qquad \int \left[\sum_\rho \delta J_\rho \, \delta X_\rho \right] dV \geq 0 \tag{10}$$

where the δJ_ρ, δX_ρ are excess quantities evaluated around the reference steady state. Then, as a consequence of inequality (9), asymptotic stability with respect to small deviations is ensured. Far from equilibrium, however, negative terms in the sum (10) may become dominant and, as represented in Fig. 2, an instability of the thermodynamic branch may appear. As a result, beyond some threshold value of the deviation from equilibrium, there may be a spontaneous evolution toward a new state of organization with properties completely different from those of the thermodynamic branch. This opens a new field, nonequilibrium physical chemistry. These transitions to a new type of dynamic structure are very similar to the usual equilibrium transitions. However, let us stress again that these new types or organization of matter, called dissipative structures, can persist only if the coupling with the outside world through flows of matter and energy is sufficiently strong.

Let us mention a few relevant examples:

One of the best understood kinds of chemical instability that leads to

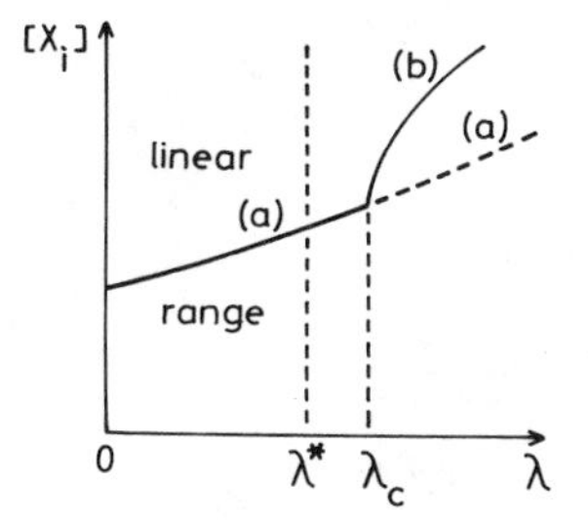

Fig. 2. Stability of the thermodynamic branch as a function of some parameter λ that measures the chemical system's distance from equilibrium. In the linear range (i.e., for $0 \leq \lambda \leq \lambda^*$) the steady states belong to the thermodynamic branch (a) and are stable. Beyond this domain there may exist a threshold point λ_c at which a new stable nonequilibrium branch of solutions (b) appears while the thermodynamic branch becomes unstable.

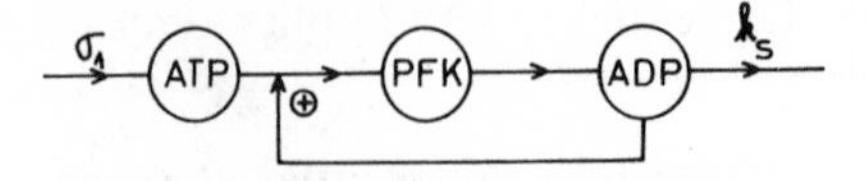

Fig. 3. Regulation of phosphofructokinase (PFK) by a product of reaction.

temporal order involves allosteric enzymes regulated by a positive feedback of the reaction product.[24] The enzyme phosphofructokinase is a representative system of this type and its catalytic properties can be summarized in Fig. 3.[26] The enzyme is part of the glycolytic chain; σ_1 and k_s stand, respectively, for the rate of supply of ATP and the rate of product decay (assumed to take place through a monomolecular consumption step). The product ADP binds preferentially to the active form of the enzyme; as a result, its accumulation displaces the conformational equilibrium in favor of the active form and promotes a positive feedback effect. When some conditions on the kinetic constants are satisfied, this feedback can appropriately be described by a trimolecular step.[11,26]

A vast class of instabilities in enzymatic systems arises from the pH dependence of enzymatic activity. In general, protein molecules contain a number of ionizing groups, such as (—COOH), which can be ionized to give the negatively charged ($—COO^-$) ion, and ($—NH_2$), which can add on a proton to give ($—NH_3^+$). The active enzyme may be represented as in Fig. 4; then addition of acid or base to the active enzyme may be depicted as in Fig. 5, or schematically

$$EH_2 \underset{H^+}{\overset{OH^-}{\rightleftarrows}} EH \underset{H^+}{\overset{OH^-}{\rightleftarrows}} E \tag{11}$$

The enzymatic activity versus pH is thus generally a bell-shaped curve as in Fig. 6. An autocatalytic effect may appear when the reaction products have an acid-base effect (often the case). A simple example is the glucose oxidase reaction, shown in Fig. 7. Notably, the rate versus product (H^+) curve indicates an autocatalytic effect on the alkaline branch, that is, for $pH > pH_0$ (see Fig. 6). Systems presenting analogous properties have been studied by R. Caplan et al.[27] and we also learn more about them from D. Thomas in this volume.

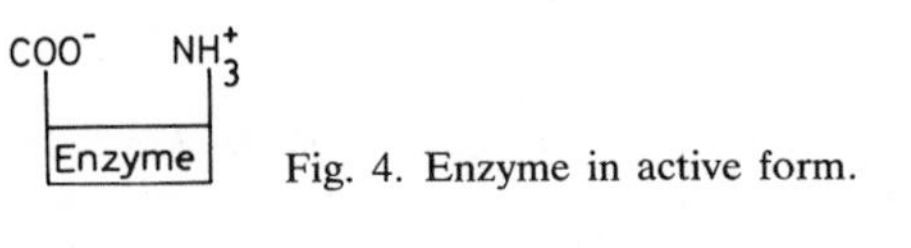

Fig. 4. Enzyme in active form.

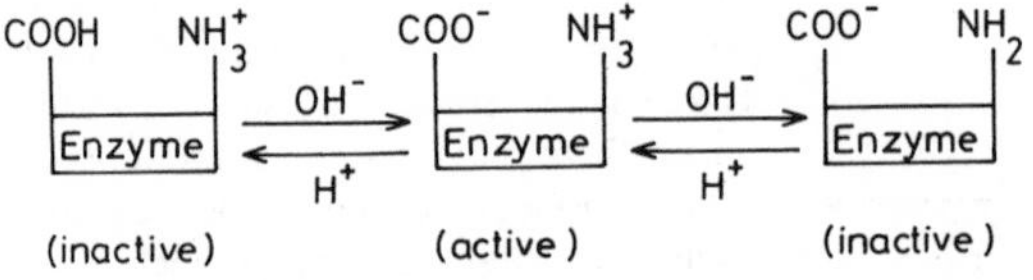

Fig. 5. Schematic representation of the acid-base equilibrium of enzymes.

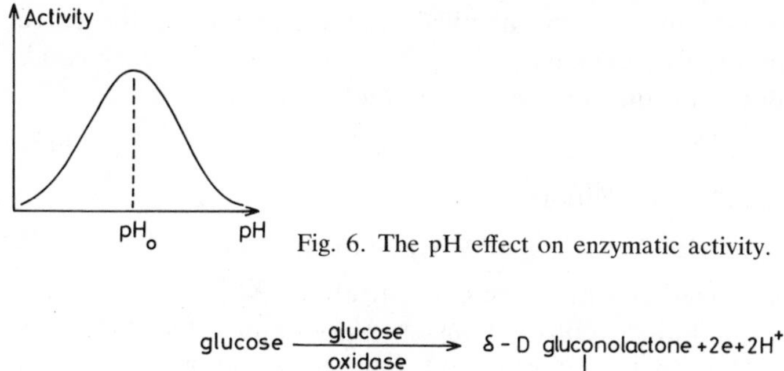

Fig. 6. The pH effect on enzymatic activity.

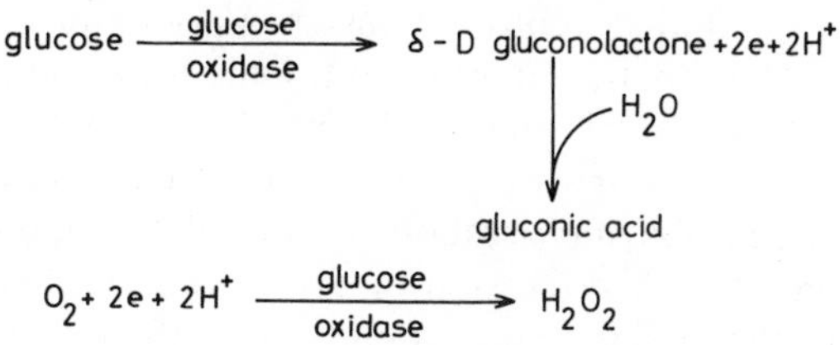

$$\text{glucose} \xrightarrow[\text{oxidase}]{\text{glucose}} \delta\text{-D gluconolactone} + 2e + 2H^+$$

$$\delta\text{-D gluconolactone} \xrightarrow{H_2O} \text{gluconic acid}$$

$$O_2 + 2e + 2H^+ \xrightarrow[\text{oxidase}]{\text{glucose}} H_2O_2$$

Fig. 7. Glucose oxidase reaction.

III. GENERAL FORMULATION

The systems considered here are isothermal and at mechanical equilibrium but *open* to exchanges of matter. Hydrodynamic motion such as convection are not considered. Inside the volume V of Fig. 8, N chemical species may react and diffuse. The exchanges of matter with the environment are controlled through the boundary conditions maintained on the surface S. It should be emphasized that the consideration of a bounded medium is essential. In an unbounded medium, chemical reactions and diffusion are not coupled in the same way and the convergence in time toward a well-defined and asymptotic state is generally not ensured. Conversely, some regimes that exist in an unbounded medium can only be transient in bounded systems. We approximate diffusion by Fick's law, although this simplification is not essential. As a result, the concentration of chemicals X_i ($i=1, 2, \ldots, r$ with $r \leqslant N$) will obey equations of the form

$$\frac{\partial X_i}{\partial t} = \sum_\rho v_{\rho i} v_\rho(\{X_j\}) + D_i \frac{\partial^2 X_i}{\partial z^2} \qquad (i=1, \ldots, r; \quad j=1, \ldots, N) \quad (12)$$

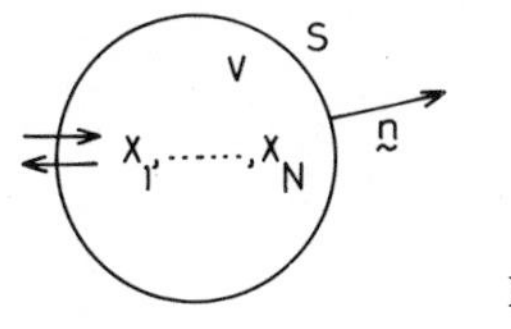

Fig. 8.

where v_ρ is the rate of the ρth chemical reaction and $\nu_{\rho i}$ is the corresponding stoichiometric coefficient of X_i. Two types of boundary conditions are considered on the surface S: Dirichlet conditions

$$\{X_1, \ldots, X_r\} = \{\text{const}\}_S \tag{13a}$$

or Neumann boundary conditions

$$\{\mathbf{n} \cdot \nabla X_1, \ldots, \mathbf{n} \cdot \nabla X_r\} = \{\text{const}\}_S \tag{13b}$$

These conditions together with those concentrations X_l $(l = N - r, \ldots, N)$ whose value is maintained constant inside V constitute the constraints applied to the system by the environment. Only for some particular set of values of these constraints is an equilibrium state realized between V and its external world. Although we refer here only to chemical systems, the class of phenomena obeying parabolic differential equations of the form (12) is much broader. A discussion of or references to self-organization phenomena in other fields (e.g., ecology, laser theory, or neuronal networks) can be found in Ref. 2.

In fact, the variety of phenomena that may be described by this sort of reaction diffusion equation is properly amazing. Some order can be brought into the results by considering as the "basic solution" the one corresponding to the thermodynamic branch. Other solutions may then be obtained as successive bifurcations from this basic one, or as higher-order bifurcations from a nonthermodynamic branch, taking place when the distance from equilibrium is increased. Investigations along such lines of thought have been made by Ortoleva and Ross[18] and inside our group.[2] On the other hand, the relation between chemical instabilities and the nature of the catalytic properties appearing in the reaction scheme has been investigated by Tyson.[28]

As we have mentioned, in the last two years much work has been done on the reaction scheme (1) from the point of view of the mathematical study of bifurcations and numerical computations. For this reason we start with a review of these results.

IV. DISSIPATIVE STRUCTURES IN A SIMPLE MODEL SYSTEM

Let us go back to the Brusselator (1). In a one-dimensional medium of unit length the reaction-diffusion equations (12) become

$$\begin{aligned} \frac{\partial X}{\partial t} &= A + X^2 Y - (B+1)X + D_1 \frac{\partial^2 X}{\partial z^2} \\ \frac{\partial Y}{\partial t} &= BX - X^2 Y + D_2 \frac{\partial^2 Y}{\partial z^2} \end{aligned} \tag{14}$$

We take the boundary conditions

$$X(0, t) = X(1, t) = A; \qquad Y(0, t) = Y(1, t) = \frac{B}{A} \tag{15}$$

These values correspond to the homogeneous steady-state solution of (14), which belongs to the thermodynamic branch. We first determine the conditions under which this solution becomes unstable and subsequently the properties of the new regimes that branch off at the point of instability.

A. Linear Stability Analysis

Setting

$$\begin{pmatrix} X \\ Y \end{pmatrix} = \begin{pmatrix} A \\ \dfrac{B}{A} \end{pmatrix} + \begin{pmatrix} x(z, t) \\ y(z, t) \end{pmatrix} \tag{16}$$

with

$$\left|\frac{x}{A}\right|, \quad \left|\frac{y}{\dfrac{B}{A}}\right| \ll 1 \qquad \text{and} \qquad x(0) = x(1) = y(0) = y(1) = 0$$

the linearization of (14) yields

$$\begin{pmatrix} \dfrac{\partial x}{\partial t} \\ \dfrac{\partial y}{\partial t} \end{pmatrix} = L \begin{pmatrix} x \\ y \end{pmatrix} \tag{17}$$

where L is the differential matrix operator

$$L = \begin{pmatrix} D_1 \dfrac{d^2}{dz^2} + B - 1 & A^2 \\ -B & D_2 \dfrac{d^2}{dz^2} - A^2 \end{pmatrix} \tag{18}$$

The solutions of (17) satisfying the boundary conditions are of the form

$$\begin{pmatrix} x(z, t) \\ y(z, t) \end{pmatrix} = \begin{pmatrix} x_0 \\ y_0 \end{pmatrix} e^{\omega_n t} \sin n\pi z \tag{19}$$

where ω_n is the eigenvalue of (18) corresponding to the wave number n. In order for the homogeneous steady-state solution to be unstable it is sufficient that the real part of some particular ω_n be positive. The critical point at which the exchange of stability takes place is thus determined from the condition

$$\mathrm{Re}\,\omega_n = 0 \tag{20}$$

This exchange of stability may correspond to the branching of a time periodic solution when

$$A^2+1+(n\pi)^2(D_1+D_2)<B<A^2+1-(n\pi)^2(D_2-D_1) \\ +2A[1-(n\pi)^2(D_2-D_1)]^{1/2} \tag{21}$$

or of a new steady-state solution when

$$B>B(n)=\left[1+\frac{D_1}{D_2}A^2+\frac{A^2}{D_2(n\pi)^2}+D_1(n\pi)^2\right] \tag{22}$$

These conditions have been plotted in Figs. 9 and 10 for two values of the

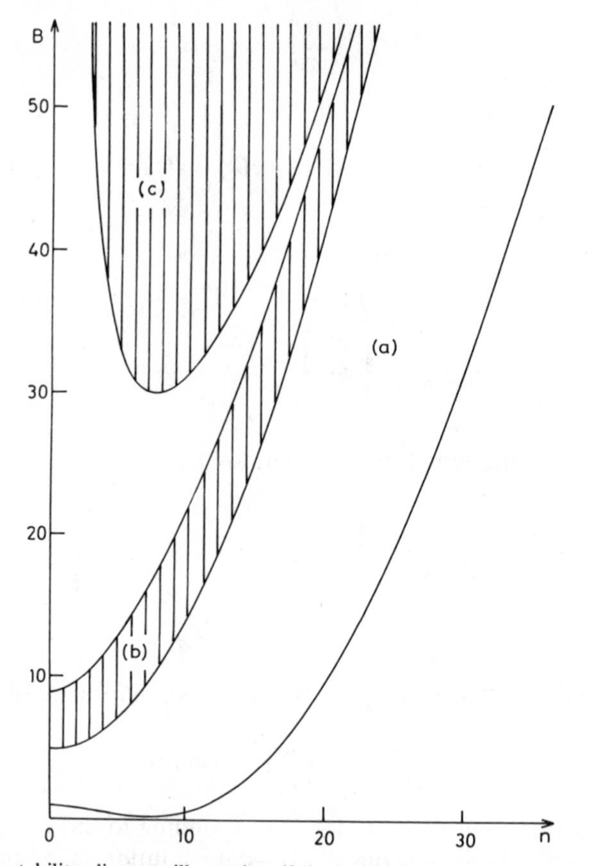

Fig. 9. Linear stability diagram illustrating (21). The bifurcation parameter B is plotted against the wave number n. (a) and (b) are regions of complex eigenvalues ω_n; (b) shows region corresponding to an unstable focus; (c) region corresponding to a saddle point. The vertical lines indicate the allowed discrete values of n. $A=2$, $D_1=8\cdot10^{-3}$; $D_2=1.6\cdot10^{-3}$.

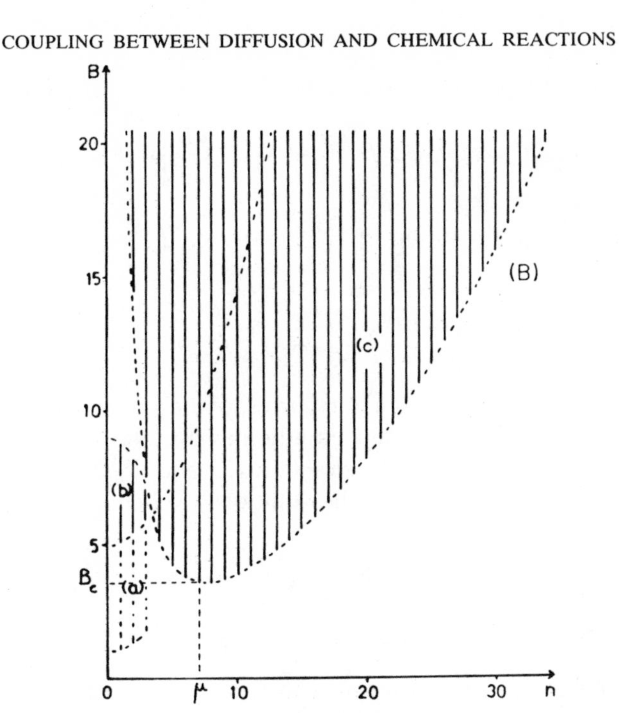

Fig. 10. Linear stability diagram illustrating the branching of a new inhomogeneous steady-state solution. The regions (*a*), (*b*), (*c*) are defined as in Fig. 9. $A=2$; $D_1 = 1.6 \cdot 10^{-3}$, $D_2 = 8 \cdot 10^{-3}$. The critical mode μ is the integer that gives to $B(n)$ its minimal value B_c.

ratio of diffusion coefficients D_2/D_1. It is seen that when this ratio increases (Fig. 10), we first encounter (for increasing values of B) the critical point (22) suggesting the appearance of an inhomogeneous pattern of concentration. This situation as well as the one corresponding to Fig. 9 has been studied analytically by the methods of bifurcation theory.[13,14] We shall limit our analysis here to the branching of steady-state solutions and report in the next section some interesting numerical results on time periodic solutions in a one- and two-dimensional reacting medium.

The surprising result we would like to emphasize here is that quite different properties arise depending solely on the symmetry characteristics of the critical mode n. It is clear from Fig. 10 that there exists a critical mode μ for which $B(n)$ assumes its minimal value $B_c(n)$. We shall proceed to evaluate the steady-state solutions branching off near this critical point. Inserting the decomposition (16) into the rate equations and keeping the nonlinear contributions in x and y, we obtain

$$L_c \begin{pmatrix} x \\ y \end{pmatrix} = \begin{pmatrix} -h(x, y) \\ h(x, y) \end{pmatrix} \tag{23}$$

where L_c is the operator (18) evaluated for the critical mode μ and

$$h(x, y)=(B-B_c)x+2Axy+\left(\frac{B}{A}\right)x^2+x^2y \tag{24}$$

The vector $\begin{pmatrix} x \\ y \end{pmatrix}$ is a solution of (23), provided L_c satisfies the solubility condition

$$\left\langle (x^+, y^+)\begin{pmatrix} -h \\ h \end{pmatrix}\right\rangle \equiv \int_0^1 dz(y^+-x^+)h(x, y)=0 \tag{25}$$

where (x^+, y^+) is the eigenfunction corresponding to a null eigenvalue of the adjoint of L_c. Both L_c and its adjoint have the same critical wave number μ whereas the amplitudes (x_0^+, y_0^+) and (x_0, y_0) are in general different. In order to calculate the solutions explicitly we introduce the expansions

$$\begin{pmatrix} x \\ y \end{pmatrix}=\epsilon\begin{pmatrix} x_0 \\ y_0 \end{pmatrix}+\epsilon^2\begin{pmatrix} x_1 \\ y_1 \end{pmatrix}+\cdots \tag{26}$$

$$B-B_c=\gamma=\epsilon\gamma_1+\epsilon^2\gamma_2+\cdots \tag{27}$$

into (23) and on identifying terms with equal powers of ϵ, we obtain the system of equations

$$L_c\begin{pmatrix} x_k \\ y_k \end{pmatrix}=\begin{pmatrix} -a_k \\ a_k \end{pmatrix} \qquad 0\leqslant k\leqslant\infty \tag{28}$$

The validity of the solubility condition (25) to each order [i.e., for each $a_k(z)$] leads to the determination of the coefficients γ_j and permits us to express ϵ in terms of $B-B_c$ from (27). The explicit form of the bifurcating solution is then obtained by substituting ϵ as well as $\begin{pmatrix} x_i \\ y_i \end{pmatrix}$ of the successive approximations of (23) into (26). The results can then be summarized as follows.

If the critical wave number μ is *even*, one finds

$$x(z)=\pm\left(\frac{B-B_c}{\phi}\right)^{1/2}\sin \mu\pi r-\frac{B-B_c}{\phi}\frac{8(\mu\pi)^2}{\pi^5 D_1 A}$$
$$\times\ [2[D_1(\mu\pi)^2+1]-B_c]\sum_{n:\text{odd}}\frac{n}{(n^2-\mu^2)^2}\frac{1}{n^2-4\mu^2}\sin n\pi z \tag{29}$$

and a similar expression for $y(z)$. The function ϕ involves A, D_1, B_c, and μ. Depending on its sign, two types of bifurcation diagram are obtained.

When $\phi>0$ the amplitude of the new solution grows in the supercritical region $B>B_c$ as $(B-B_c)^{1/2}$. Furthermore, these solutions possess the

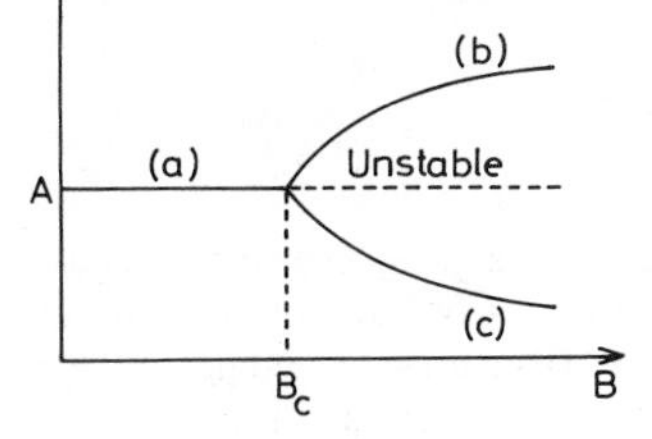

Fig. 11. Bifurcation diagram for an even wave number and $\phi > 0$. (*a*) Uniform steady-state solution; (*b*) and (*c*) stable dissipative structures extending supercritically with respect to the bifurcation point B_c.

following remarkable properties:

1. Because of the "critical exponent" $\frac{1}{2}$ the solutions are degenerate. The emergence of a spatial organization is associated with a *symmetry-breaking* transition at the point of instability. The choice of a particular solution is determined by the initial condition of the system.
2. As a direct consequence of the nonlinearity of the equations and of the boundary conditions there appears in (29) an infinite series introducing subharmonic terms that are a source of *spatial asymmetry* (see Fig. 12). The dominant contribution to this distortion is given by the term whose n is the odd closest to the value $n \simeq \sqrt{3}\mu$.
3. In general, the transition of a dissipative structure is associated with a variation of the total amount of X and Y present in the system.

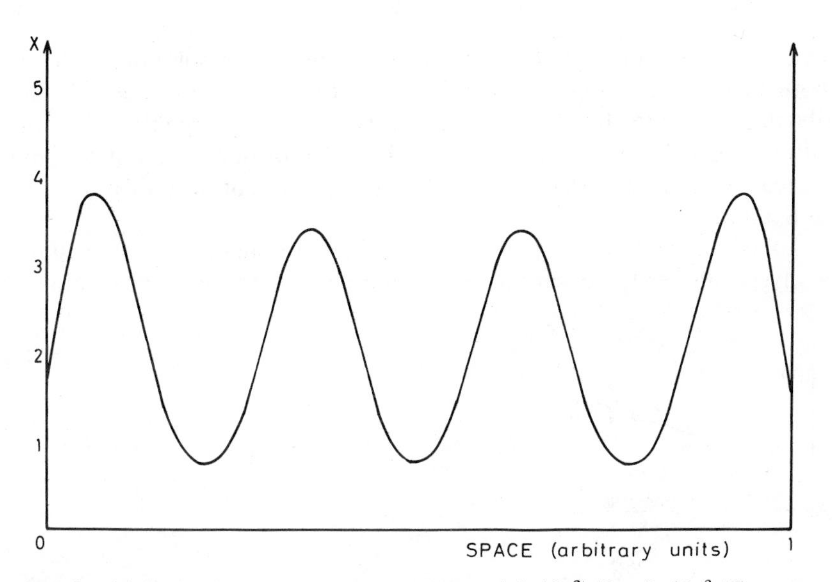

Fig. 12. Spatial distribution of X. $A = 2$, $B = 4.6$, $D_1 = 1.6 \cdot 10^{-3}$, $D_2 = 8 \cdot 10^{-3}$. The greater amplitude of the two extreme peaks is due to the subharmonic terms in (29).

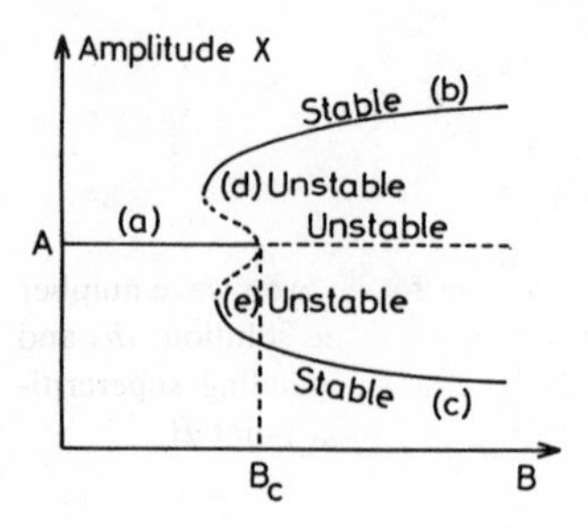

Fig. 13. Bifurcation diagram for an even wave number and $\phi<0$. (*a*): Uniform steady-state solution; (*b*) and (*c*): stable dissipative structures extending on both sides of the bifurcation point; (*d*) and (*e*): subcritical unstable dissipative structures.

When $\phi<0$, the bifurcation diagram is as in Fig. 13. There exists a subcritical region in which three stable steady-state solutions may coexist simultaneously: the thermodynamic branch and two inhomogeneous solutions. It must be pointed out that the latter are necessarily located at a finite distance from the thermodynamic branch. As a result, their evaluation cannot be performed by the methods described here. The existence of these solutions is, however, ensured by the fact that in the limit $B\rightarrow 0$, only the thermodynamic solution exists whereas for $B \gg B_c$ it can be shown that the amplitude of all steady-state solutions remains bounded.

If the critical wave number μ is *odd*, one finds

$$x(z)=-\tfrac{3}{8}(B-B_c)(\mu\pi)^2A\{2[D_1(\mu\pi)^2+1]-B_c\}^{-1}\sin\mu\pi z+O[(B-B_c)^2] \tag{30}$$

and a similar expression for $y(z)$. The new branch of solutions is defined both below and above B_c (see Fig. 14). Below it is unstable, whereas above it is stable. In this case there is no symmetry breaking associated with the instability as only one branch passes through the critical point. Instead we have a *hysteresis* type of behavior. We consider this case again in Section VI.

To conclude this section we would like to mention that a similar analysis can easily be carried out in the case of the zero flux boundary

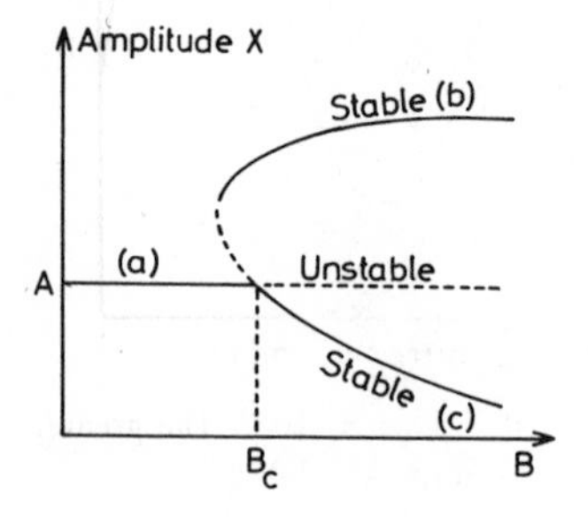

Fig. 14. Bifurcation diagram for an odd critical wave number.

conditions (13b). To the dominant order the steady-state solutions bifurcating from the homogeneous state are of the form

$$\begin{pmatrix} X \\ Y \end{pmatrix} = \begin{pmatrix} A \\ \dfrac{B}{A} \end{pmatrix} + \begin{pmatrix} x_0 \\ y_0 \end{pmatrix} \cos \mu\pi z \tag{31}$$

The bifurcation diagrams are similar to those of Figs. 11 and 13.[14] The expressions for $x(z)$ are qualitatively the same whether μ is even or odd. However, a remarkable feature is the spontaneous appearance of a *macroscopic gradient* along the system each time that μ is odd, the value of the concentration at the two boundaries $z=0$ and $z=1$ being different.

V. CONCENTRATION WAVES IN ONE AND TWO DIMENSIONS

We have discussed the case in which the unstable homogeneous steady state is a saddle point, that is, when a real eigenvalue ω_n changes sign. Let us now consider what happens when the uniform steady state is an unstable focus and a time periodic regime sets in.

One may describe analytically this behavior by the same principles as the stationary case, but we do not go into the procedure here. The manipulations are somewhat longer, as one more quantity (the frequency of oscillation in time) needs to be evaluated, and they may be found in the original paper.[13b,2] A presentation of the technique as well as the study of other systems is also given elsewhere.[2] Our discussion illustrates these behaviors on the model (1) and emphasizes some of their properties.

When D_1 is sufficiently larger than D_2, the first instability of the system according to (21) is at $B=B_0=1+A^2$. The first bifurcating solution at B_0 is homogeneous in space. Those bifurcating at $B>B_0$ are all the more inhomogeneous that B is greater.

A. Fixed Boundary Conditions in One Dimension

The solution at B_0 is ruled out. The first solution that satisfies the boundaries arises at $B=B_1=1+A^2+(\pi)^2(D_1+D_2)$ and to the dominant approximation, its structure is of the form

$$X(z,t) \simeq A + x_0 e^{i \operatorname{Im} \omega_\mu t} \sin \mu\pi z \tag{32}$$

At each point in space the oscillations are similar to the limit cycle type of behavior in ordinary differential equations. The various regions of the medium oscillate on phase and one has a kind of standing wave as in a vibrating string. This synchronization of the phase at each point arises from the coupling of the oscillations through diffusion. The effect of this coupling on the chemical reactions becomes weaker when B increases or when the size of the medium increases. The superposition of standing waves corresponding to different modes in space corresponding to different time frequencies may then lead to an *apparent* wave propagation phenomenon as shown in Fig. 15. It is believed that these behaviors are similar to the trigger waves described by Winfree[8a] and Kopell and Howard[33] in the context of the Zhabotinski reaction. The properties of

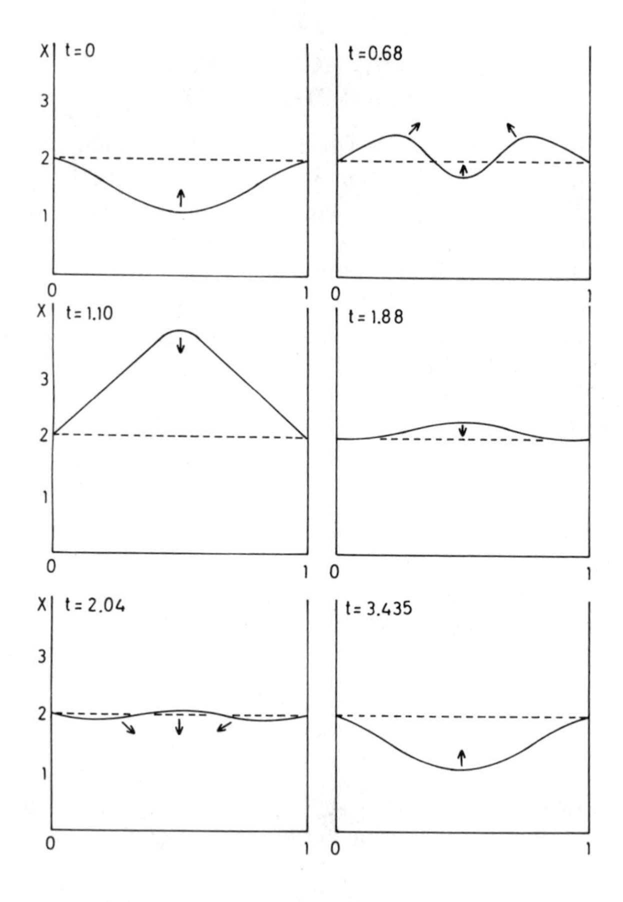

Fig. 15. Spatiotemporal dissipative structure in a one-dimensional system with fixed boundary conditions. dashed line: unstable homogeneous steady state. Full line: stable periodic regime. $A=2$, $B=5.45$; $D_1 = 8 \cdot 10^{-3}$; $D_2 = 4 \cdot 10^{-3}$.

these waves may further be summarized as follows:

1. The propagation velocity $v(z, t)$ is a function of space. At $z=0$ and $z=1$ one has

$$v(0, t)=v(1, t)=0$$

at $z=\frac{1}{2}$

$$v(\tfrac{1}{2}, t)\simeq \pm\infty$$

This shows that the usual ideas associated with propagating waves in electromagnetism or fluid dynamics do not describe the behaviors found here. These differences could be expected because of the mathematical structure of reaction-diffusion equations, which owing to their parabolic character propagate information with infinite velocity. On the contrary, in the case of classical wave equations or hyperbolic equations there is a well-defined domain of influence and a characteristic velocity of propagation of information.*

2. These waves correspond to a transport of matter in the system. The flux of X at a given point in space on the average over one period is of the form

$$\overline{J_X(z)}=D_1\pi\frac{(B-B_1)}{\phi}\sum_{m:\mathrm{odd}} ma_m \cos m\pi z \tag{33}$$

where ϕ and a_m are functions of the parameters. As a result, one sees that

$$\bar{J}_X(\tfrac{1}{2})=0$$

whereas

$$\overline{J_X(0)}=\overline{J_X(1)}=D_1\pi\frac{(B-B_1)}{\phi}\sum_{m:\mathrm{odd}} ma_m \tag{34}$$

3. The bulk properties of these temporal dissipative structures also differ from those of the thermodynamic solution. The time average of the concentrations at a given point in space is only a second-order correction (but spatially dependent). The spatial average, as in the stationary dissipative structures, varies even to the dominant order.

B. Zero Flux Boundaries and Two Dimensions

The first bifurcating solution always corresponds to a bulk oscillation of the whole medium, which has no space dependence. When the initial

* One may obtain traveling wave solutions with other kinds of boundary conditions. This is, for example, the case when the reaction medium can be visualized as a closed curve in a two-dimensional space, or a closed surface in three-dimensional space (periodic boundary conditions).[2]

conditions are not homogeneous, one may observe a sharp concentration front propagating from one boundary to the other (in one dimension). It has been found numerically that these stages of propagation tend to disappear after a few repetitions and that one evolves toward a sustained regime of homogeneous oscillations.[29] In two dimensions, however, stable spatiotemporal regimes bifurcating at $B > B_0$ from the homogeneous steady state appear more common, although the homogeneous time periodic solution bifurcating at B_0 remains stable in the entire range of parameter values. The main feature when the dimensionality of the medium increases is that the explicit form of the wave number $k = (n\pi/l)^2$ of the Laplacian operator has to be decomposed into its components along the different dimensions. For example, in a rectangle of dimensions l_1, l_2, the explicit form of n would be

$$\left(\frac{n\pi}{l}\right)^2 = \left(\frac{n_{l_1}\pi}{l_1}\right)^2 + \left(\frac{n_{l_2}\pi}{l_2}\right)^2 \tag{35}$$

A a consequence, for a given n $(n > 0)$ *more than one solution* may bifurcate from the unstable steady state. Indeed, more than one eigenfunction may correspond to the same eigenvalue of the Laplacian operator. According to general theorems, these solutions branch off as unstable branches from the homogeneous steady state. However, they can undergo a change of stability for some critical $B^* > B_n$ and remain stable for $B > B^*$. This situation is represented in Fig. 16. It has been verified by Erneux and Herschkowitz-Kaufman[30] for the solutions branching off at $B = B_1$ in a system with a circular impermeable boundary. They showed that two kinds of spatiotemporal solutions qualitatively different may be found according to the initial conditions. One solution is characterized by a symmetry plane that passes through the center of the circle (see Fig. 17). The other corresponds to a periodic rotation around the center, which itself remains stationary (see Fig. 18). The stability of these

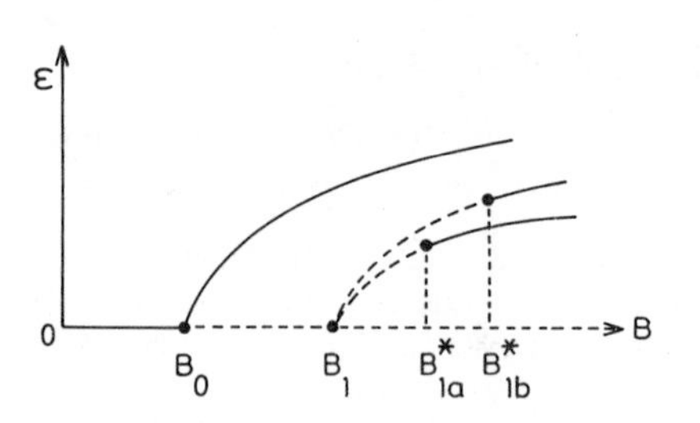

Fig. 16. Bifurcation diagram of temporal dissipative structures. ϵ (maximal amplitude of the oscillation minus the homogeneous steady-state value) is sketched versus B for a two-dimensional system with zero flux boundary conditions. The first bifurcation occurs at $B = B_0$ and corresponds to a stable homogeneous oscillation. At B_1 two space-dependent unstable solutions bifurcate simultaneously. They become stable at B^*_{1a} and B^*_{1b}. Notice that as it is generally the case: $B^*_{1a} \neq B^*_{1b}$.

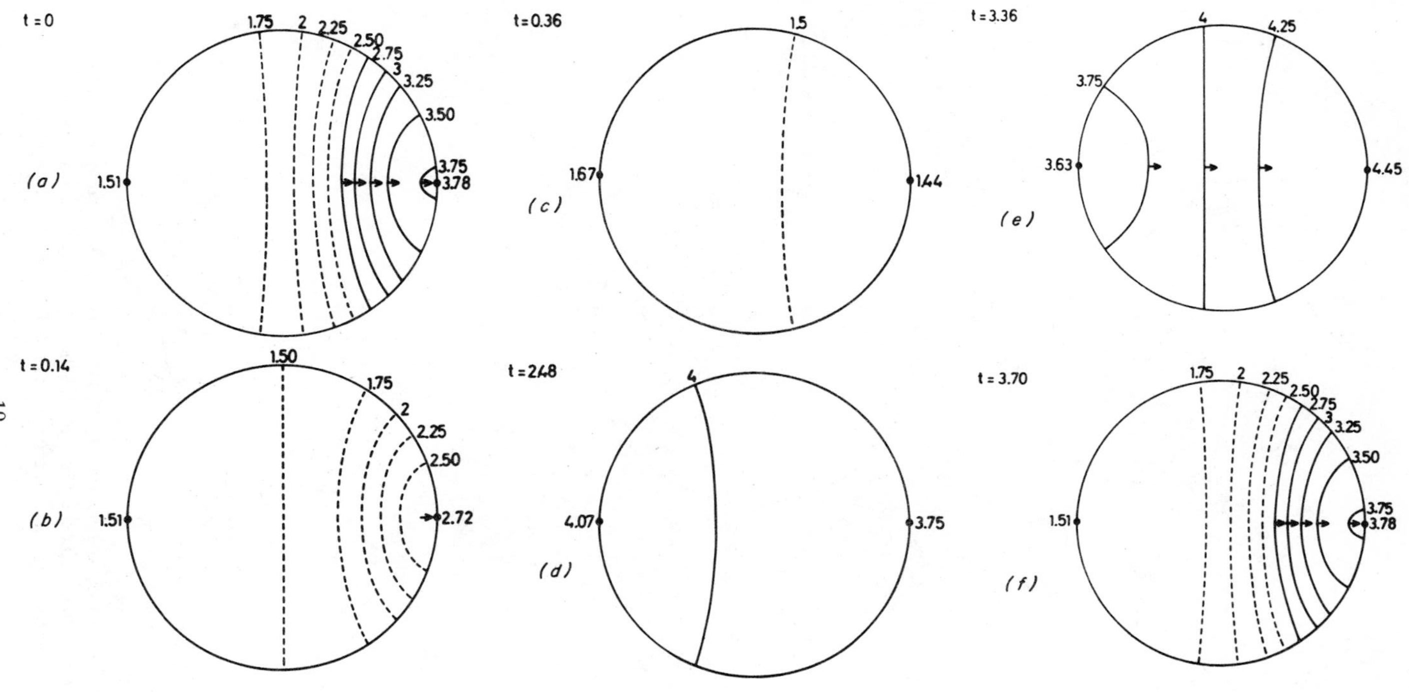

Fig. 17. Temporal dissipative structure after various time intervals during the period of oscillation. The reaction medium is a circle with zero flux boundary conditions. The lines correspond to isoconentrations. $A=2$, $B=5.4$, $D_1=8\cdot10^{-3}$, $D_2=4\cdot10^{-3}$. Curves of equal concentration for Y are represented by full or broken lines when the concentration is, respectively, larger or smaller than the unstable steady state. The radius of the circle $r_0=0.5861$.

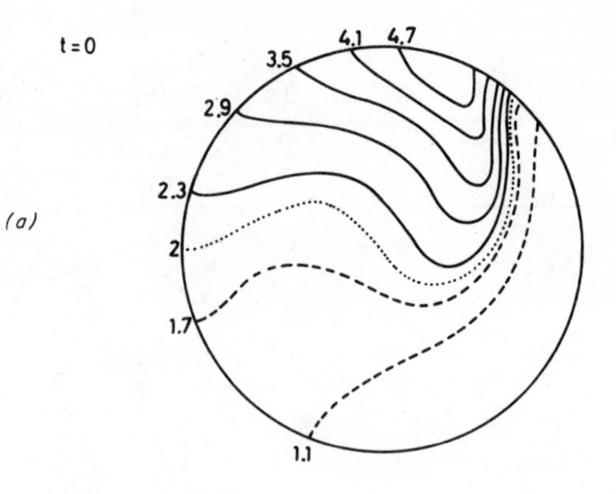

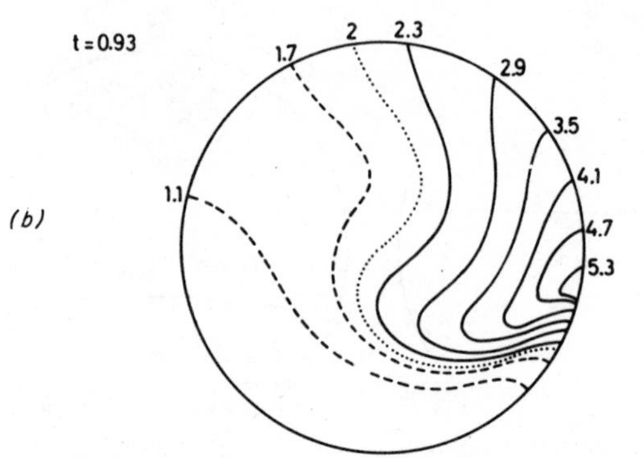

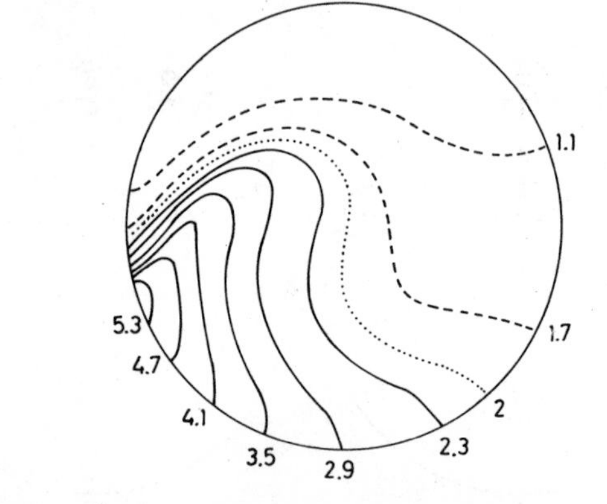

Fig. 18. Rotating wavelike solution at various times during the period of oscillation, $B = 5.8$. Curves of equal concentration for X. The other parameters are as in Fig. 17. The stabilization of the rotating wave solution occurs at $B > 5.4$.

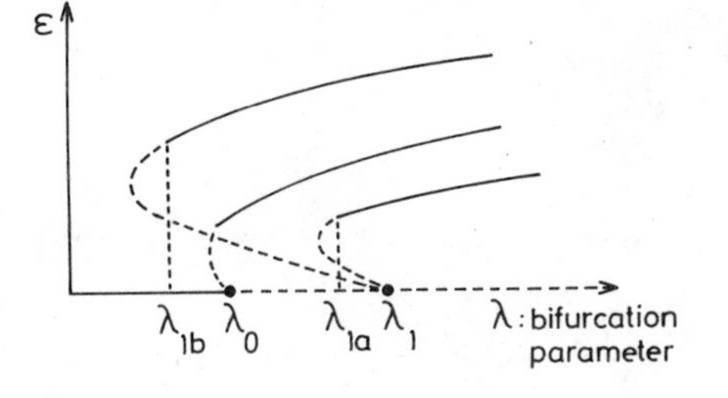

Fig. 19. Bifurcation diagram in which a rotating wavelike solution and the homogeneous steady state would be stable simultaneously.

patterns has been tested numerically, and at least to small perturbations (~1% of the concentrations value) they seem stable. One may conjecture that this behavior arises through a secondary bifurcation, as in Fig. 16. These rotating waves present some qualitative similarities with the behavior of "excitable" reaction-diffusion systems described by Winfree.[31] There is, however, one important difference; The solutions described here are stable beyond the point of instability of the homogeneous thermodynamic branch solutions; in the systems considered by Winfree the homogeneous and rotating wave solutions may be stable simultaneously, suggesting a bifurcation diagram as in Fig. 19.[30]

VI. SOLUBLE MODEL FOR DISSIPATIVE STRUCTURES

In the preceding sections we have analyzed the new solutions that appear at a point of instability and have shown that they can be calculated by the methods of bifurcation theory as long as their amplitude is small. In this section we consider a system of the form (2) whose steady-state solutions can be evaluated straightforwardly without implying any restriction on the parameters' value. This allows a complete analysis of these branches of solutions.[32]

Putting $c=-1$ and $f(x, y)=x^2y-Bx$ in (2), we obtain

$$\frac{\partial x}{\partial t}=x^2y-Bx+D_1\frac{\partial^2 x}{\partial z^2} \tag{36a}$$

$$\frac{\partial y}{\partial t}=-x^2y+Bx+D_2\frac{\partial^2 y}{\partial z^2} \tag{36b}$$

In this simplified version of the Brusselator model, the trimolecular autocatalytic step, which is a necessary condition for the existence of instabilities, is, of course, retained. However, the linear source-sink reaction steps $A \to X \to E$ are suppressed. A continuous flow of X inside the system may still be ensured through the values maintained at the boundaries. The price of this simplification is that (36) can never lead to a homogeneous time-periodic solution. The homogeneous steady states are now either

$$x = 0; \qquad y = 0 \tag{37}$$

which is always stable, or

$$x = \xi; \qquad y = \frac{B}{\xi} \tag{38}$$

where ξ is a parameter. We consider fixed an equal boundary condition

$$x(0) = x(l) = \xi; \qquad y(0) = y(l) = \frac{B}{\xi} \tag{39}$$

Adding (36a) and (36b), one has

$$D_1 x(z) + D_2 y(z) = D_1 \xi + \frac{D_2 B}{\xi} \tag{40}$$

which permits us to calculate the steady-state solutions by solving a closed-form equation for the new variables

$$w = x - \xi \qquad \text{and} \qquad z = \frac{z}{D_2^{1/2}} \tag{41}$$

We consider equation (36a) for time-independent situations and use (39) to (41). We then obtain by performing an integration over z

$$\left(\frac{dw}{dz}\right)^2 = K - F(w) \tag{42}$$

with

$$F(w) = -\frac{w^4}{2} - \frac{2}{3}\left(2\xi - \frac{\rho}{\xi}\right)w^3 + (\rho - \xi^2)w^2 \tag{43}$$

and

$$\rho = \frac{BD_2}{D_1} \tag{44}$$

where K is the constant of integration, which has a simple meaning as it determines $(dw/dz)^2$ when $F(w) = 0$. Formally, (42) presents a striking analogy with the equation expressing the conservation of total energy H of a particle moving in a potential field $V(x)$

$$m\left(\frac{dz}{dt}\right)^2 = H - V(x) \tag{45}$$

In (42), w and z play, respectively, the role of the spatial coordinate and time in (45). We may therefore use the classical analytical methods to discuss the bifurcation arising in our model system. Depending on the form of the "potential" function $F(w)$, three situations may be encountered and are schematized in Fig. 20.

As the right-hand side of (42) must be positive for the equation to admit a solution physically acceptable, the values of K must be chosen positive. As a result, in the case of Fig. 20*a* no periodic solution $w(z)$ satisfying the boundary conditions exists. The only steady-state solution of the system is the homogeneous one. In case *b* there exist besides the homogeneous steady state two other inhomogeneous steady-state solutions corresponding to downward profiles of concentration for x. Case *c* has a finite range of values for K such that spatially oscillating solutions exist. One may notice that if the cubic and quartic terms of $F(w)$ were equal to zero, one would have a parabolic potential function as in the case of the harmonic oscillator. The period of the spatial oscillation would be independent of its amplitude. The bifurcation diagram of amplitude versus length would simply be constituted by a series of vertical lines located at $l = T/2, T, 3T/2, \ldots$ (where T is the period). Let us consider these three possibilities in further detail. We fix the value of ξ and vary ρ.

1. For $\rho < \xi^2/2$ the only stationary solutions possible is the homogeneous state in which the concentrations are constant and equal to the value at the boundaries. This corresponds to the domain of stability of the thermodynamic branch.

2. For $\xi^2/2 < \rho < \xi^2$ and $0 \leq K \leq F(-\xi)$, (42) has inhomogeneous steady-state solutions, which require that the system be of length l equal to

$$L(K) = 2\int_0^{w(K)} \frac{dw'}{[K - F(w')]^{1/2}} \tag{46}$$

where $w(K)$ is the value of w at which the integral of (46) becomes singular. Note that for any $0 < K < F(-\xi)$ the integrand behaves as

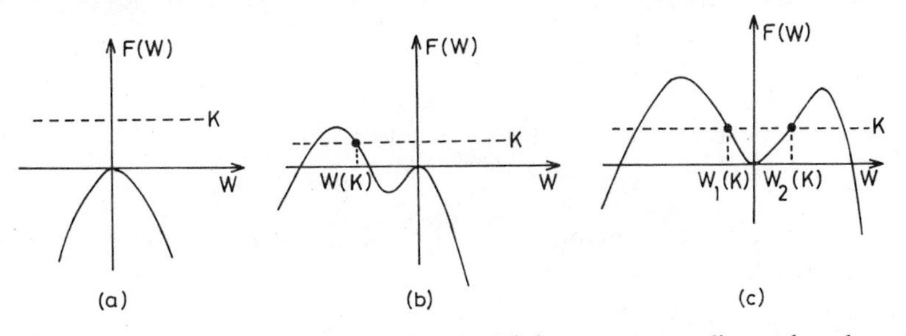

Fig. 20. Sketch of the various graphs possible for $F(w)$ versus w according to the values of ρ and ξ.

$[w(K)-w']^{-1/2}$ near $w(K)$ so that $L(K)$ is finite in this range. However, when $K=0$ or $K=F(-\xi)$, the integrand in (46) diverges as $[w(K)-w']^{-1}$, so that $L(K)$ becomes infinite. Thus there exists a value K_m such that $L(K_m)$ is minimal and two situations may arise, depending on the length l of the system considered. If it is smaller than $L(K_m)$ the only steady-state solution possible is again the homogeneous state, and if l is larger than $L(K_m)$ there exist two inhomogeneous steady-state solutions that satisfy the boundary conditions and correspond to different values of K. Necessarily these profiles of concentration lie entirely under the reference state $w(z)=0$; between $z=0$ and $z=l$ the concentration w first decreases, passes through a minimal value at $z=l/2$, and then increases again up to the value $w=0$ in $z=l$. The profile whose concentration at $z=l/2$ is the smallest and the homogeneous state $w(z)=0$ are stable; the other one is unstable. Thus in this region two steady states are stable simultaneously: the homogeneous thermodynamic solution and an inhomogeneous nonequilibrium type of solution. In the latter the concentrations can only pass through one extremal value and the multiplicity of solutions does not increase with the length l as in the dissipative structures described in Section IV; rather, the concentration of x tends to values close to zero over the entire interval l [whereas y tends to $(D_1/D_2)x(0)+y(0)$] except for some narrow region near the boundaries. The evaluation of these solutions cannot be carried out by the methods of bifurcation theory described in Sections IV and V.

3. For $\rho>\xi^2$, $F(w)$ presents two maxima above the w-axis. They are located at $w'=-\xi$ and $w''=-\xi+B/\xi$, whereas $w=0$ has become a minimum. Depending on the value of ρ, the highest peak is either at $w'(\rho<2\xi^2)$ or at $w''(\rho>2\xi^2)$. In either case there exists a range of values for K—respectively, $0\leqslant K^{1/2}\leqslant F^{1/2}(w'')=K''$ and $0\leqslant K^{1/2}\leqslant F^{1/2}(w')=K'$] for which the equation $K-F(w)$ has four real roots for positive values of K (α, β, γ, δ, in decreasing order). Under these conditions (42) has solutions of the same nature as the spatial dissipative structures described in Section IV. The half-period of these inhomogeneous solutions is given by

$$L_i(K)=\frac{4\sqrt{2}}{\sqrt{(\alpha-\gamma)(\beta-\delta)}}F(\varphi_i\backslash\epsilon) \qquad i=1,2 \tag{47}$$

where $F(\varphi_i\backslash\epsilon)$ is the incomplete elliptic integral of the first kind with modular angle ϵ.

$$\sin^2\epsilon=\frac{(\beta-\gamma)(\alpha-\delta)}{(\alpha-\gamma)(\beta-\delta)} \tag{48a}$$

The slope at the origin of the half-period being positive or negative, its amplitude is given by

$$\sin^2 \varphi_1 = \frac{(\alpha-\gamma)\beta}{(\beta-\gamma)\alpha} \qquad \text{or} \qquad \sin^2 \varphi_2 = \frac{(\beta-\delta)\gamma}{(\beta-\gamma)\delta} \tag{48b}$$

respectively. Only when the two maxima of $F(w)$ are equal [i.e., when there is no cubic term in (41), $\rho = 2\xi^2$] are the two half-period solutions for a given K completely symmetrical. For a system of length l the only admissible solutions are those that correspond to a half-period $L_i(K) = l$ or such that

$$n_1 L_1(K) + n_2 L(K) = l \qquad \text{with} \quad |n_1 - n_2| = 0 \text{ or } 1 \tag{49}$$

where n_1, n_2 are positive integers. Developing K and F around $w = 0$, it is easily seen that $L_1(K)$ and $L_2(K)$ tend to the minimal value l_m predicted by the linear stability analysis:

$$l_m = \frac{\pi}{(\rho - \xi^2)^{1/2}} \tag{50}$$

Thus as usual there is a minimal size beyond which only the dissipative structure may appear. For $l < l_m$ only the homogeneous state is a steady-state solution. Equation (49) also implies that K can only assume a discrete set of values and that there is a maximum to the number of periods and solutions that can be contained in a system of given length. In other words, in a finite system there is a quantization of the acceptable values of K and a multiplicity of solutions that increases rapidly with the size of the system. In general, the lengths L_1 and L_2 being unequal, the spatial averages of the concentrations differ from their value in the homogeneous reference state.

In Fig. 21 we have drawn the bifurcation diagram of the fundamental steady-state solutions for three values of ρ [$\pm K^{1/2}$ is plotted versus $L(K)$ as the bifurcation parameter]. There is a subcritical region in the upper or lower branch, depending on the relative height of the peaks in Fig. 20*c*. The asymptotes K' and K'' of these branches correspond to half-period solutions of infinite length. When $\rho \leq \xi^2$ the asymptote K' merges with the w-axis; therefore situation 2 above can be viewed as a particular case of situation 3 above, in which the bifurcation point moves to infinity.

These results are thus in agreement with those of bifurcation theory. In the case of odd wave numbers they demonstrate that in general the bifurcation diagrams have to exhibit a subcritical branch. However, there always exists even for odd wave numbers a value of the parameters such that the bifurcation is soft and this value marks the transition from an upper to a lower subcritical branch (see Fig. 21). This feature was less

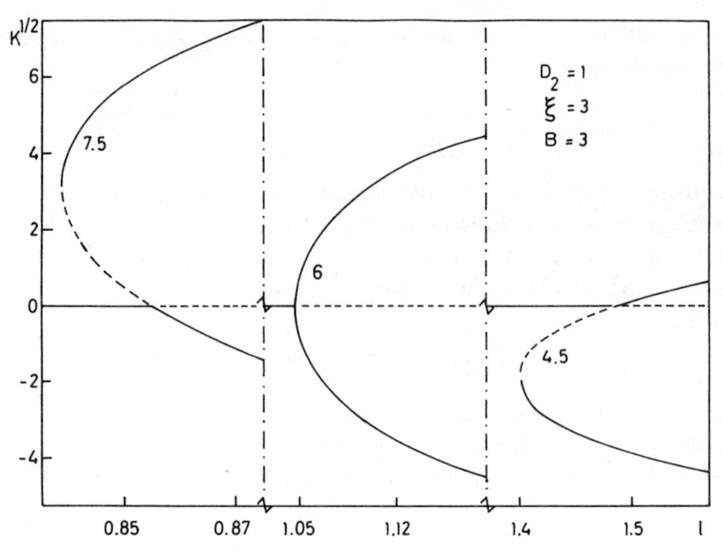

Fig. 21. Bifurcation diagram of $K^{1/2}$ for three values of the ratio of diffusion coefficients. As can be seen in Fig. 20, the amplitude of the inhomogeneous solution is proportional to $K^{1/2}$.

obvious in the case of model (1) and had not been reported. In the case of even wave numbers the bifurcation is always soft and of the type given in Fig. 11.

There are two more important advantages of these models. One is that it is possible under some conditions to carry out an exact stability analysis of the nonequilibrium steady-state solutions and to determine points of exchange of stability corresponding to secondary bifurcations on these branches. The other is that branches of solutions can be calculated that are not accessible by the usual approximate methods. We have already seen a case here in which the values of parameters correspond to domain 2. This also happens when the fixed boundary conditions imposed on the system are arbitrary and do not correspond to some homogeneous steady-state value of X and Y. In that case Fig. 20 may, for example,

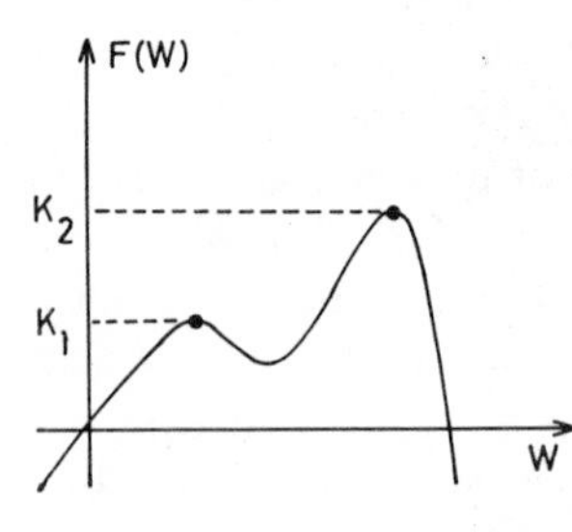

Fig. 22. A behavior of $F(w)$ when the values of x and y imposed on the boundaries do not correspond to a homogeneous steady-state solution.

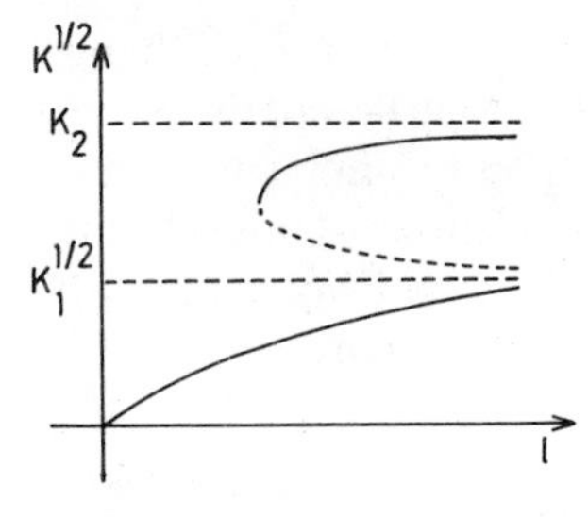

Fig. 23. Bifurcation diagram of $K^{1/2}$ in the case of Fig. 22. The lower branch of solutions corresponds to the thermodynamic branch. It tends to the asymptotic value $K_1^{1/2}$, which separates it from nonequilibrium types of solutions.

transform into the type of situation shown in Fig. 22, yielding a bifurcation diagram as in Fig. 23, from which it is clear that the branches of solutions bifurcate at infinity.

VII. GENERAL CONCLUSIONS

Instabilities and dissipative structures play a basic role in our understanding of chemical kinetics in far from equilibrium situations. Through dissipative structures the characteristics of the system become explicitly dependent on macroscopic parameters (dimensions, boundary conditions) that relate it to its environment.

The mathematical theory of dissipative structures is mainly based on approximate methods such as bifurcation theory of singular perturbation theory. Situations like those described in Section VI and that permit an exact solution are rather exceptional.

In the models discussed here we have considered primarily as bifurcation parameters the affinity of reaction as measured by the parameter B in model (1) or the length l as in Section VI. The results have illustrated that when B or l increase, the multiplicity of solutions increases. This is not astonishing, as a variation in length is a simple way through which the interactions of the reaction cell with its environment can be increased or decreased.

The use of the length or size as the bifurcating parameter is, of course, quite natural in biological processes involving growth or morphogenetic effects.

It is interesting that in the soluble case studied in Section VI a close connection appears between initial value problems of dynamics and boundary value problems for dissipative structures. In discussion of initial value problems the concept of stability plays an important role. Only for simple dynamical systems such as separable systems do we find, in general, stability in the sense that trajectories originating from neighboring points remain "close" for all times. It would be very interesting to investigate along similar lines the stability of dissipative structures, and the extent to which small changes in the boundary conditions may change the space structure drastically. It may be expected that dissipative structures arising in biological systems should have stability similar to that of the simple systems of analytical mechanics. This constitutes an interesting limitation on the type of chemical reactions that may be involved and we hope to come back to this problem in a future publication.

Acknowledgments

We thank Professor G. Nicolis, Dr. M. Herschkowitz-Kaufman, and Dr. J. Wm. Turner for numerous stimulating discussions and suggestions. The financial support of the Solvay Company, of the T. Welch Foundation (Houston, Texas), and of the CGER (Caisse Générale d'Epargne et de Retraite) is gratefully acknowledged.

References

1. P. Glansdorff and I. Prigogine, *Thermodynamic Theory of Structure, Stability and Fluctuations*, Wiley-Interscience, New York, 1971.
2. G. Nicolis and I. Prigogine, *Selforganization in Nonequilibrium Systems*, Wiley-Interscience, New York, 1977.
3. R. Lefever, Thèse de Doctorat, Université Libre de Bruxelles, 1970, I. Prigogine and R. Lefever, *J. Chem. Phys.*, **48,** 1695 (1968), R. Lefever and G. Nicolis, *J. Theor. Biol.*, **30,** 267 (1971).
4. P. Hanusse, *C. R. Acad. Sci.*, **274,** 1245 (1972).
5. J. J. Tyson, *J. Chem. Phys.*, **58,** 3919 (1972).
6. J. J. Tyson and J. Light, *J. Chem. Phys.*, **59,** 4164 (1973).
7. Faraday Symposium No. 9, Physical Chemistry of Oscillatory Phenomena, Chemical Society, Ed. R. I. London, 1975.
8. A. T. Winfree, (a) *Sci. Am.*, **230,** 82 (1974); (b) *Science*, **175,** 634 (1972).
9. A. Pacault, P. Hanusse, P. De Kepper, C. Vidal, and J. Boissonade, *Acc. Chem. Res.*, **9,** 445 (1976).
10. J. J. Tyson, The Belousov-Zhabotinskii Reaction, Lecture Notes in Biomathematics, vol **10.** Springer-Verlag, New York, 1976.
11. E. E. Sel'kov, *Eur. J. Biochem.*, **4,** 79 (1968).
12. A. Goldbeter, *Nature*, **253,** 540 (1975).
13. J. F. G. Auchmuty and G. Nicolis, (a) *Bull, Math. Biol.*, **37,** 323 (1975); (b) *Bull. Math. Biol.*, **38,** 325 (1976).
14. M. Herschkowitz-Kaufman, *Bull. Math. Biol.*, **37,** 589 (1975).
15. J. P. Keener, *SIAM*, **55,** 187 (1976).
16. T. J. Mahar and B. J. Matkowsky, to appear *SIAM J. Appl. Math.* **32,** 394 (1977).
17. P. C. Fife, *J. Chem. Phys.*, **64,** 554 (1976).
18. P. Ortoleva and J. Ross, *J. Chem. Phys.*, **63,** 3398 (1975); *J. Chem. Phys.*, **60,** 5090 (1974).
19. J. A. Boa, *Stud. Appl. Math. LIV*, 9 (1975).
20. F. Schlögl, *Z. Phys.*, **253,** 147 (1972).
21. D. Bedaux, P. Mazur, and R. A. Pasmanter, The Ballast Resistor; *Physica*, **86A,** 355 (1977).
22. D. Thomas and J. P. Kernevez, Eds., *Analysis and Control of Immobilized Enzyme Systems*, North-Holland, New York, 1976.

23. I. Prigogine, *Etude Thermodynamique des Phénomènes Irréversibles*, Desoer, Liège, 1947.
24. B. Hess and A. Boiteux, in *Regulatory Functions of Biological Membranes*, J. Järnefelt, Ed., Elsevier, Amsterdam, 1968; B. Hess, A. Boiteux, and J. Krüger, *Adv. Enzyme Regul.*, **7,** 149 (1969); B. Hess, A. Boiteux, H. G. Busse, and G. Gerish; *Adv. Chem. Phys.*, **29,** 137 (1975).
25. A. Goldbeter and R. Lefever, *Biophys. J.*, **12,** 1302 (1972).
26. R. Lefever, in *Instabilities and Phase Transitions*, T. Riste, Ed., Plenum, New York, 1975.
27. S. R. Caplan, A. Naparstek, and N. J. Zabusky, *Nature*, **245,** 364 (1973).
28. J. J. Tyson, *J. Chem. Phys.*, **62,** 1010 (1975).
29. Th. Erneux and M. Herschkowitz-Kaufman, *Biophys. Chem.*, **3,** 345 (1975).
30. Th. Erneux and M. Herschkowitz-Kaufman, *J. Chem. Phys.*, **66,** 248 (1977).
31. A. T. Winfree, in *Mathematical Aspects of Chemical and Biochemical Problems and Quantum Chemistry*, Vol. 8, *SIAM-AMS* Proceedings, D. S. Cohen, Ed. American Mathematical Society, Providence, R. I., 1974.
32. R. Lefever, M. Herschkowitz-Kaufman, and J. Wm. Turner, *Phys. Lett.* **60A,** 389 (1977).
33. N. Kopell and L. N. Howard, *SIAM*, **52,** 291 (1973).

LIST OF INTERVENTIONS

1. Intervention of Hess

Experimentally, in chemical systems a variety of dynamic states can be observed resulting from thermodynamic and kinetic conditions as defined as a prerequisite for the evolution of dissipative structures. Today we distinguish the following states: (1) the maintenance of multiple steady states with transitions from one to another, (2) rotation on a limit cycle around an unstable singular point resulting in oscillations of concentrations (several limit cycles are possible), (3) sustained oscillations coupled to diffusion resulting in chemical waves.

A typical chemical system is the oxidative decarboxylation of malonic acid catalyzed by cerium ions and bromine, the so-called Zhabotinsky reaction; this reaction in a given domain leads to the evolution of sustained oscillations and chemical waves. Furthermore, these states have been observed in a number of enzyme systems. The simplest case is the reaction catalyzed by the enzyme peroxidase. The reaction kinetics display either steady states, bistability, or oscillations. A more complex system is the ubiquitous process of glycolysis catalyzed by a sequence of coordinated enzyme reactions. In a given domain the process readily exhibits continuous oscillations of chemical concentrations and fluxes, which can be recorded by spectroscopic and electrometric techniques. The source of the periodicity is the enzyme phosphofructokinase, which catalyzes the phosphorylation of fructose-6-phosphate by ATP, resulting in the formation of fructose-1,6 biphosphate and ADP. The overall activity of the octameric enzyme is described by an allosteric model with fructose-6-phosphate, ATP, and AMP as controlling ligands.

In a series of experiments we have tested the type and range of entrainment of glycolytic oscillations by a periodic source of substrate realizing domains of entrainment by the fundamental frequency, one-half harmonic and one-third harmonic of a sinusoidal source of substrate. Furthermore, random variation of the substrate input was found to yield sustained oscillations of stable period. The demonstration of the subharmonic entrainment adds to the proof of the nonlinear nature of the glycolytic oscillator, since this behavior is not observed in linear systems. A comparison between the experimental results and computer simulations furthermore showed that the oscillatory dynamics of the glycolytic system can be described by the phosphofructokinase model.

These studies demonstrate the general mechanism of synchronization of biochemical systems, which I expect to be operative in even more complex systems, such as the mitochondrial respiration or the periodic activity of the slime mold *Dictyostelium discoideum.* As shown in a number of laboratories under suitable conditions mitochondrial respiration can break into self-sustained oscillations of ATP and ADP, NADH, cytochromes, and oxygen uptake as well as various ion transport and proton transport functions. It is important to note that mitochondrial respiration and oxidative phosphorylation under conditions of oscillations is open for the source, namely, oxygen, as well as with respect to a number of sink reactions producing water, carbon dioxide, and heat.

Some time ago we recorded an oscillatory state of a population of cells

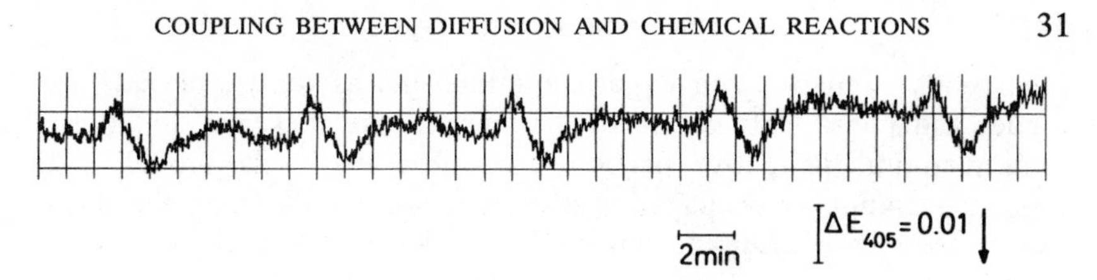

Fig. 1. Oscillation of light scattering in *Dictyostelium discoideum* cell suspension.

of the slime mold *Dictyostelium discoideum* as shown in Fig. 1. This state is part of the overall life cycle of this species and initiates a process of cellular aggregation. In a center of an aggregation territory, cells begin to produce periodically cyclic AMP, which is ejected into the extracellular space, generating chemical pulses. Then a pulse of cyclic AMP is bound by a receptor site of a neighbor cell, which triggers a chemotactic response of this cell in the direction of the incoming signal and furthermore the synthesis of an amplified cyclic AMP pulse, which again is ejected toward the periphery of the territory. Recently, a variety of periodic functions, such as the periodic activity change of the membrane-bound adenylate cyclase were identified. Furthermore, a model was suggested to describe these phenomena on the basis of a limit cycle state of the enzyme adenylate cyclase coupling to various cellular functions.

References

1. B. Hess and A. Boiteux, *Ann. Rev. Biochem.*, **40,** 237–258 (1971).
2. A. Boiteux and B. Hess, *Faraday Symp.*, **9,** 202–214 (1975).
3. B. Hess, A. Boiteux, H. G. Busse, and G. Gerisch, in *Membranes, Dissipative Structures, and Evolution*, G. Nicolis and R. Lefever, Eds., Interscience-Wiley New York, pp. 137–168.
4. G. Gerisch and B. Hess, *Proc. Nat. Acad. Sci.* (U.S.), **71,** 2118–2122 (1974).
5. A. Boiteux, A. Goldbeter, and B. Hess, *Proc. Nat. Acad. Sci.* (U.S.), **72,** 3829–3833 (1975).

2. Intervention of Goldbeter

I would like to comment on the theoretical analysis of two systems described by Professor Hess, in order to relate the phenomena discussed by Professor Prigogine to the nonequilibrium behavior of biochemical systems. The mechanism of instability in glycolysis is relatively simple, as it involves a limited number of variables. An allosteric model for the phosphofrucktokinase reaction (PFK) has been analyzed, based on the activation of the enzyme by a reaction product. There exists a parameter domain in which the stationary state of the system is unstable; in these conditions, sustained oscillations of the limit cycle type arise. Theoretical predictions compare with experimental results as to the period, amplitude, and phase shift of the oscillations. In the presence of diffusion, spatiotemporal dissipative structures are observed in the model, in the form of standing or propagating concentration waves, when the dimension of the system is supracellular [for a review see A. Goldbeter and G. Nicolis, *Progr. Theor. Biol.*, **4,** 65 (1976)].

A similar model has been proposed for the cyclic-AMP (cAMP) signaling system in the slime mold *Dictyostelium discoideum* [A. Goldbeter and L. A. Segel, *Proc. Nat. Acad. Sci.* (U.S.), **74,** 1543 (1977)]. Here the model is based on the activation of adenylate cyclase by extracellular cAMP. As in the PFK reaction, it can be shown that a parameter domain corresponds, far from equilibrium, to sustained oscillations in both intra- and extracellular cAMP. Near the oscillatory domain, there exists a region of parameter values for which excitability in the adenylate cyclase reaction can be demonstrated: as in the experiments with *D. discoideum* suspensions (G. Gerisch et al., in *Cell Interactions in Differentiation*, L. Saxen and L. Weiss, Eds., Academic, New York, 1977), the system responds to a small extracellular cAMP change by synthesizing a large pulse of intracellular cAMP. Both the oscillation and the relay of cAMP pulses play an important role in *D. discoideum* differentiation.

3. Intervention of Babloyantz

Professor Prigogine showed us the wide variety of phenomena that may appear in a nonlinear reaction-diffusion system kept far from thermodynamic equilibrium. The role of diffusion in these systems is to connect the concentrations in different parts of space. When the process of diffusion is approximated by Fick's law, this coupling is linear in the concentration of the chemicals.

It is possible to show that when the different parts of a system are connected by nonlinear interactions, one can again obtain oscillation in concentrations, patterns of chemical substances in space, and wave propagation. These phenomena are important in some biological problems when the reaction-diffusion mechanisms cannot give an adequate description of the system. Morphogenetic fields and neural networks are examples of such systems.

Theories that account for pattern formation in a morphogenetic field, as a result of reaction-diffusion processes, must assume the existence of at least two small diffusable molecules throughout the field. These hypotheses can be relaxed if one considers that the concentration of morphogenetic substances is altered in each cell via nonlinear interactions between cell surface receptors.

In neural networks the interactions between neurons are highly nonlinear. Models based on the concept of "dissipative structures" can be constructed that account for spatiotemporal patterns in epileptic seizures.

3.1 Intervention of Ter Minassian

Kamo et al. [*Biochim. Biophys. Acta*, 367, 1 and 11 (1974)] have shown that nonionic sugars modify the zeta potential of slime mold cells. Aggregation of colloids is related to their surface charge and their surface potential. This fact shows evidence of long-range electrostatic interactions controlled by metabolic reactions taking place at the membrane and able to modify the composition of the membrane medium interface. In this process the diffusion is not relevant, as indicated by Mrs. Babloyantz.

4. Intervention of Ubbelohde

Several mathematical features in the analysis presented by the authors are particularly notable:

1. The boundary conditions used make no reference to membrane characteristics.
2. An evident analogy emerges between their equations with a time variable and those in which various transformations replace this by a space variable. This prompts the suggestion that when the latter indicate a "node in space," this could be analogous to a membrane with respect to neighbor regions of space.
3. What are the specific *functions* of membranes for systems behavior; do they mainly serve to maintain "nodal" boundaries?

4.1 Intervention of Prigogine

Boundary conditions in the systems I have considered primarily ensure the maintenance of a state of nonequilibrium with respect to the environment; their relation with the size and form of the reaction medium, with the chemical and transport processes, determines the nature and properties of the dissipative structures that occur. These boundary conditions are imposed once and for all and do not couple with surface effects or electrostatic interactions. Such a coupling is likely to be the source of self-organization processes also, but was not the object of my talk.

4.2 Intervention of Ubbelohde

Do the equations point to any minimum size of system capable of showing critical behavior?

4.3 Intervention of Prigogine

Yes, generally there is such a minimal size. Its value depends on the specificities of the system considered, i.e. more precisely on the values of some relevant kinetic constants and diffusion coefficients appearing in the kinetic equations.

4.4 Intervention of Hess

The boundary conditions for the development of spatial ordering depend on the critical length of the system as well as its chemical activities. Here it is interesting to remember the actual cellular size and the density of its active enzyme components. With an average concentration of the total soluble enzyme protein in the order of 10^{-3} *M* in yeast cells such as *Saccharomyces carlsbergensis* I computed an average distance between protein units between 40 and 50 Å. With this spatial distribution at hand it is possible to estimate the transit time describing the time of transfer of the product of one enzyme from its product-releasing site to the substrate-binding site of the next enzyme in a metabolic pathway as glycolysis. For a small molecule like pyruvate a transit time between collision on the active site in the range of a few μsec is computed. This time is short in comparison to the turnover numbers of the glycolytic enzyme system in the range of 100 per second per binding site. Thus the packing of the glycolytic enzymes in this cellular type is sufficiently dense to avoid any time delays by diffusion of intermediates (see B. Hess, "Organization of Glycolysis: Oscillatory and Stationary Control," in *Rate Control of Biological Processes*, Vol. 27, Cambridge University Press, Cambridge, 1973, pp. 105–131) and the specific size for instability might not be reached. I should add that some time ago [Hess, B, Boiteux, A. and Busse, H. G., *Adv. Chem. Phys.* **29,** 137 (1975)] we reported the observation of spatial ordering of glycolytic oscillation in a cell-free extract of yeast, a phenomenon still under investigation (A. Boiteux and B. Hess, unpublished observations).

In the context of this discussion it is important to consider that the dimensionality of the diffusion process is of greatest importance for an evaluation of boundary conditions. Whereas in the case of glycolysis of *Saccharomyces carlsbergensis* the occurrence of three-dimensional diffusion of molecules of small molecular size is still a valid assumption, in membrane-bound systems two-dimensional or one-dimensional diffusion should be considered, a concept that was discussed early by Bücher, T., *Adv. Enzymol.* **14,** 1 (1953) as well as by Adam, G. and Delbrück, M., in *Structural Chemistry and Molecular Biology*, A. Rich and N. Davidson eds., Freeman, San Francisco (1968). In membrane-bound transport processes, depending on the number of interacting molecules, the rate of diffusion, and their diffusion coefficient as well as their chemical reactivity, a spatial distribution of molecules not only in the direction through the membrane, but also at the surface of the membrane is possible. Here a great number of parameters such as different surface tensions of lipids and proteins, charge distribution, and location of proteins on or within the membrane are important. One would guess that a hexagonal structure would be the most likely one. Also it is important to note that the size of the space unit (e.g., a cell or a mitochondria) has to be much greater than the average protein diameter.

Recently we investigated the problem of the occurrence of two-dimensional diffusion of cytochrome *c* on the mitochondrial membrane surface. It is suggested that cytochrome *c* will be attracted by the negative electrostatic potential on the surface of the phospholipid membrane. During its residence time on the surface, two-dimensional diffusion will occur with a relatively high diffusion coefficient, owing to the mobility of the phospholipids in the membrane, as discussed by Dr. McConnel. Statistically, a certain area of the membrane can be covered by the substrate, producing a highly efficient sink for substrate molecules round each target oxidase molecule. The encounter distance for the reaction between the enzyme and the nonspecifically bound substrate may then be represented by the diffusional distance, which is related to the two-dimensional diffusion coefficient and the dissociation constant of the bound substrate. A radius of diffusion of 126 Å was well fitting the kinetic reaction velocity constants and in accord with the distance between cytochrome *c* oxidase molecules in the order of 400 Å [Roberts, H. and Hess, B., *Biochim. Biophys. Acta* **462,** 215 (1977)]. Finally, I would like to add that the mechanism of the energy-dependent spatial distribution of ions and protons across a membrane and on the surface of membranes—even periodically changing in opposite direction in oscillation mitochondria (see above)—is under investigation in many laboratories.

5. Intervention of Lefever

Concerning the characteristic order of magnitude of inhomogeneities in spatial stationary dissipative structures, the estimates in the case of glycolysis range between 10^{-2} and 10^{-4} cm. This suggests that if spatial patterns due to glycolysis exist and are biologically relevant phenomena, they are likely to be of importance at a supracellular rather than cellular level. From a practical point of view, this characteristic length of inhomogeneities is rather small and makes experimental observations difficult compared with nonbiological systems in which spatial patterns have been reported (like the Belousov–Zhabotinski reaction). I may also stress that the conditions that allow for stationary spatial patterns are, in general, more demanding than those that lead to chemical oscillations or wave propagation phenomena. For example, it is clear that in a one-dimensional medium (as some artificial membrane systems), oscillations may result from the coupling of a single chemical process with diffusion. Spatial patterns would require at least one other distinct chemical process. If we take the case of glycolysis, all the ingredients required on the kinetics for time order are found in the reaction of PFK with ATP and ADP. To see a spatial pattern it would be necessary to include at least one more step, possibly the adenylate-kinase reaction, into the system.

6. Intervention of Thomas

1. The equations written by Dr. Prigogine, by taking Dirichlet conditions, are by themselves equations of a membrane system.

2. In the described systems the length is not a critical point by itself. The critical parameter is the Thiele modulus (including length, enzyme density, and diffusion properties). This parameter is a dimensionless one.

7. Intervention of Williams

I wish to ask about the concept of a unit of length in a biological cell. My concern is with the numerical value to be used for the diffusion coefficient. Most of biological space is heavily organized even in a single cell, and therefore a diffusion constant is not a simple property. To put my question simply, it is easiest, however, to consider enzyme sites that are otherwise identical in two situations (1) with water between them, (2) with a biological membrane between them. Is it not the case that the "unit of length" is quite different, for the diffusion in (1) is virtually free diffusion whereas in (2) the diffusion is constrained most probably as a series of activated hops?

8. Intervention of Caplan

It may be worthwhile discussing the papain oscillator in some detail, since it was quoted as an example by Professor Prigogine and has the advantage of simplicity. This will show directly how "membrane boundary conditions" can be introduced and will also relate to the point of Dr. Lefever that more than one chemical reaction is required to give rise to a dissipative structure. The rate of hydrolysis R of an ester substrate by papain and similar proteolytic enzymes, giving rise to an acid as one of the products, is well known to have a bell-shaped pH dependence (see Fig. 1); clearly the high-pH limb of the curve corresponds to a typical autocatalytic region. A membrane containing immobilized papain presents a one-dimensional diffusion-reaction problem, which can be solved

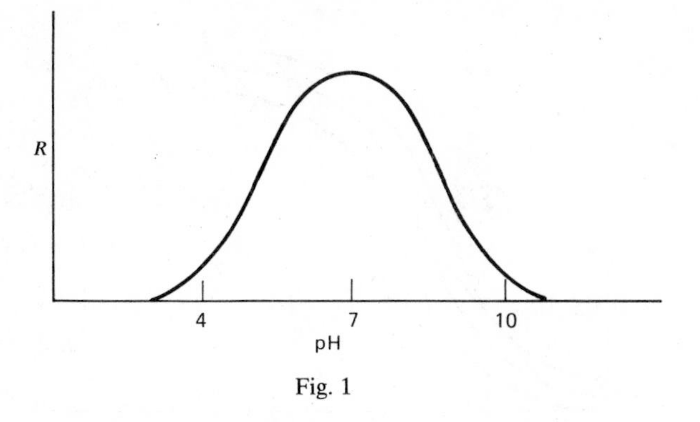

Fig. 1

completely since the kinetic parameters are all known [Caplan, S. R., Naparstek, A. and Zabusky, N. *Nature*, **254,** 364 (1973)]. It is useful, however, to consider a "zero-dimensional" system in which diffusion and reaction are spatially separated (Hardt, S., Haparstek, A., Segel, L. A. and Caplan, S. R., in *Analysis and Control of Immobilized Enzyme Systems*, D. Thomas and J. P. Kernevez, eds., North-Holland, New York, 1975, p. 9). Computer simulations of the behavior of the enzyme membrane are in fact based on such a procedure. Suppose diffusion of substrate as well as hydrogen ions takes place through a membrane that separates an outer reservoir from a well-stirred inner compartment containing the enzyme. The boundary conditions are given by the substrate and hydrogen ion concentrations in the reservoir, S_0 and H_0, respectively (see Fig. 2). In the autocatalytic pH range ($pH_0 \simeq 10$) one obtains the type of behavior shown in Fig. 3 (the diagram is entirely schematic). The curves corresponding to a series of S values in the inner compartment represent rate of reaction as a function of hydrogen ion concentration in that compartment, whereas the straight line represents rate of diffusion of hydrogen ion J_H *out* of the enzyme

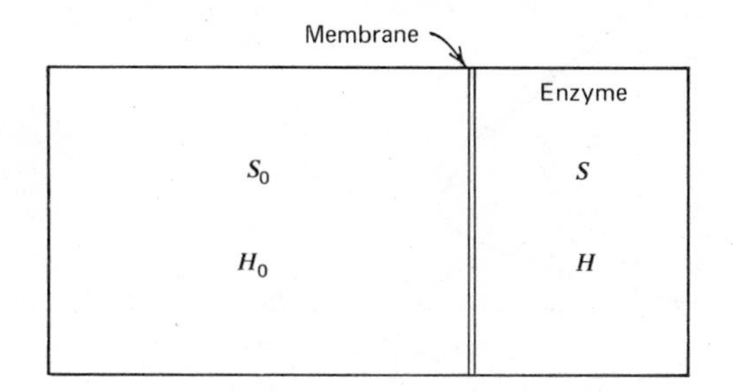

Fig. 2.

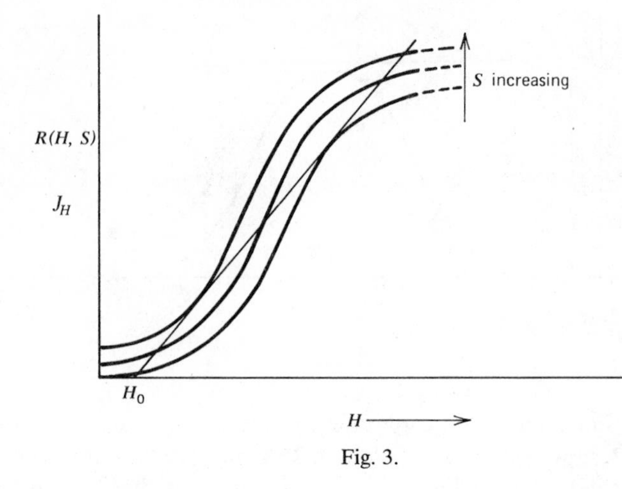

Fig. 3.

compartment into the reservoir. In a certain range of S three steady states are possible. Plotting all such steady-state reaction rates as a function of substrate concentration in the enzyme compartment yields a sigmoid relationship, as shown in Fig. 4. We superimpose on the sigmoid curve lines representing the rate of diffusion of substrate *entering* the inner compartment J_S for three different values of S_0. It is seen that when S_0 is low (S_0^1), stable "reaction-controlled" states are achieved; when S_0 is high (S_0^3), stable "diffusion-controlled" states are achieved. In an intermediate

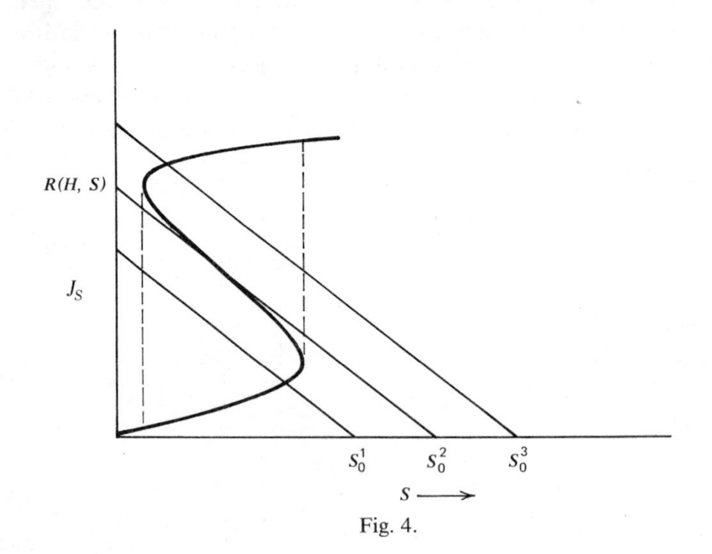

Fig. 4.

range (S_0^2) no stable state is achieved and the system describes a limit cycle. Now the "one-dimensional" or continuous system may be considered analogous to a series of infinitesimally narrow enzyme-loaded compartments separated by diffusion barriers through which substrate and hydrogen ion must pass en route between the outer surface and the interior. A convenient practical arrangement is that devised by Dr. Thomas for experimental purposes [Naparstek, A., Thomas, D. and Caplan, S. R. *Biochim. Biophys. Acta,* **323,** 643 (1973)], in which the membrane is coated on a glass pH electrode. In this configuration the "inner" region of the membrane is bounded by an impermeable wall (see Fig. 5). At low values of S_0 all the hypothetical compartments of the membrane remain in reaction-controlled steady states, whereas at higher values of S_0 the inner compartments may pass into diffusion-controlled steady states with a drastic decrease in local pH accompanied by the formation of a sharp pH "front" within the membrane. At intermediate values of S_0 this front oscillates back and forth through the thickness of the membrane, as the local compartments describe individual limit cycles shifted in phase with respect to one another. The oscillation constitutes a spatiotemporal dissipative structure. The participation of a second chemical reaction in this phenomenon is readily seen on writing down the equations of the system. Electrical effects are ignored on the basis of an excess of supporting electrolyte. For the substrate we have

$$\frac{\partial S}{\partial t} = -R(\mathrm{H}, S) + D_S \nabla^2 S$$

However, on writing the corresponding relation for hydrogen ion the formation of water must be taken into account. The equilibrium between water, hydrogen ion, and hydroxyl ion ($K_w = \mathrm{H} \cdot \mathrm{OH}$) may be considered to be established instantaneously,

$$\frac{\partial \mathrm{H}}{\partial t} = R(\mathrm{H}, S) + D_{\mathrm{H}} \nabla^2 \mathrm{H} - \frac{\partial \mathrm{H_2O}}{\partial t}$$

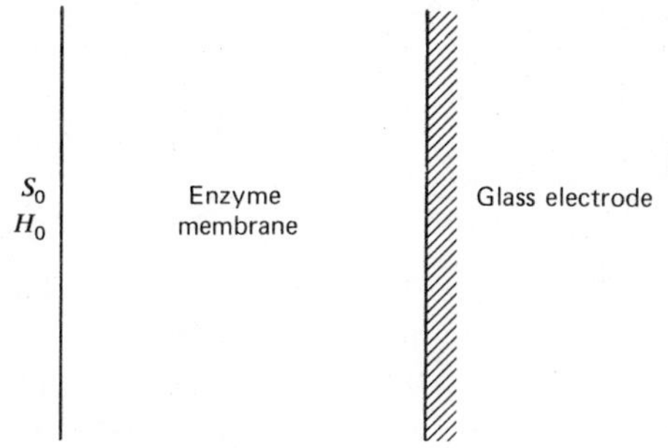

Fig. 5.

In addition,

$$\frac{\partial \text{OH}}{\partial t} = D_{\text{OH}} \nabla^2 \text{OH} - \frac{\partial \text{H}_2\text{O}}{\partial t}$$

Combining the second and third equations by assuming that the diffusion coefficients are approximately equal and introducing the quantity $\Delta = \text{H} - \text{OH}$, we have

$$\frac{\partial \Delta}{\partial t} = R(\text{H}, S) + D_{\text{H}} \nabla^2 \Delta$$

The first and last equations then constitute the pair governing the system.

It may also be useful to note that for simple Michaelis–Menten kinetics the Thiele modulus we have been discussing is given, in conventional notation, by

$$\frac{\Delta x}{2} \left(\frac{V_{\max}}{K_M D_S} \right)^{1/2}$$

where D_S is the diffusion coefficient of the substrate and Δx is the thickness of the membrane.

9. Intervention of Ross

I would like to make some further remarks on the topic of spatial and temporal structures in chemical instabilities. There are three topics I wish to discuss briefly, topics of possible interest to this conference. First, let me say that Professor Prigogine introduced me to the field of instabilities in reaction-diffusion systems more than ten years ago by discussing with me Turing's pioneering work and the plans for his incisive research. This is an appropriate occasion to thank him for that and many helpful talks on the subject since then.

As a first example let me discuss a particularly simple system. [A. Nitzan and J. Ross, *J. Chem. Phys.*, **59,** 241 (1973); C. L. Creel and John Ross, *J. Chem. Phys.*, **65,** 3779 (1976).] Consider a glass cell containing an equilibrium mixture of NO_2 and N_2O_4. The equilibrium is maintained by the simple reaction

$$N_2O_4 \underset{k_2}{\overset{k_1}{\rightleftarrows}} 2NO_2 \tag{1}$$

Now irradiate this glass cell with light of a frequency absorbed by NO_2 but not by N_2O_4 (such as the green light of an argon ion laser). Photons absorbed by the NO_2 are turned after about 100 μsec into heat, which increases the temperature. However, since the rate coefficients k_1 and k_2

in (1) are temperature dependent, as the temperature increases there is a shift to a higher NO_2 concentration. Thus we have a simple, positive feedback mechanism among the variables of concentration and temperature. The kinetic equations for the system

$$\frac{dx}{dt} = 2k_1(C - \tfrac{1}{2}x) - 2k_2x^2$$

$$\frac{dT}{dt} = \alpha I_0 x - \beta(T - T_e) - \lambda\left(\frac{dx}{dt}\right)$$

where x is the NO_2 concentration, C is the concentration of N_2O_4 if all the gas were in the form of N_2O_4, I_0 is the incident light intensity; T is the temperature of the system, and T_e is the temperature of the surroundings. The remaining symbols are constants and are related to the parameters of the system. Notice that we are displacing the system far from equilibrium by the imposition of light, that is, a flow of photons across the boundary of the system. At steady state this energy input is balanced by a flow of heat across the boundary of the system. The energy input is proportional to the concentration of NO_2 and as that energy input is increased we increase the density of photon absorbers. The two kinetic equations are nonlinearly coupled because of the exponential dependence of the rate coefficients and temperature. There are three constraints imposed on the system and under our experimental control: the intensity of the laser light, the boundary temperature, and the total amount of gas in the system.

We obtain calculated values of steady-state concentrations from solutions of the kinetic equations with the terms dx/dt and dT/dt set equal to zero (see Fig. 1). If the boundary temperature is more than 250°K, the calculated "isotherm" (used here to refer to constant boundary temperature) is monotonic and for each value of the light intensity there is only one stationary stable state. If the boundary temperature is lowered below a critical value we find the possibility of multiple stationary states, two of which are stable and one unstable. The sets of steady stable states form branches that have positive slopes. For such a situation if we vary the light intensity but maintain at each point steady state, then on variation of light intensity we expect chemical hysteresis. This is shown in Fig. 2, where on increasing the light intensity we expect to trace out the path *AFBCD* whereas on decreasing the light intensity we expect to trace out the path *DCEFA*. We need to note that so far we have considered only a homogeneous system and only the results of the deterministic kinetic equations for such a system (i.e., in the absence of fluctuations).

Experiments on this system have been completed and confirm the

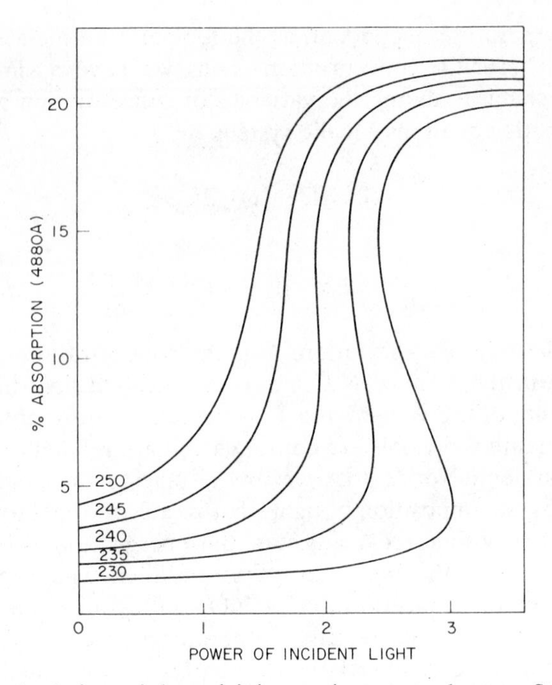

Fig. 1. Numerical solutions of deterministic equations at steady state. Steady-state NO_2 concentration is given by percent absorption and plotted versus incident laser light intensity for various values of the boundary temperature T_e (degrees Kelvin).

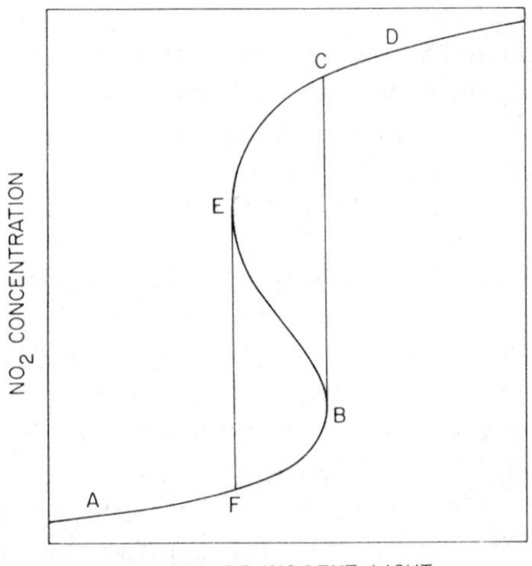

Fig. 2. Plot similar to Fig. 1. According to the deterministic equations on increasing the light intensity, the curve *AFBCD* is traced; on decreasing the light intensity, the curve *DCEFA* is traced.

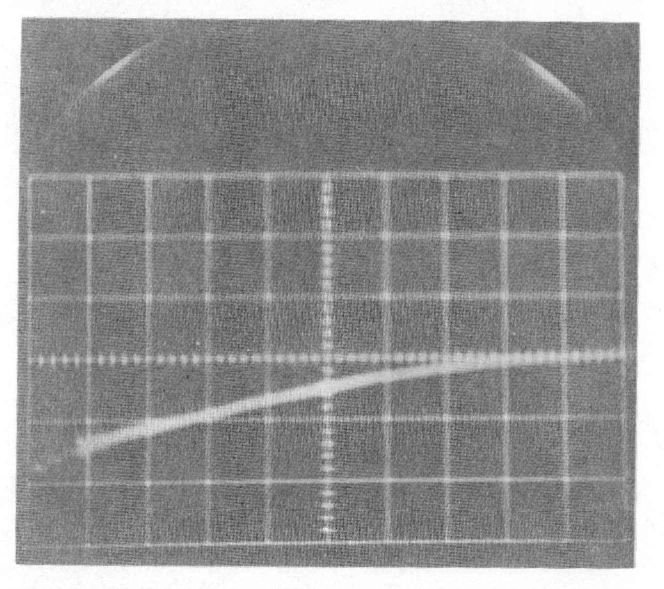

Fig. 3. Percent absorption (10% per division), a measure of NO_2 concentration, versus incident laser light intensity (0.254 W per division) for the boundary temperature of 253°K. The light intensity is increased from 0 to 2.5 W and back again to 0 Watt in 2 sec. Note the superposition of the up and down sweep of the laser light intensity. (From Creel and Ross, op. cit.)

predictions of the presence of multiple stationary states and chemical hysteresis. In Fig. 3 there is shown a plot of percent of absorption that is proportional to the steady-state concentration of NO_2 versus light intensity. The boundary temperature is 253°K and the single trace results from both increasing and decreasing the light intensity. The experiment in Fig. 3 also shows that there are no artificial sources possibly contributing to a hysteresis effect. In Fig. 4 the boundary temperature is 240°K and now on increasing the light intensity along the abscissa the lower curve is traced out, whereas on decreasing the light intensity the upper curve is obtained. The time necessary for scanning up and down in both Figs. 3 and 4 is about 2 sec. On increasing the scanning time to more than a minute the hysteresis loop disappears. The time necessary to attain steady state is about 40 to 50 ms. The time necessary for transition from one stable branch to another is longer than that but certainly less than a few seconds.

The observation of hysteresis depends on the time scale of the fluctuation around a stable branch as compared to the time scale of the transition from one branch to the other. So far we have only talked about

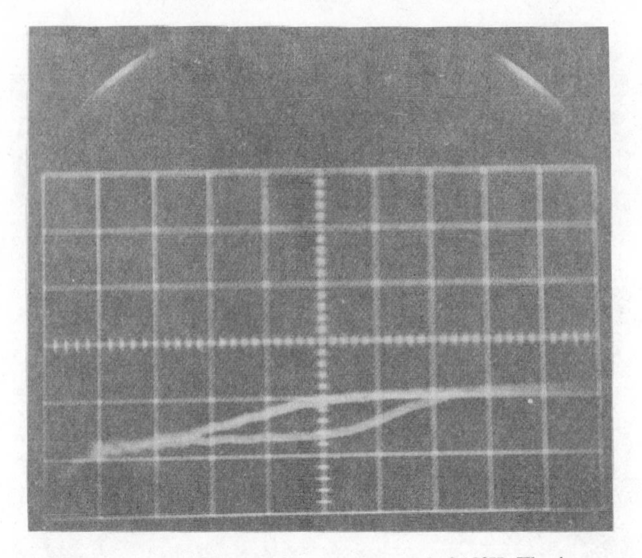

Fig. 4. Same as Fig. 3 except that boundary temperature is 243°K. The increase of the laser light intensity sweeps out the lower curve; on decreasing the laser light intensity the upper curve is traced. (From Creel and Ross, op. cit.)

deterministic equations. Let us see what the calculated hysteresis loop looks like for the solutions of the deterministic equations to which have been added stochastic terms corresponding to fluctuations. The calculated results are shown in Figs. 5 to 7, for small, medium, and large fluctuations. Note in Fig. 5 the increased amplitude of the fluctuations near the marginal stability point. In Fig. 6 the calculations were performed three times for the same external constraints and same mean strength of fluctuations. The results for the three calculations are quite different and the difference arises from the imposed fluctuations.

It is very likely that the physical system does not make a transition from one branch of stable states to another in a homogeneous way. Much more likely, inhomogeneous fluctuations lead to nucleation centers of a stationary state of one branch within a region of a state of the other branch. The observation made then constitutes an average over the volume of the system, and from the calculations shown in Figs. 5 to 7 we expect a gradual transition from one branch to another as the light intensity is varied through the marginal region, and this is confirmed by experiment. The observation of the gradual transition in the experimental system allows the inference that, as expected, fluctuations become larger as the marginal stability point is approached. The fluctuations in far-from-equilibrium systems are of fundamental importance in understanding the

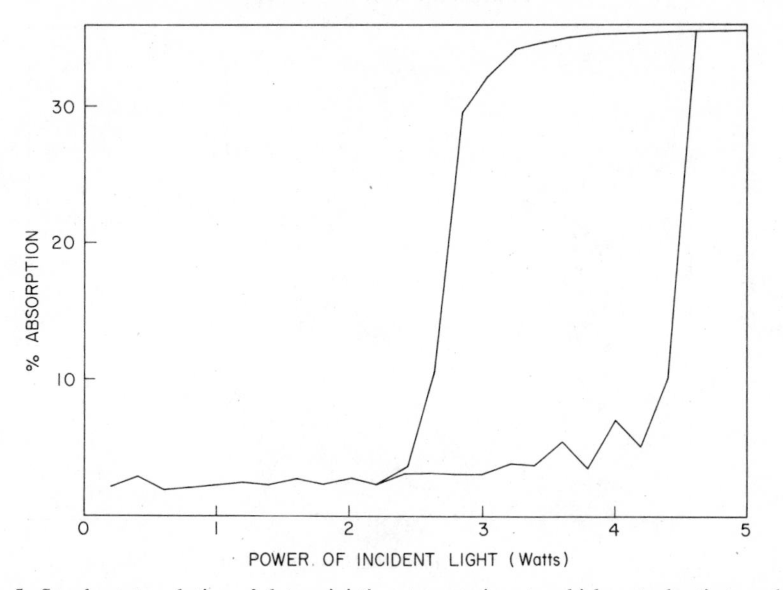

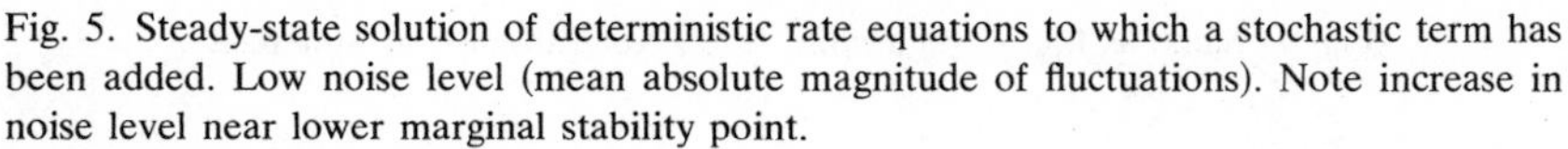

Fig. 5. Steady-state solution of deterministic rate equations to which a stochastic term has been added. Low noise level (mean absolute magnitude of fluctuations). Note increase in noise level near lower marginal stability point.

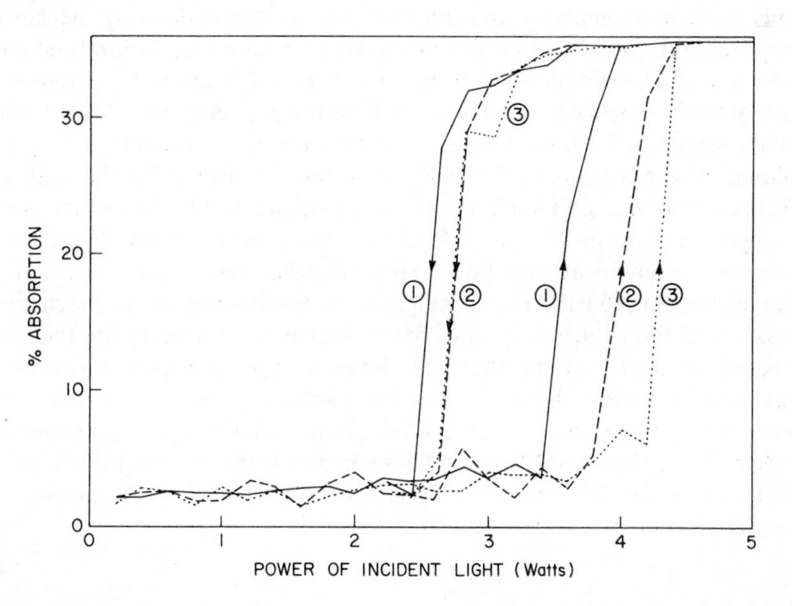

Fig. 6. Same as Fig. 5: intermediate noise level. Traces 1, 2, and 3 are for identical constraints and noise level.

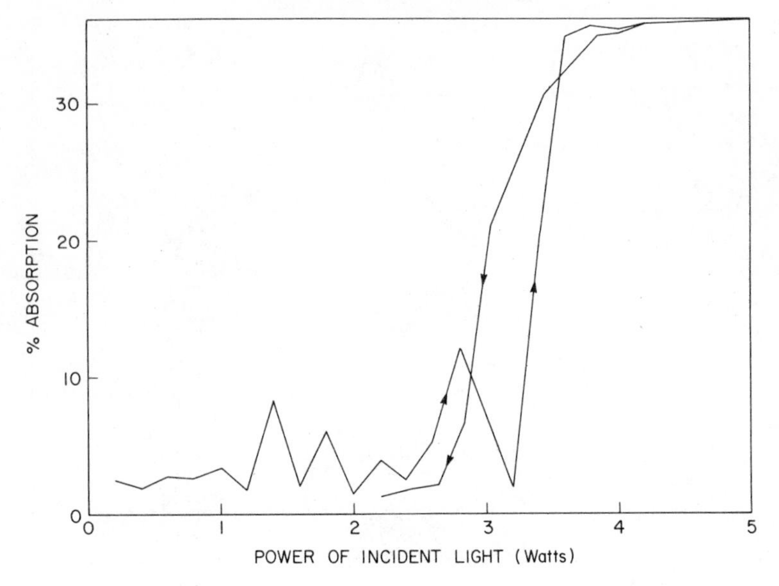

Fig. 7. Same as Fig. 5: high noise level. Note absence of hysteresis.

behavior of such systems. We are presently designing experiments to study both homogeneous and inhomogeneous fluctuations by measuring temporal and spatial concentration and temperature correlation functions.

As a second example, which may be of possible interest to studies of autocatalytic reactions occurring on immobilized enzymes, let me consider a system that consists of a set of bulk reactions occurring in a given volume. The reactions in the bulk are stable. Immersed in the bulk are localized catalytic sites such as electrodes or membranes on which occur reaction mechanisms that may lead to various types of instabilities. For the case of a single site one finds expectedly that oscillations, for instance, depend not only on the rate coefficients of the homogeneous mechanism but also on the diffusion coefficients of reactants coming to the localized site and products leaving that site. Imagine now, however, an array of such localized sites; the array may be regular or random. Then we find interesting possibilities of cooperative phenomena depending on the site density [E. K. Bimbong-Bota, A. Nitzan, P. Ortoleva, and John Ross, *J. Chem. Phys.*, **66,** 3650 (1977)]. Consider the following simple example of reactions occurring in the bulk of the system:

$$X \xrightarrow{k} D$$

$$B \xrightarrow{k} Y \longrightarrow C$$

At each of the localized sites there occurs a product-activated enzyme catalysis

$$X + E'' \rightleftharpoons E'$$
$$X + E' \rightleftharpoons E$$
$$Y + E \rightarrow E + X$$

The system can now be driven far from equilibrium by increasing the concentration of the species B. In Fig. 8 we plot the concentration of B necessary for obtaining an instability versus the catalytic site density (a^{-1}). This calculation can be carried out analytically for the case of the diffusion coefficient of one species being quite different from that of the other. The maximum in the curve as shown corresponds to a cooperative effect of the catalytic sites at a given density. Note furthermore that if the density of sites is changed at constant concentration of species B (the line 1–2), then we traverse three regions. The region around point 1 has a single stable stationary state, the region from 1′ to 2′ has three stationary states, and the region around point 2 has again one stationary stable state.

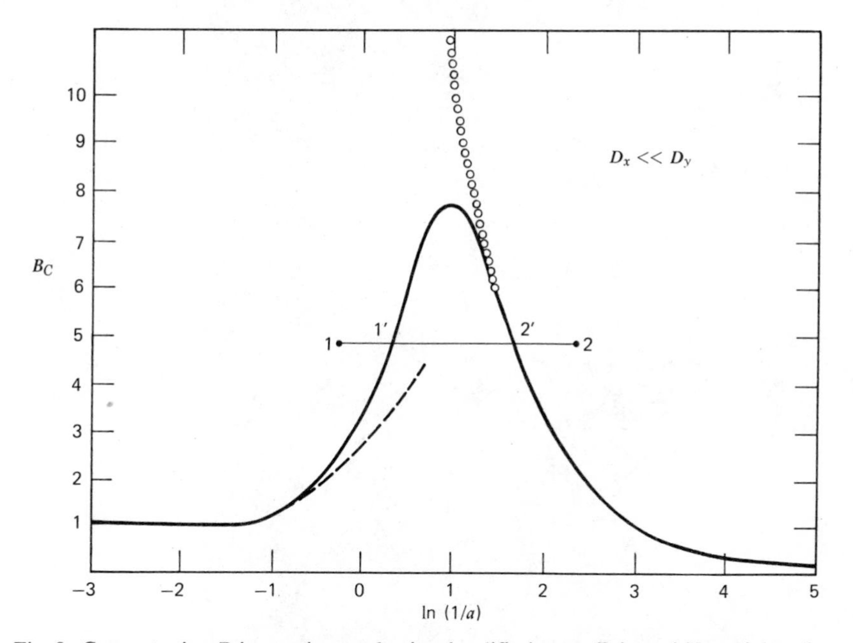

Fig. 8. Concentration B in reaction mechanism for diffusion coefficient of X much less than that of Y, necessary to achieve instability in a system in which the autocatalytic mechanism occurs on a one-dimensional array of local sites, versus logarithm of site density a (solid line). Dashed and dotted lines are for the isolated and continuum site limit, respectively.

Finally, I wish to turn to spatial periodic structures as exemplified by Liesegang rings. Such a structure is shown in Fig. 9 [taken from M. Flicker and John Ross, *J. Chem. Phys.*, **60,** 3458 (1974)]. A solution usually in gel) of one soluble electrolyte, say $Pb(NO_3)_2$, is put in a test tube and above it is placed a solution of another electrolyte, say KI. Periodic rather than continuous precipitation of PbI_2 occurs as shown in the figure (the dark rings). There have been many explanations of this phenomena, all of which require the existence of gradients of concentrations of Pb^{++} and I^- in opposite directions. However, a number of experimental facts do not fit these explanations. One of these experiments was performed by M. Flicker, following a suggestion of P. Ortoleva. She made a sol of PbI_2, a colloid in which the colloidal particles are initially small (the sol is nearly colorless), added gelatin, and placed the solution in a Petri dish (Fig. 10). After a short period the gel sets. Although the sol is

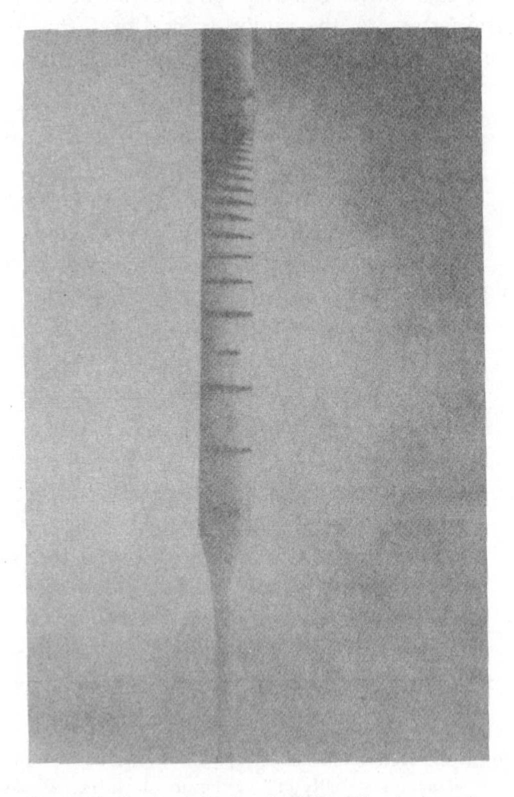

Fig. 9. Liesegang ring periodic precipitation in PbI_2. The dark regions are yellow rings of PbI_2.

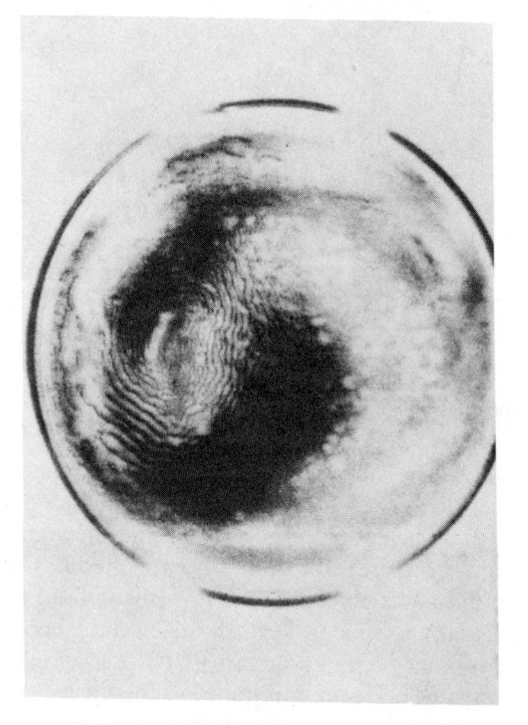

Fig. 10. Inhomogeneous growth of colloid in initially homogeneous PbI_2 sol.

initially homogeneous and care is taken to assure the absence of gradients of temperature and other variables, the homogeneous colloidal growth becomes unstable, and macroscopic inhomogeneities are formed. Note the existence of an inherent characteristic "wave length" in the wavelike patterns. In the cited reference Flicker and I proposed an autocatalytic reaction mechanism (involving the electric double layer around each colloidal particle) that, when coupled with diffusion, leads to the possibility of inhomogeneous instabilities and the formation of spatial structure.

A more general analysis is possible of the Liesegang ring problem as well as other similar kinetics of phase transitions (R. Lovett and John Ross, unpublished; P. Ortoleva, private communication). Suppose we mix a solution of $Pb(NO_3)_2$ and KI. After a very short time interval we have a colloid with most of the colloidal particles containing small numbers of PbI_2 molecules. Let c_1 be the concentration of monomer of PbI_2, which we choose as one quantity of interest. The concentration of c_1 is in quasi-equilibrium with ions and higher polymers during the slow growth of colloidal particles. The growth may be described crudely by a single

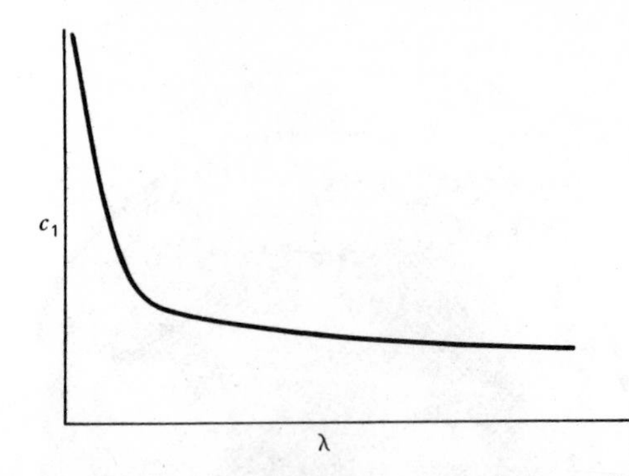

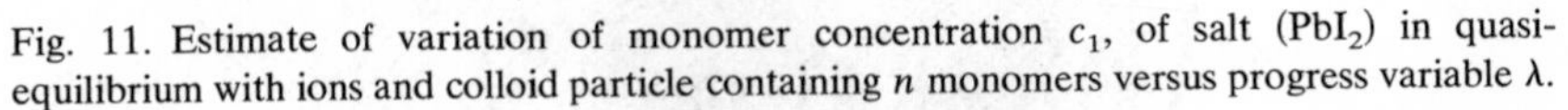
Fig. 11. Estimate of variation of monomer concentration c_1, of salt (PbI_2) in quasi-equilibrium with ions and colloid particle containing n monomers versus progress variable λ.

progress variable λ. The variation of c_1 with the progress variable is then as shown in Fig. 11, where the initial rapid decrease in c_1 (or ions) corresponds to the nucleation process and the slower decrease to the aging process. Now picture an initially homogeneous situation in which the progress variable λ is constant in space (the colloid has grown to the same extent uniformly in space). Perturb this homogeneous growth by the addition of a monomer at one point so that the monomer concentration c_1 is altered from the uniform condition as shown by the dotted line in Fig. 12. On a time scale very short compared to the time scale of growth of the colloid the added monomer concentration disappears rapidly to increase the average colloidal size, that is, the progress variable (Fig. 13). This *increase* in λ at one point *decreases* c_1 in quasi-equilibrium at that point (Fig. 14). That decrease brings about diffusion of c_1 from the neighboring regions (indicated by arrows in Fig. 14), thus further advancing λ.

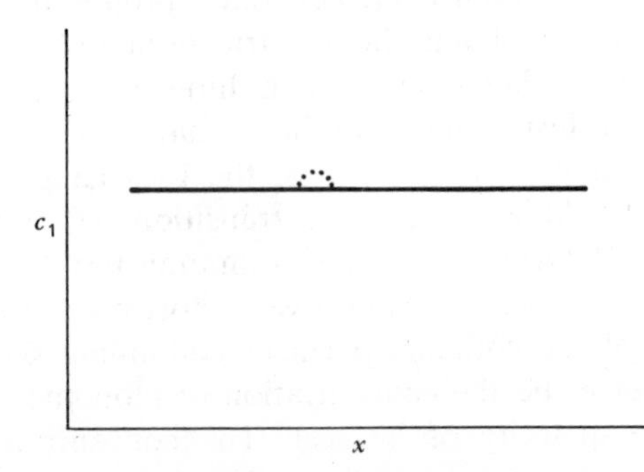

Fig. 12. Variation of monomer concentration versus spatial coordinate x. The dotted line indicates an external inhomogeneous perturbation (addition) of monomer.

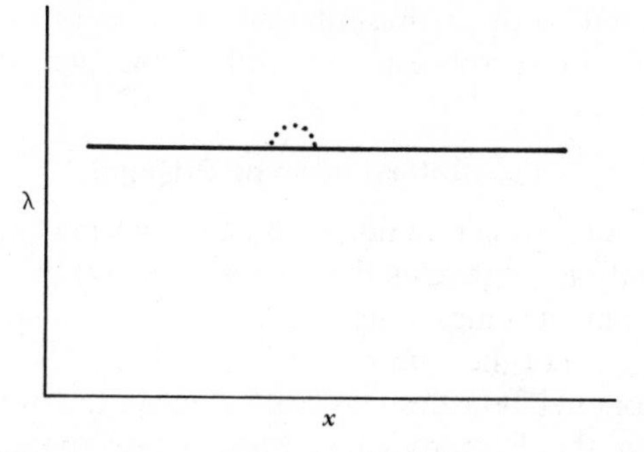

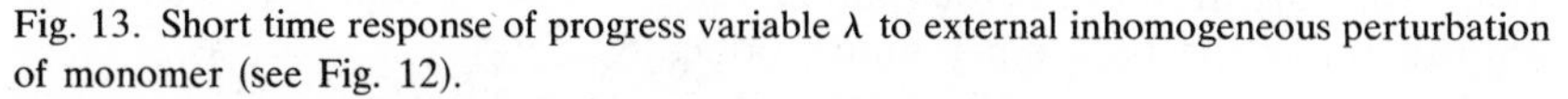
Fig. 13. Short time response of progress variable λ to external inhomogeneous perturbation of monomer (see Fig. 12).

This simple, nonlinear mechanism predicts spatial instabilities under quite general conditions for a variety of precipitation and phase change processes.

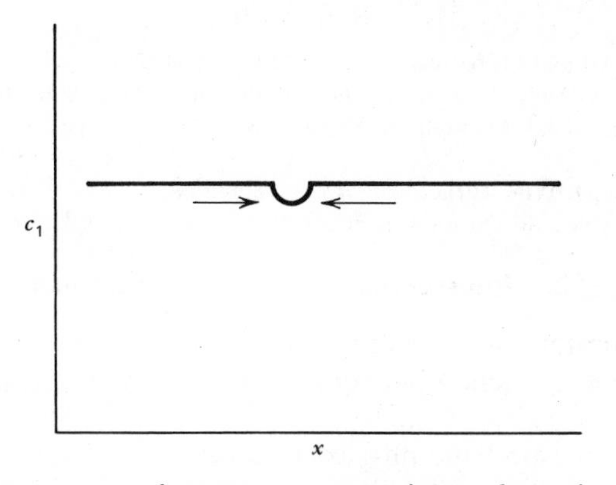

Fig. 14. Short time response of monomer concentration to change in progress variable shown in Fig. 13.

10. Intervention of Mayer

My impression is that most in vitro experimentally observed periodic reactions have recurrence times of the order of minutes, seldom more than an hour. This may be an artifact of choice by the experimenter to choose conditions to have this convenient range; I do not know. Those of us here who have just arrived from California are suffering from our built-in 24 hour clock. There is at least one well-known perodicity of four weeks affecting a large fraction of humans. Probably the periodicity of one year observable in most temperate-zone organisms is a reaction to external stimuli such as temperature or relative day-night lengths, but I have been assured, without presentation of the evidence, that there are unexplained built-in period responses with recurrence times of order of a year.

11. Intervention of Prigogine

In the vicinity of bifurcation points, fluctuations play an important role. A stochastic analysis [1–3] reveals that as the instability is approached from below, the system develops long-range spatial correlations and a concomitant divergence of the variance of fluctuations. The calculation of the divergence exponents indicates classical values in the presence of fluctuations varying in the hydrodynamic range.[4] For spatial fluctuations of shorter range the problem remains open.

Numerical simulations illustrating the role of diffusion in the onset of instabilities have been carried out by Hannusse.[5] They confirm the modification of macroscopic behavior by local fluctuations that introduce such phenomena as delays or metastable states.

References

1. H. Lemarchand and G. Nicolis, *Physica*, **82A,** 521 (1976).
2. C. Gardiner, K. McNeil, D. Walls, and I. Matheson, *J. Stat. Phys.*, **14,** 309 (1976).
3. G. Nicolis, M. Malek-Mansour, K. Kitahara, and A. Van Nypelseer, *J. Stat. Phys.*, **14,** 414 (1976).
4. G. Nicolis and J. Wm. Turner, *Physica A*, submitted.
5. P. Hanusse, Thèse de Doctorat ès Science, University of Bordeaux.

12. Intervention of Delmotte and Chanu

In our laboratory we perform experiments adapted to the study of diffusion chemical reaction phenomena under maintained nonequilibrium conditions.

In order to interpret the physicochemical steps of retinal transduction as well as membrane excitability, we analyze macroscopic properties of membranes within biological components. Such membranes separate two aqueous ionic phases the chemical compositions of which are kept constant separately. The total flux through the membrane is directly deduced from the counterbalance quantities in order to maintain the involved thermodynamical affinities constant. From such measurement, we calculate the dynamical membrane permeability. This permeability depends not only on membrane structure but also on internal chemical reactions.

It must be emphasized that all the flux measurements are performed in steady state *strictly*—in other words, without any time evolution of clamped constraints. This method was especially chosen to avoid any delay between the setup of new constraints and stationary regimes, and above all to *separate* perfectly possible constraint variations from intramembranar kinetics. The membranes used are from 0.07 to 0.09 mm thick and geometrical coefficient V/S of our apparatus is about 70 mm (V being the half-cell volume and S the effective area of the membrane). The experimental conditions are precisely controlled and permit to realize: optimalized homogenization, initial input of solutions of each reference composition, imposed stationarity with accuracy of better than 1% of concentration levels. In addition, we underline the necessity of pursuing the time observation during five to eight hours while keeping the transmembrane fluxes quite constant.

This approach has been used with an ATPase of mammal cerebral cortex under sodium ionic gradient. In a gradient range corresponding to the mean biological values, we observed two stable flux regimes. The latter are laying between two spaces: one of high dynamical membrane permeability with low gradient and the other one of low permeability against higher gradients.

In the case of rhodopsin membrane under calcium gradient, the technique shows a nonlinearity of fluxes with respect to imposed gradients. We also notice some permeability variations just after rhodopsin bleaching and at last a calcium adsorption consequent to lighting performed in situ.

We can say, following these experiments, that the structural permeability loses its physical meaning. In fact, such nonlinearities between fluxes and forces and permeability variations establish the very existence of a "chemical effect," leading us to the concept of the *dynamical membrane permeability.*

References

1. *Membranes à permèabilité sélective,* Colloque DGRST, Paris, February, 1967.
2. M. Delmotte, Thèse d'Etat, *Approche dissipative des transports membranaires en biologie et mécanismes de conduction et transduction nerveuses. Méthode d'investigation ou interprétation?* Université Paris VII, 1975.
3. M. Delmotte and J. Chanu, *Ionic Transport in Artificial Biomembranes, Bioelect. Bioenerg.,* **3,** 474 (1976).
4. M. Delmotte and J. Chanu, *On the measurement of membrane potential in biology.* Colloque C.N.R.S. *Echanges ioniques transmembranaires chez les végétaux,* Rouen, July, 1976.

STRUCTURES AND ENERGETICS OF PROTEINS AND THEIR ACTIVE SITES

I. D. CAMPBELL, C. M. DOBSON, AND R. J. P. WILLIAMS

Inorganic and Biochemistry Laboratories
Oxford University
Oxford, England.

I. INTRODUCTION

It has been the custom in the last 20 years to look on protein structures through the detailed electron density maps produced by X-ray diffraction techniques. The interested biologist has to accept the models produced as he is unable to engage in the exercise of structure determination himself. Unfortunately, there is now reason to believe that although these models may give the gross features of protein structure, they are not necessarily to be trusted when structure is to be used *in detail* to discuss function. In this article we wish to draw attention to different views of the nature of a protein structure when it is examined by using different methods. We wish to discuss two aspects of protein structure. The first is the general distribution of energy in the whole protein, including rotational and vibrational states. The second is the more localized imposition of steric and electronic bond strain (the entatic state) that can arise from the protein fold but be concentrated in a small group of atoms within the whole structure.

The newly evolving ideas of a protein differ from those that can be generated from the X-ray structure mainly because they are dynamic and not static in nature. Nuclear magnetic resonance spectroscopy (nmr) is the technique that is revealing the most detail about the dynamic aspects of protein structure, and much of this article is concerned with the results of nmr studies. It is unfortunately the case that dynamic models are much the more difficult to describe and to use in a precise way, but it is very important to realize that quite a new outlook on a protein has to be developed. The dynamics of a protein structure are now known to be specific to a given protein, and the response of the protein to external changes, either physical or chemical, is therefore specific also.

The discussion of activity from X-ray data in conjunction with kinetic data is also difficult because, apart from considerations of dynamics, these techniques do not provide the essential knowledge about the energy states of given atoms or groups. It is necessary to inspect the electronic structure of at least certain regions of the protein. Methods exist for this inspection, and these include electron paramagnetic resonance, ultraviolet, circular dichroism, Raman and Mössbauer spectroscopies. The full understanding of activity can only come when the information derived from all available methods is assimilated and rationalized.

A. The Protein Fold

A protein consists of a single chain of peptide residues (or of several chains in the case of multi subunit proteins) that are covalently linked. Rotation about certain of the peptide covalent bonds is allowed, but in the folded state this rotation must be severely restricted. In this folded state new (noncovalent) types of bonding—hydrophobic interactions, electrostatic interactions, and hydrogen bonds—are generated. The spontaneous folding of a protein clearly shows that the folded state is thermodynamically favored over the unfolded state. Moreover, there is often, though not always, a *unique* fold, and all other folds are therefore considerably less stable. The folding energy is given by the sum of a large number of interactions between side chains of amino acids and between main-chain segments, as well as between side chains and main chains and between both these elements and water. This sum must have a negative total free energy change relative to the same sum in the unfolded form. It is not essential to the folding problem that some parts of the sum give a negative and other parts a positive contribution so long as the whole sum is decidedly negative. It is naive to suppose, for example, that just because a negatively charged center such as $-CO_2^-$ and a positively charged center such as $-NH_3^+$ have a high electrostatic interaction in a vacuum, they have a negative (favorable) free energy of interaction in a folded protein relative to their individual interactions with water in the unfolded form. It may well be the case that they are so heavily solvated that the reaction

$$\text{(unfolded)}\ -CO_2^- + -NH_3^+ \rightarrow [-CO_2^- \cdots -NH_3^+]\ \text{(folded)}$$

has a positive (unfavorable) free energy change. After all, Na^+Cl^- crystals dissolve in water. Perhaps the best assessment of the partial free energy (ΔG) values that contribute to folding is that involved in the association of very hydrophobic groups, for here analysis has been possible in comparable small nonpolar model systems. When we turn to polar group association all that can be said is that unlike charge association is known not to be invariably favorable and that it is usually more favorable than the association of like charges. Thus the folding of a protein may be controlled more by energy of charge repulsion than by that of charge attraction.

The overall result of the folding of a globular protein is that a tightly packed structure is produced. Examination of many X-ray structures shows that, in general, hydrophobic groups are found in the center of the folded structure, whereas polar groups are found on the outside of the structure. However, this is not always the case, and there are often specific regions (e.g., those for binding or catalysis) where the arrangement of groups appears to be thermodynamically unfavorable. Most proteins possess certain regions for binding other molecules, and enzymes possess active sites for catalysis.

So far the rigidity of the protein structure has not been considered. Although globular proteins are tightly packed and appear to crystallize in specific and unique structures, the forces holding the folded structure intact are not necessarily strongly directional. Thus the concept of a protein fold by no means excludes the possibility of a degree of motion within the structure.[1–5] Thus the molecule may undergo breathing or deforming motions resulting from the possession of thermal energy. The definition of the extent and nature of these motions is a major aim of recent studies of protein structure. Thus the full concept of a structure requires not only knowledge of spatial coordinates for each atom, but also definition of the time dependence of these coordinates. In addition, it is well known that conformational changes may be induced by binding of molecules (both large and small) to the protein, by the change in the charge of specific groups in the molecule by ionization of side chains, or by oxidation and reduction of redox centers in the molecule.

Despite these considerations it is conventional to represent proteins by rigid ball-and-stick models derived from X-ray studies. A somewhat different approach to the mechanics of a protein has been developed by Hopfield,[2] who likens the fold to a series of springs connected by arms. The arms are relatively nonadjustable, whereas the springs are adjustable according to their loading (binding) and can take up strain either locally or extensively. This physical image is related to the idea of the entatic state of Vallee and Williams[3] (see below), but it provides a clear, dynamic mechanical picture of energy distribution, and it is easy to build working models. These models are useful in illustrating the properties of proteins that are quite hidden in the rigid ball-and-stick structures. However, a protein is very versatile, and the spring model does not consider other types of motion. For example, one could picture protein mechanical motions as sheets of structure slipping over one another (cf. friction-driven action) or twisting around one another (cf. screw gears).

B. Energetics of Reactions

Even though there is limited accumulation of protein structures available at present, it is clear that protein folding does force some groups into unfavorable relationships. There are clusters of positively or negatively charged residues in some proteins (e.g., lysines on the surface of cytochrome *c* or carboxylates in the active site of lysozyme). Because such groups have a lowered thermodynamic stability, they have a greater drive toward a change of state than is conventional for such side chains in unfolded proteins. If this instability in a local region of a folded protein differentiates between the ground state of a group and a form of it that is required in any one of its reaction paths, in favor of the latter, then the group will be anomalously reactive in the protein. Now such an adjustment in reactivity is not just a property of the interaction of the side chains or other groups of the protein in the immediate vicinity of the reacting group, but a consequence of the fold to a minimum total and cooperative free energy of the whole protein. Thus a center can be activated by the fold energy and, conversely, alteration of its energy status (e.g., its charge) alters the energy of the whole molecule, the fold stability. The state of a group placed in an active site that is so activated has been called an entatic state.[3,6] Note that this term is best used to emphasize that the strain is due to folding (in which the microenvironment is only a part), reactivity is then usually due to a destabilized ground state. The effect is then such that it can be observed in the *destabilized ground state* by spectroscopic or other means. This concept throws some light on the requirement of high molecular weight for an enzyme, as a considerable entatic energy can arise only from a favorable fold overall; therefore a large number of favorable interactions elsewhere in the enzyme are required to support the raised activity of the active site. Small peptides cannot achieve this activation except through very rigid, chemically enforced, covalent cross-linking. Such cross-linking may not be desirable in enzymes, however, because, in addition to a special attacking site, they must have mobility in the active site region, for reasons that are elaborated later but that can be summarized as follows.[1,4] An enzyme is a reactant during the catalytic steps (not just a catalyst) and therefore must flow through a reaction energy pathway. This pathway requires adjustment of the chemical groups involved (protein side chains usually), and therefore great rigidity that would permit only one of the many required steps would hinder the overall reaction. The very fact that such conformational mobility is required in catalysis may have allowed the evolution of certain allosteric effects, where information is passed through a protein via internal bond adjustment, which can regulate not only the structural states of proteins, but also the rates of interconversion of the states, so modulating catalysis and thence metabolism. Note that this modulation can act on any step in the reaction and need not be seen either in the ground state with bound reactants or in the final state with bound products. Later in this article we look again at allosteric changes and ask to what degree running changes in *internal* structures of proteins are required (see under cytochrome *c*).

From the preceding discussion it is clear that for some functions a relatively high degree of rigidity, especially of the main chain fold, is required, for activity is associated with a relatively well-defined spatial distribution of side chains. In other regions of the enzyme such definition is not required, and in the case of proteins that are not enzymes, their very function may require mobility (e.g., consider contractile proteins). Again, modulation of activity by allosteric mechanisms is often considered in terms of through-bond conformational changes over wide regions of a protein (e.g., in hemoglobin). Thus limited mobility may be required over small regions of some proteins, but considerable mobility over wide regions may be needed in others.

Thus the energetics of a protein are intimately connected with structure and structural mobility. It is therefore important to investigate these structural features. In the following section we consider the type of structural information that different techniques can provide. Then general conclusions from these techniques are used to discuss the nature of protein structure as it is now understood. Finally, the structure energetics and functions of a few specific proteins are briefly discussed.

II. METHODS FOR STUDYING PROTEIN STRUCTURE

Even a small protein consists of many hundreds of atoms. The primary aim of a structure determination is to place each of these atoms in space by determining the coordinates of each atom relative to a fixed coordinate system. The only techniques that can provide detailed information of this type are based on diffraction methods, and X-ray diffraction is the primary method. We must consider the meaning of the structures derived from these studies. Now X-ray diffraction can only be used to determine structures in crystals. One is, however, really concerned with the structure of proteins in solution, and it is therefore necessary to examine the difference between structure in the solid state and the solution state. We consider differences in general between these two states, and then differences in specific cases. To perform structural studies in solution, spectroscopic methods must be used. These methods are quite different from diffraction methods, being concerned with specific absorption or emission of energy between energy levels within the molecule. Depending on the method chosen, different aspects of the molecule may be studied. Only one method, nuclear magnetic resonance, can provide information that relates to the overall detailed structure of the molecule. Signals can be detected from virtually every atom in the molecule, and techniques exist for relating the nmr spectrum of the molecule to its three-dimensional structure. Thus this method is capable of allowing direct and detailed comparison of the overall fold of a protein in the solid and solution states.[5] In addition, nmr can provide information about the time dependence of the protein structure, which is not possible with diffraction methods. It is also possible to use nmr to follow both large and small conformational changes caused by chemical or physical perturbations to the protein.

Other spectroscopic methods cannot provide the same overall picture of protein structure or dynamics. However, they can give information about specific atoms or groups in the protein. In order to gain detailed information from these techniques, it is generally necessary to study metal atoms, which in some cases are a natural part of the protein and in other cases may be specifically introduced. Techniques such as UV, visible, Raman, and epr spectroscopies provide information about the metal atom and its environment, which is concerned both with structural features and with energetic features.

For different reasons all these physical techniques for studying proteins are most powerful when the protein contains metal atoms. We shall therefore consider metal ion probes and isomorphous replacement in general before turning to individual techniques.

A. Metal Ion Probes and Isomorphous Substitutions

In crystallography, heavy atom derivatives are required to solve the phase problem before electron density maps can be obtained from the diffraction patterns. In nmr, paramagnetic probes are required to provide structural parameters from the nmr spectrum. In other forms of spectroscopy a metal atom itself is often studied. Now many proteins contain metal atoms, but even these metal atoms may not be suitable for crystallographic or spectroscopic purposes. Thus isomorphous substitution has become of major importance in the study of proteins. Isomorphous substitution refers to the replacement of a given metal atom by another metal that has more convenient properties for physical study, or to the insertion of a series of metal atoms into a protein that in its natural state does not contain a metal. In each case it is hoped that the substitution is such that the structural and/or chemical properties are not significantly perturbed.

TABLE I. Isomorphous Replacements

Metal	Protein	Replacement
Zn(II)	Carboxypeptidase Carbonic anhydrase Alcohol dehydrogenase	Co(II),[a] Mn(II), Cd(II), Cu(II)
Ni(II)	Urease	Co[II], Fe(II)
Ca	Troponin, parvalbumin	Ln(III) Mn(II)
Mg	Kinases (e.g., phosphoglycerate kinase)	Mn(II)[a] Ln(III)
K	Many enzymes	Tl(I)
Na		None
Cu(II)	Blue proteins, etc.	Co(II), Cd(II) Zn(II)
Fe(III)	Hemeproteins, e.g., cytochromes Myoglobin, hemoglobin, peroxidases	Co(III), Cu(II), Co(II), Mn(II) or (III)

[a] The underlined element is the one of choice.

Table I lists isomorphous replacements for various metalloproteins. Consider zinc enzymes, most of which contain the metal ion firmly bound. The diamagnetic, colorless zinc atom contributes very little to those physical properties that can be used to study the enzymes. Thus it has become conventional to replace this metal by a different metal that has the required physical properties (see below) and *as far as is possible* maintains the same activity. Although this aim may be achieved to a high degree of approximation [e.g., replacement of zinc by cobalt(II)], no such replacement is ever exact. This stresses the extreme degree of biological specificity. The action of an enzyme depends precisely on the exact metal it uses, stressing again the peculiarity of biological action associated with the idiosyncratic nature of active sites. (The entatic state of the metal ion is an essential part of this peculiarity.) Despite this specificity, the replacement method has provided a wealth of information about proteins that could not have been obtained by other methods. Clearly, there will often be a compromise in the choice of replacement. Even isomorphous replacement that should retain structure will not necessarily retain activity at all. However, it has become clear that substitutions can be made for structural studies where the substituted protein is inactive (e.g., in the copper proteins and the iron-sulfur proteins). It is also possible to substitute into metal coenzymes. Many studies have been reported of the substitution of iron in hemoglobins, peroxidases, and cytochromes. There is the distinct possibility of following the properties of chlorophylls containing metal substitutions such as copper or cobalt for magnesium. It would be useful to study vitamin B_{12} where cobalt(III) has been substituted by Cr(III) or even Mn(II).

Similar probes can be used in the study of proteins that do not require a metal ion. One example has been the study of lysozyme, where the probes used have been the lanthanide(III) cations and various anions (Table II). Additionally, for X-ray crystallographic purposes, a whole range of metal atoms has been incorporated into proteins. For these X-ray studies, it is only necessary that the metal atom has high atomic weight and that major structural perturbations are not caused.

TABLE II. Examples of Nmr Probes for Nonmetalloproteins

	Type of perturbation		
Class of probe	Relaxation	Shift	Blank
Cationic	Gd(III)	Ln(III)[a]	La(III), Lu(III)
	Mn(II)	Co(II)	Ca(II), Zn(II)
	$[Cr(NH_3)_6]^{3+}$	$[Ru(NH_3)_6]^{3+}$	$[Co(NH_3)_6]^{3+}$
Anionic	$[Cr(CN)_6]^{3-}$	$[Fe(CN)_6]^{3-}$	$[Co(CN)_6]^{3-}$
	$Gd(edta)^-$	$Ln(edta)^-$	$La(edta)^-$
			$Lu(edta)^-$
Neutral	Nitroxide spin label	—	Reduced spin label
	Gd(nta)	Ln(nta)	La(nta)
			Lu(nta)

[a] Ln refers to paramagnetic lanthanide cations other than Gd.

B. Physical Techniques

In this section we do not wish to discuss in detail the operation of any given physical technique. Rather, we try to summarize the nature of the procedures that specific techniques require and to make general comments on the type of information each gives.

1. *X-ray Crystallography*

The techniques of X-ray crystallography are well known. The method has been applied to proteins of molecular weight in excess of 100,000, but the difficulties in signal-to-noise ratios and in resolution of the diffraction patterns increase considerably as the molecular weight increases (cf. nmr below). In fact, the procedures and results of X-ray diffraction studies of proteins differ considerably from those of small molecules. In small molecules the bond lengths and angles can be defined closely by direct determination of the electron density map, in which the constituent atoms can be identified. In a study of a protein, electron density maps can generally be obtained only by use of heavy atom derivatives to interpret the observed diffraction patterns. The resolution of protein electron density maps is generally not at the atomic level. Thus the procedure is to fit the known primary sequence of the protein to the electron density map in such a way that the best fit of the electron density to a single structure is obtained. If any part of the molecule exists in more than one conformation or if molecular motion is occurring more rapidly than the time for data collection (which may be several days), then the single structure obtained is not strictly correct but represents some form of mean or average conformation. Nevertheless, it is undoubtedly the case that X-ray crystallography can define the overall fold of the protein correctly and that a model derived from these studies is likely to be a good description of the atomic positions in the crystal, allowing for the preceding provisos. Clearly, the goodness of fit depends on the degree of motion and existence of different conformers, which cannot yet be defined easily in the crystal. The most difficult regions are undoubtedly those of the surface, and the electron density for some surface groups, chiefly arginine and lysine, is often diffuse, indicating an ill-defined structure for these groups.

A common procedure in crystallography is the soaking of crystals in solutions containing inhibitors or other molecules known to bind to the protein in solution. These molecules can diffuse to binding sites in the crystal. Two difficulties can arise here. First, the packing of the crystals may restrict binding in certain regions of the protein. This is the case in at least one crystal form of lysozyme where part of the active site cleft is blocked by an adjacent protein molecule. Second, if binding requires or produces a conformational change, this may be resisted by forces in the crystal. This sometimes results in cracking the crystals, but in other cases results in binding the molecule in a different mode or even in a different position in the crystal to that which occurs in solution. Because of these objections, binding studies in the crystalline state must be shown to correlate well with similar studies in solution.

The procedure of crystallography is concerned with the determination of the overall shape of the molecule by a low resolution study. Then the structure is refined to produce a high resolution structure. How far this refinement procedure can proceed is not certain. It is important to discover to what extent the solution structure resembles the X-ray structure. This comparison can be done at two levels, that of the overall fold and that of the high resolution structure. At the highest resolution it is clearly only an approximation to describe molecular structure even in the crystal by a single structure. What is not clear is the level of approximation and how this differs from crystal to solution states.

2. *Nuclear Magnetic Resonance*

The nmr spectrum of a molecule consists of a plot of the absorption of radiofrequency radiation against energy. The energy of a transition, which occurs between nuclear energy levels, is dependent on the environment, both chemical and physical, of the nucleus. The advantage of nmr over other forms of spectroscopy (e.g., UV) lies in the very-high-resolution, narrow linewidths and in the fact that the area under each peak is strictly proportional to the number of nuclei giving rise to the absorption at that energy. Thus nmr is directly analytical for virtually every atom in the protein.

It is inappropriate here to discuss details of nmr spectroscopy. However, the full possibilities for structural studies arise from the use of probes that bind to the molecule under study and perturb the nmr spectrum. These probes are generally paramagnetic species, in particular the lanthanide cations. In the nmr spectrum separate signals arise from each nucleus in the molecule, provided that the nucleus possesses a nonzero nuclear spin (e.g., ^{1}H, ^{13}C, ^{14}N). The extent of the spectral perturbations of a given signal depends on the relative geometries of the paramagnetic species and the nucleus in question. Thus structural parameters can be obtained, in principle, for most atoms (nuclei) in a protein molecule.[5]

The use of lanthanide probes for structure determination in small molecules has been extensively emphasized. In practice, bond lengths and angles are taken from crystallographic data, and the rotation angles about single bonds are determined from the nmr studies. The results of small molecule studies have been reviewed elsewhere.[7,8]

As with X-ray crystallography, as the molecular weight of the molecule increases, the difficulties in performing structural studies increase. The lanthanide probe method has so far been applied in great detail to only one protein, lysozyme.[5] The nuclei studied were protons. Because of the complexity of the protein nmr spectrum, it was necessary to apply techniques to resolve and simplify the spectrum and to assign the observed resonances to specific nuclei in the molecule. Table III lists some details of these methods, with references.

At present the nmr methods are not sufficiently advanced to allow structural studies in solution to be carried out in the detailed manner in which X-ray diffraction studies are carried out. In fact the nmr procedure differs in principle from the X-ray method in that an overall picture of

TABLE III. Techniques Used in Nmr Studies of Lysozyme

Resolution and simplification of spectra	Reference
(a) Paramagnetic probes	9, 10
(b) Convolution difference (resolution enhancement)	9
(c) Pulse methods	11, 12
(d) Difference spectroscopy	9, 10
(e) Deuteriation	
Assignment of spectra	
(a) To type of nucleus	10, 13
(b) To type of residue	10, 14
(c) To specific residue in sequence	5, 9, 10, 13, 14, 15

the structure is not obtained as a first stage in the structure determination. Rather, information about specific groups is obtained, and the nmr studies advance by obtaining information about an increasingly large number of groups. Thus the approach we adopted in the study of lysozyme was to compare the nmr structural data with the X-ray structure in order to discover how closely the two structures (solid and solution) correspond. As we see later, already very close similarities in the protein fold are certain, and close similarities for the side-chain conformations of many groups, particularly in the active site region, have been demonstrated. Other probes, as well as the lanthanide cations, have been used to aid these studies (see Table II). It is of course the case, and we consider this in the next section, that the nmr data cannot be fitted to a single structure in the presence of molecular motion and conformational equilibria and in this respect suffer from the same difficulties as X-ray crystallography. However, the nmr methods do allow the *detection* and in favorable cases the quantitative definition of the motion.

Thus, although the nmr methods are important for comparison of the structure of proteins in the solid state and in solution, they are also of importance in areas where X-ray crystallography can provide little information. One of these areas concerns the time dependence of protein structure, for molecular motion over a wide range of time scales can be detected. Table IV indicates the methods and references to these studies. The range of the nmr technique is from about 10^{-10} s to slower than 10 s. Thus, although more restricted than X-ray crystallography in direct structure determination, the nmr studies can check the solution structure and complement the diffraction studies once the overall similarity between the solid and solution structure is proved. In this way nmr relates the static picture of a protein structure to the kinetic data of solution chemistry.

TABLE IV. Nmr Methods for Detecting Rate Processes

Method	Approximate time scale (sec)	Reference to protein studies
Relaxation parameters	10^{-9}	11, 15–19
Exchange effects	10^{-3}	14, 20–25
Cross-saturation	10^{-1}	23, 26
Successive spectra	>10	13, 27

Nmr methods are also valuable in other areas of protein structure. For example, information about the role and extent of induced conformational changes (caused, for example, by inhibitor-substrate binding, by ionization of side chains, or by redox changes) can be obtained.

3. *Other Spectroscopic Methods*

It is not necessary here to give details of other spectroscopic methods. However, Table V summarizes the information that some of these methods can give about proteins. Valuable contributions have been made by electronic (UV) spectroscopy in determining the site symmetry and spin state of transition metal ions, often isomorphous substitutions for naturally occurring ions such as Zn^{2+}. However, epr, fluorescence, resonance, Raman, and so on, have also been used extensively, as Table V shows. All in all it is these methods that have revealed the most about the electronic states of atoms and groups in proteins.

TABLE V. Information about Metals in Proteins from Spectroscopic Techniques Other than Nmr

Type of information	Techniques
(1) Oxidation state (2) Spin state	UV/visible spectroscopy (absorption, CD, MCD, ORD) Infrared, Raman X-ray absorption (EXFAS) Mössbauer EPR
(3) Nature of ligand (4) Symmetry (ligand field parameters and distortion-strain)	UV/visible spectroscopy X-ray absorption EPR
(5) Metal—ligand distances (6) Metal—long-range distance	X-ray absorption Fluorescence EPR

III. THE NATURE OF PROTEIN STRUCTURE

In this section we wish primarily to draw conclusions about the nature of protein structure in solution. In discussing this topic in detail, we look at some differences between structures in solids and in solutions, for we wish to bring together data obtained in the two states. First, however, we look at some results for small molecules.

A. Structures of Small Molecules in Solids and Solutions

Crystallization is a specific process, and it is generally the case that a given molecule is seen in only one form in a particular crystal, although this is not universally true. The energy barrier to conformational change from one crystal form (one conformation) to another crystal form (another conformation) is a large cooperative lattice energy, and only small motions are likely to be seen in the crystals of one molecular form. The observed conformation is usually the form in which the molecule is least soluble in the mother liquor from which crystallization took place. However, once again this is not necessarily true. The value and danger of assessing conformation in solution directly from solid-state conformations has been discussed for small molecule systems,[28,29] and we look at some of these arguments later.

In solution the barriers to conformational change are often small, even when the molecule has a built-in restriction on motion. Conformational barriers calculated for isolated molecules in the gas phase that reveal the nature of some of these barriers are likely to be good reflections of the real barriers in solvents such as chloroform. There usually are many conformations present at any time in such solutions and they are in equilibrium. The equilibria are likely to be much more restricted in polar media. It is very important for us to discover the extent of the equilibrium, that is, the number of conformations involved, the relative proportion of each, and the rate of transformation between them. Such a task is virtually impossible from theoretical considerations, and two major approaches using physical techniques, mainly nmr, are possible. These have been discussed in some detail for small molecules and can be summarized as follows.

1. *Method* (*a*)

The physical data from measurements on solutions are used to generate a "family" of single molecular conformations in which each single conformation reflects the physical observations well. This family covers a larger volume than a molecular model of any single conformation would give. For example, a methyl group becomes a ball and an ethyl group a cone.

This derivative of an averaged shape over a range of bond angles is the one we have adopted in conformational determinations of several small molecules.[7,8] It attempts to give an overall impression of the space occupied by a molecule that is a direct representation of the data obtained by experiment. If the physical data arise from combinations of very different individual conformations,[28] then this procedure is not expected to be possible.[7,8]

2. *Method* (*b*)

The physical data plus theoretical and/or empirical knowledge of the most likely combination of conformations are taken, and the properties of the different conformations are variables. This method of fitting is likely to give a closer reproduction of the experimental data than method (a), as more adjustable parameters are available. We consider that method (b) should be applied only if method (a) fails to satisfy the physical data using reasonable bond lengths and angles. The objection to procedure (b) is that it is impossible to inspect a very large number of contributing conformations, and therefore it is necessary to choose a limited number of conformations on the basis of crystal structures or theoretical calculations. The use of both solid state and theoretically derived conformations is hazardous, as the role of the solvent cannot be readily included. There is no need to stress that solvents can bias conformational equilibria, especially polar solvents, for this has been observed.

It seems clear that in solution conformational equilibria exist for small molecules. However, in the solid state single conformations are generally found. This undoubtedly arises from the fact that crystal forces are often greater than the differences in energy between the different conformers. Thus crystal structures of a number of penicillins differing in molecular formula only in the side chain have different puckers of the five-membered thiazolidine ring (which is remote from the side chain). Nmr studies using lanthanide ion probes show[29] that in solution there are no differences in the conformation of the different molecules. It appears from the nmr data that in solution there is rapid equilibrium between the different ring puckers, and the equilibrium populations are the same for the different molecules.[29] As the different molecules crystallize, different puckers are selected by the crystal forces in each case. Thus crystallography can define closely the conformations in the solid state, but it is incorrect to assume that these are the only conformations in solution. Nmr studies cannot define conformations as precisely as crystallography, but give a more realistic answer to the solution situation.

B. Structures of Proteins in Solids and Solutions

Having seen that differences between crystal and solution conformations can exist for small molecules, we now turn to the consideration of proteins. First, we must differentiate between those proteins that crystallize and those that do not. Proteins that crystallize are usually globular, whereas those that do not are often nonglobular and are unlikely to conform to a single conformational family. These may be compared to industrial organic polymers such as those used in the plastics industry.

For those proteins that crystallize, there are differences between their crystals and those of small molecules. For example,

1. The solvent from which the proteins are obtained usually has a very high salt content, often about 3 molar
2. The crystals contain large quantities of salt solution, sometimes greater than 50% of the total volume.
3. The derived X-ray data indicate that the solvent-salt region in the crystals is in part not crystalline, and this is consistent with fast diffusion of small molecules into the crystals.

Thus X-ray diffraction studies of proteins are likely to be more difficult than those of small molecules because the motion of molecules that we know occurs in solution could hardly be entirely damped out in the crystals of proteins. By contrast, just from this point of view, solution studies are likely to be easier for proteins than for small molecules because in a globular protein the extent of gross freedom of movement of groups is less than that found in small molecule systems. In other words, the problem of defining a molecular structure in the solid state has become more like the problem of defining the structure in solution. Thus neither nmr data in solution nor X-ray crystallography in the solid state can define a real single conformation of a protein. Because of the complexity of protein structure, it is virtually impossible to consider all the possible interconverting conformers, as method (b) of the previous section requires. Thus method (a) must be used even by X-ray diffraction to define the structure in outline; then attempts must be made to discover the extent of motion that is occurring in the molecule. Clearly, crystallography can define the outline structure better than any other technique. It is then necessary to show whether this outline structure is preserved in solution. After this, nmr and other spectroscopic methods are clearly best for examining the extent of motion in the molecule.

There is good evidence in a number of systems that the overall folding of a globular protein (the outline structure) is in fact very similar in the crystal and in solution. This evidence includes estimates of overall molecular dimensions, exposure or nonexposure of specific groups in the structure to reagents in solution, binding properties of the molecules in the two states, and measurements of activity in crystals and in solution. However, evidence of this type gives no quantitative details of the similarity of the crystal and solution folds. In order to attempt this comparison we have used nmr methods to study in particular the protein lysozyme in solution. The progress of these experiments has been reported elsewhere,[5,10] and may be summarized as follows. *Within the experimental limitations of the present state of both X-ray diffraction and nmr spectroscopy, no gross differences in the protein fold can exist.* In the nmr experiments a lanthanide metal ion was shown to bind in the active site of lysozyme to two carboxylate groups, one of which could be proved to be glu 35 (see Fig. 1). In the crystal, lanthanide cations were also found to bind in the active site, close to two carboxylate groups, one of which

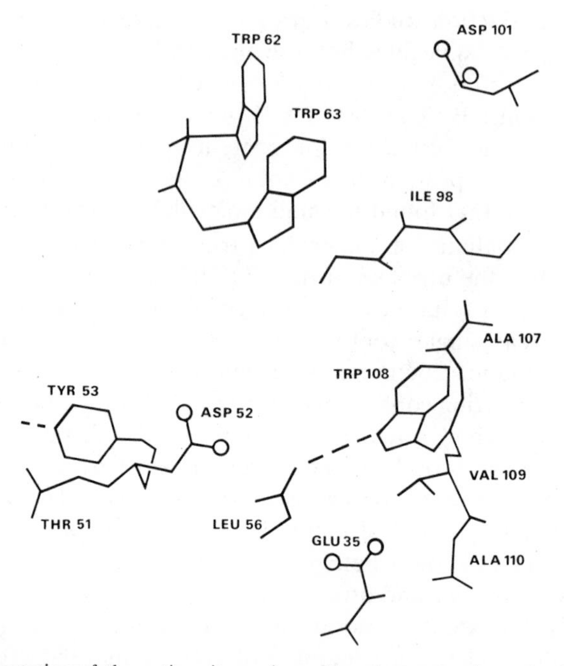

Fig. 1. Representation of the active site region of hen lysozyme, from the X-ray structure (adapted from Ref. 31). The lanthanide cations bind in the region between asp 52 and glu 35.

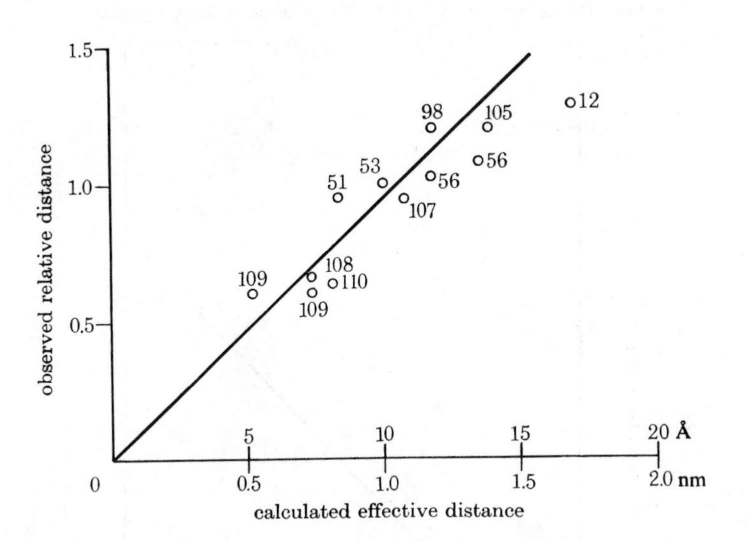

Fig. 2. Agreement between experimental distances of nuclei of lysozyme, from the bound Gd(III) (obtained from the sixth root of the observed nmr broadening in solution) and the distances calculated from the X-ray structure. The numbers in this and Figs. 3 and 4 refer to the residues in the sequence (see Ref. 5).

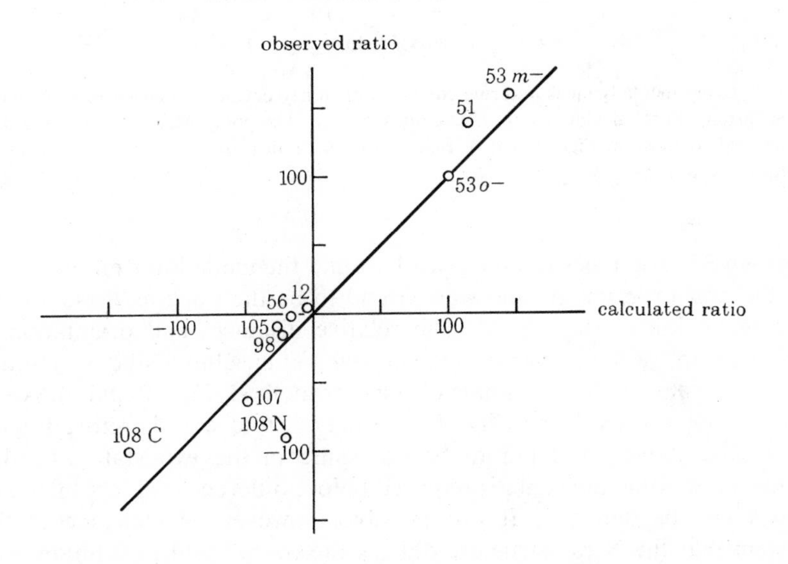

Fig. 3. Experimental dipolar shift ratios [for Pr(III)] of lysozyme nuclei plotted against values calculated from the X-ray structure, for a defined direction of the magnetic symmetry axis (see Ref. 5).

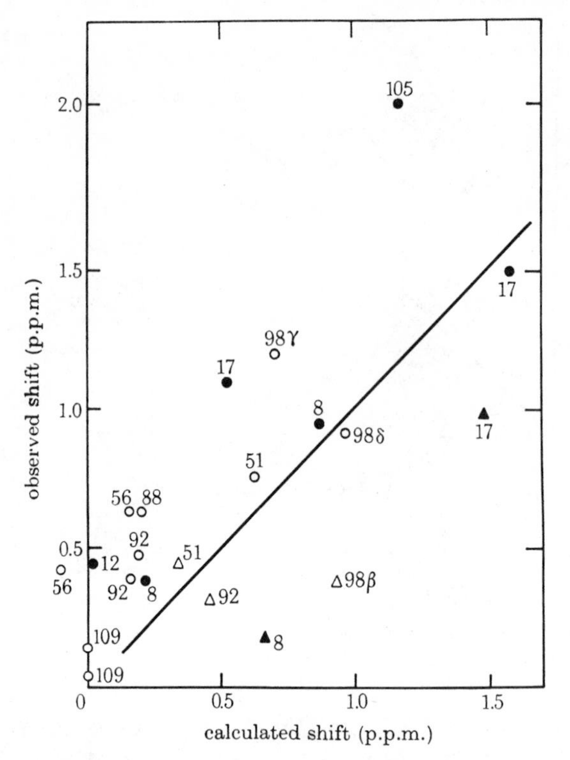

Fig. 4. Experimentally measured ring current shifts for lysozyme resonances plotted against ring current shifts calculated from the X-ray structure. The poor agreement for these data, compared to those in Figs. 2 and 3, reflects the theoretical uncertainty in the calculated values (see Ref. 5).

was glu 35. The types of amino acid around the metal ion were idenitfied in the nmr experiments and were found to be the same as those around the metal ion in the crystal. The relative distances and orientation of these groups were compared in the crystal and in solution and were found to be the same within the limits of the nmr method (Figs. 2 and 3). Along with other nmr evidence (Fig. 4), the overall fold was therefore for the first time shown in detail to be the same in the two states. Further refinement of the nmr data is required before differences of less than, say, 1.0 Å can be detected. It was possible, however, to conclude in this system that the X-ray structure defines the overall fold in solution, thus enabling further nmr studies to be made to discover the extent and rate of molecular motion in solution. We now assume that the findings for lysozyme can be generalized to a class of *globular* proteins and we therefore use the combination of nmr and X-ray data to discuss the general nature of protein structure in solution.

C. Protein Structure in Solution

It is convenient to divide the protein structure into three regions—the interior, the surface, and the active site–binding site. We look at these regions in turn.

1. *The Interior*

Let us first consider what might be expected. The internal residues of proteins are generally rather tightly packed. However, as most of the side chains are hydrophobic groups, the directional properties of the interactions are relatively poor, and the whole may be likened to a hydrocarbon solvent. The packing is quite irregular, and this part of the protein is not really constrained by lattice considerations, as would be found, for example, in the regular crystals of benzene or naphthalene. Perhaps one could liken hydrocarbon regions of a protein to molecules trapped in a foreign clathrate cage, where the cage is formed from a hydrogen bonded net of the hydrophobic residues. In clathrates there is considerable motion at 300°K, and one might expect therefore that motion exists in the interior of proteins.

Nmr data have provided quite definitive evidence for this motion (see Fig. 5). The most straightforward examples of this concerns the aromatic side chains of tyrosine and phenylalanine. In all protein studies some or all of these groups are able to flip, or to rotate, about the C_β—C_γ bond. Evidence for the flipping comes from the apparent equivalence of signals from the two ortho and the two meta protons of these groups. In a rigid molecule these signals would be distinct, because the environment of the four protons would be different. The rate of flipping varies from one group to another in each protein. In some proteins all the residues are undergoing flipping at rates that are too fast for nmr methods to follow (i.e., $>10^4\,s^{-1}$). For example, in lysozyme[5,14] all three tyrosine residues and in parvalbumin[24] at least 8 if not all 10 phenylalanine residues are in this category. In other proteins, such as cytochrome *c*[20,21] and bovine pancreatic trypsin inhibitor (BPTI),[25] the motion can be followed in detail for certain groups. In cytochrome *c* a detailed study of one tyrosine residue has been made and the activation energy measured.[23] This activation energy and those found in BPTI are too high to be due to the motion seen in a simple free phenylalanine or tyrosine residue. Thus the observation of restricted flipping means that there is further structural constraint on motion in the protein. It is then relaxation of the protein structure that permits the observed flipping.[30]

The motion of the aromatic groups is easy to detect by nmr because of the symmetry of the group, but if motion occurs for these residues it is very likely to occur for other residues of similar size (e.g., valine and leucine). Indeed, evidence for motion of these groups in lysozyme has been gathered,[5] although this is less easily interpreted.

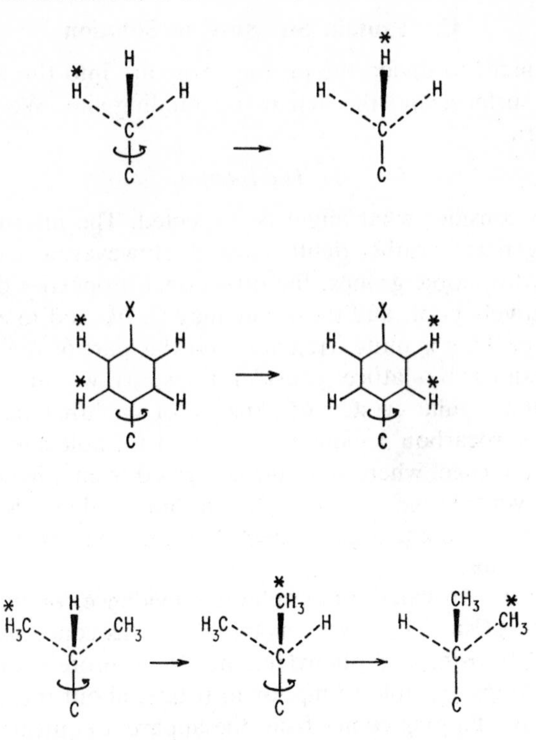

Fig. 5. Types of motion in proteins detected by nmr. Rotation about methyl groups is easily detected from threefold symmetry and is rapid. Rotation or flipping about the C_β—C_γ bonds of tyrosine or phenylalanine has been observed readily (see text) because of the twofold symmetry of the aromatic ring. Rotation of more complex side chains is more difficult to define because of the lack of symmetry.

The nmr data for this type of motion are direct and the motion clearly involves rotation about bonds in the millisecond time scale range. However, less direct evidence for motion comes from other techniques such as fluorescence depolarization, O_2 diffusion, hydrogen exchange kinetics, and nmr relaxation times (see Ref. 4). The extent of this motion is not yet easy to define, but this evidence points to motion in the nanosecond time scale range. It is tempting to see the motion in this time scale as bond oscillations rather than rotations. To put it in a different way, on this time scale the side chains have some freedom to move with respect to each other but not normally to undergo substantial bond rotation. Table IV summarizes some references for motion of different types. Additionally, nmr relaxation studies suggest that the backbone or main chain of a protein is more restricted than that of the side chains.

Putting these facts together, it is plausible to view the interior of the protein as a medium in which high-frequency but limited structural fluctuations occur. Structural fluctuations are more limited for the main chain than for side chains. Larger fluctuations of side chains occur with lower frequency, and it may be that in order for a group to be able to rotate about a bond, the high-frequency fluctuations must be in phase to allow the structure in the region of the group concerned to be relaxed sufficiently to permit rotation. Within these limitations it is known now that the structure of the interior of a protein, on time average, is quite well defined and resembles closely the X-ray structure. This arises because in large part the large motions (e.g., rotation of groups) may well involve rapid flips between different conformations, but for most of the time the molecule exists in a preferred conformation.

2. *The Surface*

The surface of a protein in solution is quite different from the interior of the molecule. The side chains are generally polar and, instead of being fully surrounded by other side chains, often project into the solvent. These projecting side chains are thus likely to resemble in some ways small molecules dissolved in solution, except that they are anchored to the bulk of the protein molecule. More extensive motion is obviously predictable for these groups, and the existence of many conformational states of similar energy. Motion or conformational variation of surface residues has even been deduced from inspection of X-ray data. Lysine and arginine groups, exposed on the surface, with their positively charged head groups not bound back to negatively charged carboxylate groups, are poorly defined in the electron density map of, for example, lysozyme.[31] The definition of the $—CH_2—$ groups becomes worse the further the $—CH_2—$ group is from the protein backbone.[31] Additionally, the X-ray structures of lysozyme crystallized in different space groups show distinct differences in the orientation of surface residues but little differences elsewhere.[32] The packing in the crystal and the likely ordering of surface groups by the salt solutions the crystals contain clearly influence the surface structure.

In solution, nmr data provide direct evidence for this high mobility. Relaxation times for surface groups are quite different from those for other groups. This has been observed in lysozyme (^{1}H studies[11,17] and also ^{15}N studies[18]) and in larger proteins. This means that at least some of the surface groups (lysine, arginine, and histidine mainly) are performing extensive and rapid motion that is independent of the motion of the protein as a whole (Table IV). It is clearly necessary to use the comparison between the solution and crystalline states with great care for these surface regions. In an aqueous medium of low salt content the lysines, for example, would be freely mobile, whereas in the solid state, where very high salt concentrations are present, the charged groups could have accumulated anions to some degree. This association would restrict mobility. The accumulation of anions could be greater the better the hydrogen bonding strength of the anion (e.g., $HPO_4^{2-} > SO_4^{2-} > Cl^-$) and the greater their concentration. Thus the surface of a crystalline protein may well show a salt dependence that is of little biological consequence.

Other groups on the surface of proteins that are known to show mobility are sugar residues. This has been observed[33] in the nmr spectral data of peroxidase, for example. The mobility of the surface is of importance in theories of protein-protein recognition.

The recognition of one surface of a protein by another is often pictured as the matching of two shapes, as in a jigsaw puzzle. In the light of the high mobility of protein surfaces this is unlikely to be true. A more plausible approach is to start from the firm ground that the binding energy is given by the following sum.

$$\Delta G(\text{binding}) = \sum \Delta G_{(1,2,\ldots,n)} \text{ (associated and hydrated)}$$
$$- \sum_{1}^{n} \Delta G_n \text{ (separated species with hydration of all separate surfaces)}$$

This equation states nothing about structure and it could well be that there is no structural match in the associated unit that would have been recognized easily by an X-ray or nmr structure of the isolated members of the pairs, even assuming that the correct hydration was maintained from solution to crystal. All that is necessary is that ΔG (binding) should be large, which can come about by several methods.

1. Precise fitting of opposite charges and hydrophobic regions, which could be recognized by structural studies of components.
2. Advantageous gain of free energy of water molecules, often a part of hydrophobic bonding, but no particular structure requirement.
3. Cross-linking by small molecules or ions such as Ca^{2+}.
4. Direct covalent linkage between the separate units, as in the formation of disulfide bridges.

Situations such as items 2 to 4 can only be understood from the analysis of associated units, and not from studies of component units. Note that under these three headings considerable changes in the structure and mobilities of the components are permitted in the adduct.

3. *Binding Sites and Active Sites*

The regions of proteins that are responsible for binding other molecules and for catalysis are somewhat intermediate between surface and internal regions of the protein. These regions must necessarily be accessible to the groups to be bound and therefore must be to some extent on the surface. (We leave until later the question concerning transport of electrons.) It is often the case that these sites are clefts in the molecule and contain groups that are not generally found on the surface of proteins. We have already touched on the problem of rigidity and mobility here. Some rigidity is clearly necessary for specificity of binding, but mobility may be necessary for an ensuing reaction to occur.[1,4] In lysozyme there are three tryptophan residues in the active cleft, and because these residues are bulky, extensive rotation of these groups is unlikely. However, one of these residues, trp 62, is not clearly defined in the X-ray structure,[31] and the nmr data shows that this region of the protein possesses mobility of some type,[5,22] probably a considerable oscillation. On binding of inhibitors in the active site, the group is seen clearly in the X-ray electron density map and gives sharp signals in the nmr spectrum. Its mobility is reduced. Again it is observed in lysozyme that ionization of one of the active site carboxylates (glu 35) results in a conformational change in solution.[5,13] In the protonated state this carboxylate group (35) interacts with the nearby tryptophan 108 residue, possibly by some charge transfer or π-bonding interaction (see Fig. 1). On ionization, however, the carboxylate moves away from the tryptophan residue, and the whole of the active site region undergoes some local conformational change.[13] Proton transfer from this carboxylate residue is a critical part of the proposed catalytic mechanism of lysozyme.[31]

Binding of inhibitors to lysozyme results in a conformational change in the active site that is observed both in the crystal[31] and in solution.[5,22] In solution, the nmr data indicate that this conformational change spreads throughout the protein, and some tightening of the structure occurs on binding. Thus binding in active sites can alter the mobility of the protein structure.

We now turn to a different aspect of active site regions or clefts. In many proteins the anomalous distribution of electronic energy in a protein can be seen by examining the structure or the physical properties of molecular groups in the active site (see introduction). The groove or active site itself may often be raised to a high energy.[3,6] For example, the hydrophobic groups that are generally present in the active site (see lysozyme) are held apart from each other by the interaction energy of the rest of the protein. Instead of the favorable energy of association of these groups, they are forced into contact with solvent and hydrophilic groups, where the interaction is less favorable. (Note the free energy of the whole protein is at a minimum.)

Apart from a general strain in the active site region, specific examples of strain have been observed. The conformation of the trypsin-trypsin inhibitor complex is one example seen by X-ray crystallography.[34] However, the clearest examples of the entatic states are observed in metalloproteins.[3,6] Spectroscopic and crystallographic data have revealed unusual (strained) coordination geometries for many metal ions that are catalytically (not just structurally) important in enzymes.[3,6] These have been described recently[35] and include hemeproteins and enzymes, iron-sulfur proteins, zinc enzymes, B_{12} requiring enzymes, and many others. It is accepted, for example, that the mechanism of triggering by spin-state changes in hemoglobin involves electronic energy strain at the iron-nitrogen bonds.[6] In some enzymes the strain is sufficient to generate local free radicals, as in the human ribonucleotide reductase.[36] In other cases the evidence for a strained environment is equivocal until substrate or inhibitor is added when the state of the enzyme groups can be seen to be quite anomalous. The vitamin B_{12} requiring enzymes show this feature very strongly.[35]

To summarize the structure of a globular protein in solution, we consider that it can be described as follows. The backbone of the protein is relatively rigid, and the main fold of the protein is the same as that observed in the crystalline state. There is mobility of side-chain groups over a range of time scales, and this occurs even in the interior of the protein. The surface groups have extensive freedom of motion. The active sites of enzymes have intermediate specific mobility that can be altered on binding other molecules. It is also the case that strain exists in active sites, which is important for catalysis (the entatic state). We stress that it is not our intention to discuss the energetics of catalysis as such; this has been discussed elsewhere.[3,6,37] We are concerned merely with the nature of the protein structure and the energetics of this structure. We must add the following reservations to these statements. The X-ray and nmr methods have been applied and can only be readily applied to *globular* proteins (i.e., the most rigid proteins). There are, however, a vast variety of other proteins in which mobility and even backbone mobility may well be high (see Table VI). All these points will be illustrated.

TABLE VI. Rigid and Flexible Proteins[a]

Relatively rigid	Flexible	Mixed
Cytochrome *c*	Phosphovitin	Insulin
Neurotoxins	Chromogranin A	DNA—binding proteins
Protease inhibitors	Histones	Phospholipase A.2
Lysozyme		
	Glucagon	Kinases
Peptidases	Myelin protein	Antibodies (?)
Nucleases	Vesiculin	
(Globular proteins)	Metallothioneins	

[a] Generally extracellular proteins are rigid and are often cross-linked by S-S bridges. Intracellular proteins are more mobile especially if they are required to equilibrate between bound and unbound states involving DNA, RNA, or membrane surfaces.

IV. DISCUSSION OF SOME INDIVIDUAL PROTEINS

We now wish to consider the structures and/or energetics of a number of specific proteins.

A. Lysozyme

We have frequently used lysozyme as an example in previous sections of this article, and only a few additional points will be made here. The use of nmr methods to compare the solution structure to the X-ray has been successful to the extent that the main fold of the protein and approximate positions of many side chains (in the interior and active site) are now known to be the same in the two states.[5] Figures 2 to 4 demonstrate some evidence along these lines. The observations of side-chain motion mean that further refinement of the solution structure cannot be made until these motions are better understood. It has also been found that binding of inhibitors results in a conformational change in the active site, and the rate of this conformational change has been measured[22] with one inhibitor (see Fig. 6.) Binding of inhibitors also appears to lead to a tightening of the structure, and the conformational change is spread through much of the structure.[17]

The mechanism of action of lysozyme has been suggested to involve strain in the substrate as a result of binding to the protein.[31] The protein requires that the substrate binds to the protein in a distorted state that is close to the transition state of the reaction. The energy for this distortion can arise from the substrate binding energy. This example differs from the entatic state hypothesis because this strain arises from substrate binding and is not a feature of the protein structure. However, the entatic state

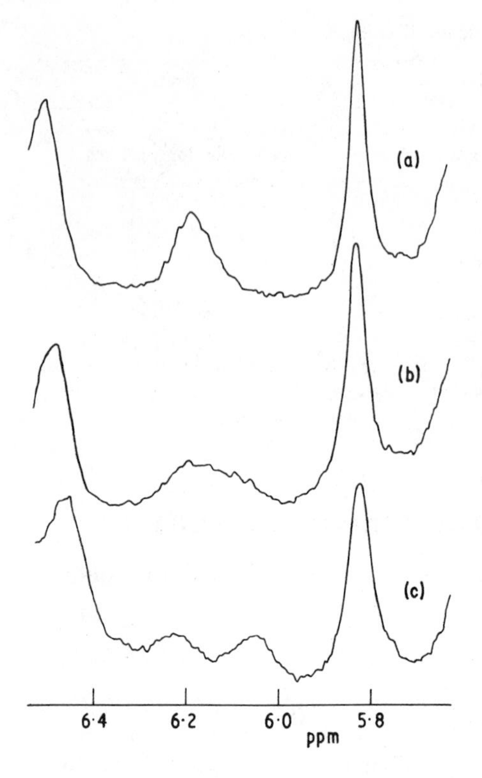

Fig. 6. Part of the spectrum of lysozyme in the presence of the inhibitor $(GlcNAc)_3$. At the concentration used, two resonances from 1 proton of trp 63 can be observed at 6.25 ppm and 6.05 ppm. These correspond to the resonance of this protein in the unbound and inhibitor-bound protein, respectively. The shift indicates that a conformational change occurs in the protein on binding $(GlcNAc)_3$. At low temperature (*c*) the rate of the conformational change is slow on the time scale of the nmr experiment, but at higher temperature (*a*) the two resonances coalesce as the rate of the conformational change increases. At 45°C (*b*) exchange broadening occurs. The rate of the conformational change was measured from these data, and at 45°C is 20 sec^{-1} (see Ref. 22).

postulates may well give an appropriate description for the state of the catalytic groups and we draw attention[5,13] to the anomalous interaction of carboxylate 35 and tryptophan 108.

B. Cytochrome *c*

Cytochrome *c* has an excellent internal probe—the heme unit. However, the characteristics of this probe cannot yet be used to give quantitative information. Thus semiquantitative studies have had to be made first, but fortunately these have given sufficient detail to enable us to state[38] with great confidence that the main-chain fold of cytochrome *c* is the same in solution as that found in the crystals. A considerable number of side-chain positions have been compared in detail, and correspondence between solution and solid state is good. As with lysozyme, we believe that the surface of the protein has considerable motion, so that no structural study of, for example, exposed lysines has much meaning. It is clear that in the interior of the protein are regions of considerable rigidity higher than that seen in lysozyme. For example, the one conserved tryptophan residue close to the heme is very specially positioned and is the same in the cytochromes *c* from different species.[36] The picture of much of the rest of the molecule is dependent on an analysis of molecular motion. For example, nmr spectroscopy shows[20,21,23] that there is differential motion of the different tyrosines and of the different phenylalanines in cytochrome *c* (Table VII).

Cytochrome *c* can exist in two forms, oxidized (ferric form) or reduced (ferrous form). After careful refinement (and removal of errors) both X-ray[39] and nmr[38] data show that in crystals and in solution there is only a very small conformation change on oxidation state change. In fact, the change of the internal conformation of the protein is far too small for this to be used as a through-bond trigger or relay for electron transport.

Nevertheless, we know[40] that cytochrome *c* in its two oxidation states has a different solubility, a different rate of movement on columns, and a different pH and temperature stability. It is also known that the protein binds anions (even chloride) or cations, or even other proteins differently on change of redox state. These binding differences would be sufficient to operate a relay.

TABLE VII. Mobilities of Aromatic Amino Acids in Cytochrome-*c*[a]

Aromatic amino acid	Mobility	
	Fe(III) state	Fe(II) state
Tryptophan 59	Not mobile	Not mobile
Phenylalanine 10	Mobile	Mobile
Phenylalanine 36	Mobile	Mobile
Phenylalanine 46	?	Only mobile above 350°K
Phenylalanine 82	?	Only mobile above 350°K
Tyrosine 97	Mobile	Mobile
Tyrosine 48	?	Only mobile above 350°K
Tyrosine 74	Not observed	Not observed
Tyrosine 67	Not observed	Not observed

[a](1) An assignment is complete only when all protons are assigned and therefore the assignment of some of the above groups is not yet certain. (2) The amino acids are placed in relation to certain channels in the protein structure in Fig. 4. (3) Mobile implies a flip rate of greater than 10^4 per second at 300°K and immobile a flip rate of less than 10^3 per second at the same temperature.

Putting both sets of data together, it appears that it is the surface state of the protein that changes on redox change of charge, rather than the interior state. Now the state of the surface is not well defined, as we have seen, and in solution is not likely to possess real order over even 10^{-8} sec. Thus we are forced to suppose that the energetics of the surface (including the concomitant fluctuating changes of conformation) are the origin of the change in properties with charge. Although the effects are not relayed from surface to surface by through-bond conformational changes (allosteric theory), they are generally concerted, as they depend on electrostatic interaction, which itself has no directional dependence but has directional significance through the variation in dielectric medium within the protein. At first sight this seems an unlikely mechanism, but it must be remembered that the change in charge is deep in the protein and occurs at a site of low dielectric constant; that is, a large change of electrostatic energy is implicated. Given that there are only very small changes in the interior of cytochrome *c*, it follows that there will be destabilization of the largely positively charged surface of the protein on increasing the oxidation state and that this energy of destabilization will be sufficient to account for the loss of pH and temperature stability. Consider the following pair of charged states in such a protein

$$\text{(Inside) } M \qquad \text{R}'\text{—NH}_3^+ \cdots \text{CO}_2^- \cdot \text{R} \qquad \text{(surface)}$$

and

$$\text{(Inside) } M^+ \qquad \text{R}'\text{—NH}_3^+ \cdots \text{CO}_2^- \cdot \text{R} \qquad \text{(surface)}$$

It is clearly advantageous to rearrange the last system to

$$\text{(Inside) } M^+ \qquad \text{R—CO}_2^- \cdots \text{NH}_3^+\text{—R}' \qquad \text{(surface)}$$

The consequences of such a conformational change could be dramatic for the physical and chemical properties of the protein and would explain the cytochrome *c* behavior. Related effects on electron transfer catalysis of the protein have been discussed elsewhere.[40]

In the case of cytochrome *c*, these electrostatic terms are due to changes in the redox states of the internally bound protein metal ion. In other cases where the charges on anions or cations are numerically fixed, the ions can dissociate (e.g., as the metal ion leaves the protein) or migrate (e.g., Na^+, K^+, Ca^{2+}, Cl^-, HPO_4^{2-}, H^+). If the exchange of these ions involves sites, especially hydrophobic sites, deep inside proteins, on the one hand, and free solution or surface sites, on the other hand, then they will be expected to have an electrostatic influence on the protein much as in a change of redox state. Thus we look next at two calcium binding proteins and later at insulin.

C. Calcium Binding Proteins

We now consider two calcium binding proteins, parvalbumin and troponin. The first, carp parvalbumin, is a small protein of known crystal structure.[41] It binds two Ca^{2+} ions per molecule. One of the Ca^{2+} ions is essential for a folded structure to be maintained; the second may be lost with a relatively small conformational change.[42]

The protein examined in detail (denoted carp pI 4.25) contains 10 phenylalanine residues, which are widely distributed throughout the structure.[42] In solution, it was possible to resolve the proton nmr resonances of 8 out of the 10 phenylalanine residues in the two Ca^{2+} form of the protein.[24] All the residues studied have equivalent pairs of ortho and of meta proton resonances, despite large ring current shifts, which are expected and calculated to affect the five aromatic protons of each residue differentially. The observed equivalence means that all these phenyl groups flip rapidly about the $C\beta$—$C\gamma$ bond. The whole of the protein must therefore be undergoing rapid fluctuations to allow this flipping to occur. Nmr studies have provided further evidence for conformational fluctuations. For example, the protein contains one sulfhydryl group that is not exposed on the surface. However, the protein can dimerize, by formation of a disulfide bridge, and to do this the sulfhydryl groups of two molecules must become exposed to the solvent to allow this reaction to occur.[24] The monomer and dimer can be observed in the nmr spectra to have slightly different conformations.[43]

The second protein we consider is troponin C.[44] This binds four Ca^{2+} ions and has no X-ray structure as yet. The binding of the Ca^{2+} ions to this protein, the calcium binding protein of muscle, occurs in two steps. The first two Ca^{2+} ions to bind cause a considerable tightening of the protein structure and a change of protein conformation of unknown magnitude. Now binding of the two weaker Ca^{2+} ions causes but a very minor conformational change, despite the fact that this binding is the trigger for muscle action. However, the nmr methods used would detect only conformational changes in the interior of the protein. The mobile surface groups—for example, lysines and carboxylate containing residues—would be affected by binding, but this would not have been detected. On general grounds of electrostatics, and given the preceding discussion of cytochrome *c*, we would expect this. It may be that muscle action is triggered by organizational changes in protein interactions from such simple electrostatic effects.

Thus we note that cations (and anions) in proteins would have a very general effect on protein stability, either through cross-linking (similar to the effects of disulfide bridges) or through a general electrostatic effect that helps to stabilize (or destabilize) a surface of the opposite (or the same) charge. Is this the reason that metal ions such as Ca^{2+} and Zn^{2+} are so frequently found in nonactive site regions of proteins (e.g., alcohol dehydrogenase)?

D. Insulin

The crystal structure of this protein has been shown to depend on the salt solution from which it is crystallized.[45] When crystallized from $(NH_4)_2SO_4$, the insulin hexamer is held together in part by two Zn^{2+} ions. These can be visualized as being at either end of a cylinder, and each Zn^{2+} has as a ligand one histidine imidazole nitrogen atom from one of three chains. Thus a histidine (his B.10) of each of the six chains is coordinated to Zn^{2+}, three at each Zn^{2+}. The Zn^{2+} ions occupy octahedral (trigonally distorted) sites overall.

However, crystallization from chloride medium gives a different structure in which one of the Zn^{2+} atoms moves to a position where it bridges his B.10 and another histidine B.5 and is now in a tetrahedral site. The creation of this tetrahedral site for Zn^{2+} involves considerable changes of the fold of the last six residues of the polypeptide chain. There are thus three alternative new Zn^{2+} sites that could be created. The structure has been called the four-zinc insulin as opposed to the two-zinc insulin described earlier. Equally, the same change could occur at the other Zn^{2+} site, making a six-zinc insulin structure, but this crystal form is not yet known.

The same transformation has recently been examined by nmr in solution.[46] In zinc sulfate solution, with two zinc atoms per hexamer, a certain nmr spectrum is observed. Addition of excess zinc and of chloride, iodide, or particularly thiocyanate but not sulfate converts this to a different spectrum. Detailed study suggests that this is of the six-zinc insulin, which is in equilibrium with the two-zinc form, but in slow exchange.

The change in conformation seen here (both in the crystal and in solution) may not be due to the zinc so much as to the anion. The X-ray study showed that the change in zinc coordination took place with uptake of an anion in a very hydrophobic pocket. This is consistent with the binding strength order observed $SCN^- > I^- > Cl^- > SO_4^{2-}$. In the solution studies two thiocyanate anions are required for the overall change. Apart from the dependence on cation and anion concentration, the equilibrium position between the conformations in solution is temperature and pH dependent.

The insulin monomer is of 6000 molecular weight, and either the monomer or the dimer is the active hormone. Thus in fact the zinc that binds the molecule into hexamers is not of physiological interest. What is of interest here is that the insulin hormone has some conformational flexibility that is greater than that seen in other small proteins (e.g., lysozyme or the trypsin inhibitors) but less than that seen[47] in the polypeptide hormone glucagon, which is quite flexible in solution. Thus it would appear that protein mobility can vary considerably even for molecules of similar size.

So far we have been concerned with specific features of protein structure and mobility. In the next two examples we look briefly at cases where the binding of small molecules has been examined.

E. Concanavalin A

Concanavalin A is an effector protein, which has been studied in forms with and without bound sugars,[48] so that the equilibrium considered is

$$\text{ConA} + \text{sugar} \rightleftharpoons \text{ConA}\cdot\text{sugar}$$

In the crystalline state these two forms are not isomorphous, and adding substrate to ConA crystals causes them to shatter.[48] (This is a common observation and occurs even in hemoglobin crystals.) The differences between the conformers in the ConA part of the complex are not fully known, but there is considerable rearrangement of the protein. The rates of the reaction are fast, and so in solution the protein must be able to fluctuate readily between its different forms. Comparison with lysozyme (earlier in this article) shows that when lysozyme binds its inhibitors, a conformational change does occur, but it is not so gross that in the solid state shattering of crystals occurs.

F. Phosphoglycerate Kinase

Phosphoglycerate kinase has been studied in both the crystal[49] and solution states.[50] When ATP binds to the enzyme in the crystal, little conformational change was observed. However, the protein was crystallized from sulfate, and the binding of ATP can perhaps be represented as

$$\text{E}\cdot\text{SO}_4^{2-} + \text{ATP} \rightleftharpoons \text{E}\cdot\text{ATP} + \text{SO}_4^{2-}$$

In the nmr solution study the sulfate-free enzyme was also studied, and it is known that SO_4^{2-} inhibits the enzyme. The evidence suggests that E differs from $E\cdot SO_4^{2-}$, but that $E\cdot SO_4^{2-}$ and E·ATP are more similar. There is a considerable conformation change on binding ATP to E.

Let us now turn to systems, where the energetics have been related to function.

G. Carboxypeptidase

In carboxypeptidase differences of opinion between crystallographic and solution studies seem to be becoming resolved into a matter of conformational differences between particular crystal states and particular solution states.[51] Here the mobility of a given tyrosine residue is in question and the fold of the protein is the same in crystals and in solutions. This residue lies either exposed on the protein surface or partially buried in the active site region, depending on the experimental conditions. This therefore appears to be a clear example of extensive surface mobility.

In carboxypeptidase, the active site contains a Zn^{2+} ion, bound by three protein side chains, two histidines, and one glutamate. This leaves the Zn^{2+} open sided, and in a distorted tetrahedral environment that, as predicted by the entatic state hypothesis,[3] is of high energy compared to Zn^{2+} in normal complexes allowing that the same coordinating centers could be used without the constraints of the protein fold. This strained site has been generated by the overall fold of the protein to be particularly suited for catalysis. The preceding coordination structure has been deduced from the X-ray structure. By using the method of isomorphous replacement (Zn^{2+} replaced by Co^{2+}) the active site can be studied by spectroscopic methods. The absorption spectrum shows the metal coordination to have low symmetry and is affected by binding of small molecules in different ways. However, neither the absorption spectrum nor the epr spectrum of a relatively mobile, low-symmetry site can provide definitive structural data. The details of the concept of the entatic state in catalysis by metalloenzymes has been described fully elsewhere.[3]

H. Hemoglobin and Heme Enzymes

The discussion of heme enzymes can be initiated[6] by an overall examination of the information that a large number of methods have given concerning hemoglobin. The X-ray structure[52] is the central part of the information. It is now clear that there are two extreme structures that we may identify with the two forms (T- and R-, indicating *tense* and *relaxed*) seen in crystals. The balance between these forms is altered by the following:

1. Binding of oxygen, carbon monoxide, and so on, at the iron.
2. Binding of allosteric modifiers, especially protons (the Bohr effect), organic phosphate (particularly diphosphoglycerate), and carbon dioxide.
3. Organic modifications, especially of the β-sulfhydryl residues.
4. Mutations—that is, substitution of a variety of types in sensitive regions of the sequence around the α,β contacts, the heme pocket, and the allosteric sites.

All these effects can be interpreted in terms of two forms, but there is much evidence that minor changes in tertiary structure within the *R*- and *T*-states occur, so that there is a variety of modified forms with roughly fixed either *R*- or *T*- main-chain folds. If we start from the iron atom, the interactions within the protein are triggered by the effect of ligand binding on the electronic states and oxidation states of the iron. There is then an adjustment of Fe—N bond lengths, an adjustment of the porphyrin, and an adjustment of the whole iron-porphyrin unit within the protein. The protein responds in many large or small ways as the contacts are adjusted. A simple change is the action of Fe—N bond length change on the FG helix, but there are undoubtedly many other changes. When added together, site cooperativity under items 1 to 4 can occur. Discussion of entatic state concepts for hemoglobin has been given separately.[6]

Now this model does not yet include the changes seen in cytochrome *c* (see earlier). Here charge effects produced a through space adjustment, especially of surface or polar groups separated from charge changes only by a medium of low dielectric. In a heme *enzyme* there are changes both of binding (of O_2, substrate and effectors) and of charge at different steps along the reaction path; for example,

$$Fe^{2+} + O_2 \rightarrow Fe^{2+}O_2 \rightarrow Fe^{4+}O^{2-} \rightarrow Fe^{3+}\cdot(OH_2)$$

We can see that there are a large number of changes within the protein at small or even large distances, and energy "flows" between different regions of the protein. This argument is relevant, for example, to peroxidase and cytochrome P-450.

I. Trypsin

The mechanism of action of serine proteases has been suggested to involve a tetrahedral intermediate in the acylation-deacylation steps in which there is a covalent bond between the enzyme and the substrate.[53] The tetrahedral intermediate is formed from the carbonyl group of the peptide bond to be hydrolyzed by the enzyme. This intermediate would be close to the transition state of the reaction and hence would result in a lowered activation energy for hydrolysis. In the normal hydrolytic reaction the lifetime of the tetrahedral intermediate would be very short and requires considerable active site conformational change. It has never been convincingly demonstrated. However, several naturally occurring peptides, terminating in arginyl peptides, are potent inhibitors of trypsin and plasmin, emphasizing the importance of the tetrahedral intermediate.[53]

In the complex of the protein BPT1 with the protein trypsin, a close approximation to a tetrahedral adduct has been observed[34] in the X-ray structure. This adduct is formed by addition of the reactive serine (ser

195) of the enzyme to the carbonyl group of a specific lysine residue of the protein inhibitor. (The reactivity of ser 195 is itself an example of the entatic state.) In this complex the tetrahedral intermediate clearly constitutes a minimum of free energy for the system in the crystals as it is the observed state. However, taken by itself the binding group of the inhibitor is very strained. Now the inhibitor is itself a protein and it is the fold of this protein that has forced the reactivity on the lysine peptide bond. Here is an example of an entatic condition in a protein substrate! This is a clear organic case where the overall energy of the protein is at a minimum, but at the reactive site strain exists.

J. Some Random Coil Proteins

A rigid protein fold may be of no great advantage if the function of the protein involves the specific recognition of amino acids close together in the sequence. Consider a protein that is required to be a scavenger for a particular metal cation. The side chains of the protein can be chosen following well-known principles of chelation: that is, carboxylate side chains of aspartate and glutamate will scavenge for Ca^{2+}, and thiolate side chains of cysteine for Cd^{2+}, Hg^{2+}. For effective scavenging it is sufficient to lower the calcium concentration to 10^{-7} *M*, but that of cadmium and mercury may have to be reduced to 10^{-20} *M*. The different levels of scavenging required in order to provide protection arise from the very different strengths of binding of these three metal ions to protein groups in general. To reduce the calcium concentration to 10^{-7} *M* requires a binding constant of the scavenger of 10^7. Four carboxylate centers placed on a chain so that they can fold around the metal give this binding strength, and it is now known that in certain proteins there are series of γ-carboxylated glutamate residues (two carboxylates each), called Gla, spaced some two or three residues apart (e.g., in the clotting factors[54] the sequences are γ-Gla.x.x.γ-Gla.x.γ-Gla). If we take two such γ-Gla residues spaced apart as in γ-Gla.x.x.γ-Gla or γ-Gla.x. γ-Gla and add 10^{-7} *M* Ca^{2+} ions, then this part of the chain will fold around the calcium. It is of no great advantage to have the protein tightly folded, for the speed of the calcium on-off reaction could be greatly impeded by the necessary readjustment that must be made against the fold energy.

A similar situation arises in the metallothioniens, which are the scavenger proteins for Cd^{2+} and Hg^{2+}. Here the protein sequence[55] contains a large number of thiolate residues, a relatively high concentration of proline and glycine (helix-breakers), but no aromatic amino acids, which usually provide centers for hydrophobic cooperativity, helping to generate chain folding. The thiolate (cysteine) residues are spaced apart cys. x.x. Cys and Cys. x.x.x. Cys and once again two (Hg^{2+}) or three

(Cd^{2+}) residues fold around the metal cation, giving very good binding. The protein is very likely to be quite disordered in the metal-free state and it can pick up some 6 to 10 Cd^{2+} ions. This type of protein can be contrasted with the proteins that use cysteinate side chains (thiolate) to bind zinc (alcohol dehydrogenase),[55] iron (ferredoxin[40]), or heme (cytochromes c[40]), where it is found that there are just two thiolates Cys. x.x. Cys in sequence but that a *distant* region of the sequence folds back to provide either other cysteinates or histidines to give a binding site for Zn, Fe, or heme. In this case the protein fold generates the chelation center and the two are cooperatively linked.

Thus although a protein can be made very specific for a particular metal, if it has a fold that generates matching properties to those possessed by the cation (e.g., carbonic anhydrase), this degree of matching is not required for the removal of heavy metals as they have such high binding constants to unconstrained sites. Clearly, there is a competition between two types of site for two metal ions:

$$M + \text{random site} \rightleftharpoons M\cdot\text{random site}$$

$$M + \text{folded site} \rightleftharpoons M\cdot\text{folded site}$$

and as heavy metals such as mercury bind better than lighter metals such as zinc in both sites the random site must bind Hg relative to Zn such that

$$\log \frac{K_{Hg}}{K_{Zn}} (\text{random}) \gg \log \frac{K_{Hg}}{K_{Zn}} (\text{folded})$$

It is then the fold that generates the specificity of metal involvement.

There are many other proteins that resemble the scavenger metal proteins, but they "scavenge" for quite other chemical groups. All the histones are proteins with multiple repeating sequences of the kind

$$(\text{basic residue x x basic residue x x})_n$$

and these proteins are designed for combination with the repeating units of DNA phosphate anions. Thus the protein must be unfolded before binding. It is not possible to examine such a protein by the methods of X-ray crystallography.

A different example involves the proteins of the interior of the chromaffin granule—the chromogranin A proteins.[57] Physical measurements using the ultra centrifuge[57] showed that these proteins were not globular, and detailed nmr studies[58] indicate that the proteins are largely in the random coil form as the nmr spectrum is relatively sharp and the spectrum is the sum of the spectra of the component amino acids. This protein has been examined in the vesicles of the adrenal medulla in vivo and it exists in a practically identical form to that observed in vitro in free aqueous solution.[58] The suggestion arises that the protein is employed to decrease the possibility of the formation of crystals from the mixture of ATP and adrenaline in the vesicles holding them in a viscous medium. The value of such a medium, which can be rapidly dispersed, differentiating it from crystals, will not be missed. It is suspected that other proteins of this kind will be found in other vesicles.[58]

There will also be proteins that have largely ordered folded regions joined by less rigid pieces. The development toward this state can be seen in lysozyme, which is known to have two highly ordered regions on the two sides of the reactive site cleft, one largely β-pleated sheet and the other largely helices, linked by long strands of the less-defined secondary structure. Such a protein region could act as a hinge. Other proteins in which there would appear to be highly constrained regions linked together by such strands have been found among kinases and antibodies. It is tempting to presume that proteins will be found with all kinds of intermediate mobilities between these well-organized, semimobile systems and disordered proteins. One example is phospholipase A.2.

V. CONCLUDING REMARKS

In this article we have considered some structural and energetic features of proteins that have been observed through physical techniques, particularly X-ray crystallography and nuclear magnetic resonance spectroscopy. We have not considered the details of the energetics of enzymic reactions, which are discussed in recent reviews,[3,6,37] but have mentioned certain features related to structural observations.

The overall fold of a *globular* protein molecule in solution is likely to be given by the X-ray structure of the molecule in a crystal. This has been shown to be the case in detail for one protein, lysozyme, by means of nuclear magnetic resonance techniques, and is highly probable for others. It is necessary, however, to be aware that the conditions under which the protein is crystallized, or studied in solution, can affect the fold, and this has been illustrated with insulin.

The *exact* disposition of the side chains in a *globular* protein is difficult to define in solution. Although it is likely that the peptide main chain (backbone) of the protein is relatively rigid, the side chains have been shown to be undergoing motion of several different types (see lysozyme, peroxidase, and carboxypeptidase). This means that the full definition of atomic positions in the structure requires a knowledge of the time dependence of their coordinates. The motion of side chains is likely to be different in the crystal and solution states, but this difference may well depend on the location of the group, internal, on the surface or at an active site.

The internal (hydrophobic) regions of proteins appear relatively fixed in crystals, and in lysozyme the time-averaged side-chain positions in solution have been shown to be the same as those found in the crystal, within present experimental limitations. However, in lysozyme and in other proteins studied in solution (examples in this article include cytochrome *c* and parvalbumin), the side chains are clearly shown to be mobile by nmr methods. The motions appear to vary from rapid oscillation about all bonds (in the nanosecond time scale range) to complete bond rotation. Complete rotation or flipping about the $C\beta$—$C\gamma$ bonds of tyrosine and phenylalanine residues occur in the millisecond time scale range. In contrast to the interior of a protein, surface regions are not always well defined even in crystals. In solution extensive motion is known to occur. Thus rotational and vibrational energy is distributed throughout the protein structure. Proteins of different types show very different degrees of motion. Cross-linked globular proteins such as lysozyme have relatively well-defined folds, but other proteins, such as the chromagranin A proteins of the interior of the chromaffin granule, are largely in random coil forms in solution.

There is a third region of a protein that is neither on the surface nor in the interior but that is in a cleft. Such regions are often associated with enzyme action and examples show they have (*1*) intermediate mobility (e.g., tryptophan 62 of lysozyme or the tyrosine of carboxypeptidase) and (*2*) unfavorable energetics of exposed groups—the entatic state.

Finally, there are many nonglobular proteins in which motion may well be very much more extensive.

Now, knowing that a globular protein is not a rigid body and that there are certain activated regions, we can consider some consequences of this. Let us first describe a protein as a small homogeneous object such as a deformable sphere. This sphere is charged, with the charges placed to a large extent on the outside. The reactions of such an object are dependent on such factors as temperature and pressure and on the pH and ionic strength of the medium in which it is immersed. It will swell or contract under the influence of any charge in the medium to which it is exposed. Its volume will effectively monitor the physical environment, like a measuring device. These effects have been seen in lysozyme.[17]

If we make a somewhat more realistic molecular model, such as that of linked springs (Hopfield),[2] then just, as the volume of the sphere is dependent on a great variety of external effects, so must this structure respond at the molecular level to these external effects, but now all effects become directional. The idea of allosteric equilibrium (e.g., in hemoglobin) between different conformational forms is then seen to be an extreme example of a rather specific response to a particular reagent. In general, proteins respond to lesser or greater degrees to all environmental changes, whether they are physical or chemical in nature. There is, however, another way in which information can be relayed.

In cytochrome *c* the redox change Fe(II) to Fe(III) at the heme affects various properties of the protein, including binding of molecules and ions to the protein surface. A change of charge in the low dielectric center of the protein has been transmitted to the surface of the protein, where polar and charged groups exist. These surface groups could just respond to the change in charge in the center of the protein, thus causing the observed changes in the molecular properties. No significant change in the conformation of the hydrophobic interior of the protein appears to occur on change of redox state. This type of through-space response of a protein to charge was also considered when charge addition or removal occurs (e.g., binding of metal ions or protons), as in Ca^{2+} binding proteins. Note the changes, although electrostatic, cause *selected* alterations in the surface.

The active sites of globular proteins often exhibit evidence for steric or electronic strain. Local strain in a region of a protein can exist, even though the total free energy of the structure is at a minimum. This can arise, for example, where the protein fold forces hydrophobic groups to be exposed to solvent, or hydrophilic groups to be in hydrophobic regions. This strain can produce unusual pK values of groups (e.g., glu 35 in lysozyme has a pK of 6.2), unusual geometries of groups (e.g., around metal ions; see carboxypeptidase), and unusual reactivity of groups (see ser 195 in trypsin). A strained environment in the active site of a protein has been called an entatic state. In this situation, where the catalytic groups are energized, the energy profile for a reaction *starts* from a ground state nearer to the transition state than would occur for nonenergized groups, resulting therefore in a lowered activation energy. The strained or energized ground state can be observed by physicochemical examination of the groups of the protein itself. Examples are given for carboxypeptidase and hemoglobin.

Note that the entatic state differs from the situation where the *substrate* itself is strained by virtue of binding to the protein. Such a case is also found in lysozyme, where the energy of holding the substrate (a saccharide) is used partly to distort the substrate toward its transition state, thus lowering the activation energy.

A different case occurs for the trypsin-trypsin inhibitor complex, where protein steric strain occurs in the free inhibitor.

Let us consider next a complex reaction path in which the protein must cycle through many conformational states even in the active site region. Again it may well be that the active site must be controlled through the mobility of the protein that links active and allosteric sites. Thus a compromise must be reached between the most effective static structure for a given reaction step (the entatic state) and two other requirements. The first is that the enzyme reaction may well require a series of steps and one conformation cannot accommodate all the steps. The second is that the remote control of kinetic parameter, k_{cat} or K_M, of a given step has to be achieved through mobility. Thus evolution, which is concerned with *controlled* catalysis, must have generated not only special structures in proteins but also special mobilities of the structures allowing connection between the different intermediates required along the reaction path, and allowing modulation of the reaction by distant effectors. It is readily shown that evolution has provided some remarkable active sites for single-step reactions (electron transfer), but it is much harder to understand the purposes of evolution as expressed in the active sites of enzymes at which reaction involves a series of steps. A very extensive comparative study of mobility of proteins is now under way, as illustrated previously.

Acknowledgment

This paper is a contribution from the Oxford Enzyme Group and we wish to acknowledge the help of very many members of that group.

References

1. L. A. Blumenfeld, *J. Theor. Biol.*, **58,** 269–284 (1976); R. Lumry, *Ann. N.Y. Acad. Sci.*, **227,** 46–73 (1974); R. J. P. Williams, *Pure Appl. Chem.*, **38,** 249–265 (1974).
2. J. J. Hopfield, *J. Mol. Biol.*, **77,** 207–222 (1973).
3. B. L. Vallee and R. J. P. Williams, *Proc. Nat. Acad. Sci.* (*U.S.*), **59,** 498–510 (1968).
4. G. K. Radda and R. J. P. Williams, *Chem. Br.*, **12,** 124–129 (1976).
5. I. D. Campbell, C. M. Dobson, and R. J. P. Williams, *Proc. R. Soc.*, **A345,** 41–59 (1975).
6. R. J. P. Williams, *Cold Spring Harbor Symp. Quant. Biol.*, **36,** 53–68 (1971).
7. B. A. Levine and R. J. P. Williams, *Proc. R. Soc.*, **A345,** 5–22 (1975).
8. C. M. Dobson and B. A. Levine, in *New Techniques in Biophysics and Cell Biology*, Vol. 3, 1976, pp. 19–91, Eds. R. H. Pain and B. J. Smith, Wiley, London.
9. I. D. Campbell, C. M. Dobson, R. J. P. Williams, and A. V. Xavier, *J. Magn. Res.*, **11,** 172–181 (1973).
10. I. D. Campbell, C. M. Dobson, and R. J. P. Williams, *Proc. R. Soc.*, **A345,** 23–40 (1975).
11. I. D. Campbell, C. M. Dobson, R. J. P. Williams, and P. E. Wright, *FEBS Lett.*, **57,** 96–99 (1975).
12. I. D. Campbell and C. M. Dobson, *JCS Chem. Comm.*, 750–751 (1975).
13. I. D. Campbell, C. M. Dobson, and R. J. P. Williams, *Proc. R. Soc.*, **B189,** 485–502 (1975).
14. I. D. Campbell, C. M. Dobson, and R. J. P. Williams, *Proc. R. Soc.*, **B189,** 503–509 (1975).

15. I. D. Campbell, C. M. Dobson, R. J. P. Williams, and A. V. Xavier, *Ann. N.Y. Acad. Sci.*, **222,** 163–174 (1973).
16. I. D. Campbell, C. M. Dobson, and R. J. P. Williams, *JCS Chem. Comm.*, 888–889 (1974).
17. I. D. Campbell, C. M. Dobson, G. Ratcliffe, and R. J. P. Williams, to be published.
18. D. Gust, R. B. Moon, and J. D. Roberts, *Proc. Nat. Acad. Sci.* (*U.S.*), **72,** 4696–4700 (1975).
19. W. E. Hull and B. D. Sykes, *J. Mol. Biol.*, **98,** 121–153 (1975); E. Oldfield, R. S. Norton, and A. Allerhand, *J. Biol. Chem.*, **250,** 6368–6379 (1975).
20. C. M. Dobson, G. R. Moore, and R. J. P. Williams, *FEBS Lett.*, **51,** 60–65 (1975).
21. G. R. Moore and R. J. P. Williams, *FEBS Lett.*, **53,** 334–338 (1975).
22. C. M. Dobson and R. J. P. Williams, *FEBS Lett.*, **53,** 362–365 (1975).
23. I. D. Campbell, C. M. Dobson, G. R. Moore, S. J. Perkins, and R. J. P. Williams, *FEBS Lett.*, **70,** 96–100 (1976).
24. A. Cave, C. M. Dobson, J. Parello, and R. J. P. Williams, *FEBS Lett.*, **65,** 190–194 (1976).
25. K. Wüthrich and G. Wagner, *FEBS Lett.*, **50,** 265–267 (1975); G. Wagner and K. Wüthrich, *J. Magn. Res.*, **20,** 435–445 (1975).
26. A. G. Redfield and R. K. Gupta, *Cold Spring Harbor Symp. Quant. Biol.*, **36,** 405–416 (1971).
27. J. D. Glickson, W. D. Phillips, and J. A. Rupley, *J. Am. Chem. Soc.*, **93,** 4031–4038 (1971).
28. A. S. V. Burgen, G. C. K. Roberts, and J. Feeney, *Nature*, **253,** 735–755 (1975).
29. C. M. Dobson, L. O. Ford, S. E. Summers, and R. J. P. Williams, *JCS* (*Faraday II*), **71,** 1145–1151 (1975).
30. B. R. Gelin and M. Karplus, *Proc. Nat. Acad. Sci.* (*U.S.*), **72,** 2002–2006 (1975).
31. T. Imoto, L. N. Johnson, A. C. T. North, D. C. Phillips, and J. A. Rupley, in *The Enzymes*, 3rd ed., P. D. Boyer, Ed. Vol. VII, Academic, New York, 1972, pp. 665–868.
32. L. H. Jensen, private communication.
33. R. J. P. Williams, P. E. Wright, G. Mazza, and J. R. Ricard, *Biochim. Biophys. Acta*, **412,** 127–147 (1975).
34. W. Bode, P. Schwager, and R. Huber, Proc. 10th FEBS Meeting, 1975, pp. 3–20.
35. R. J. P. Williams, *Inorg. Chim. Acta Rev.*, **5,** 137–154 (1971).
36. A. Ehrenberg and P. Reichard, *J. Biol. Chem.*, **247,** 3485–3488 (1972).
37. W. P. Jencks, *Catalysis in Chemistry and Enzymology*, McGraw-Hill, New York, 1969.
38. G. R. Moore, R. J. P. Williams, and P. E. Wright, in *Biological Aspects of Inorganic Chemistry*. Eds. A. W. Addison, W. R. Cullen, D. Dolphin and B. R. James. Wiley Interscience, New York (1977), pp. 369–401.
39. R. E. Dickerson, personal communication.
40. G. R. Moore and R. J. P. Williams, *Coord. Chem. Rev.*, **18,** 125–165 (1975).
41. P. C. Moews and R. H. Kretsinger, *J. Mol. Biol.*, **91,** 201–225 (1975); R. H. Kretsinger and C. E. Nockolds, *J. Biol. Chem.*, **248,** 3313–3326 (1973).
42. A. Cave and J. Parello, personal communication.
43. A. Cave, C. M. Dobson, J. Parello, and R. J. P. Williams, to be published.
44. B. A. Levine, D. Mercola, and J. M. Thornton, *FEBS Lett.*, **61,** 218–222 (1976); also unpublished data.
45. G. Bentley, E. Dodson, D. Hodgkin, and D. Mercola, *Nature*, **261,** 166–168 (1976).
46. G. Bentley, K. Williamson, and R. J. P. Williams, to be published.
47. B. A. Levine, J. R. P. Madden, and R. J. P. Williams, unpublished data.

48. J. W. Becker, G. N. Reeke, B. A. Cunningham, and G. M. Edelman, *Nature*, **259,** 407–411 (1976).
49. C. C. F. Blake and P. R. Evans, *J. Mol. Biol.*, **84,** 585–597 (1974).
50. P. Tanswell, E. W. Westhead, and R. J. P. Williams, *Eur. J. Biochem.*, **63,** 249–265 (1976).
51. J. T. Johansen and B. L. Vallee, *Biochemistry*, **14,** 649–660 (1975).
52. M. F. Perutz and L. F. Ten Eyck, *Cold Spring Harbor Symp. Quant. Biol.*, **36,** 295–310 (1971).
53. G. P. Hess, in *The Enzymes*, 3rd ed., P. D. Boyer, Ed., Vol. III, Academic, New York, 1971, pp. 213–248.
54. L. Sottrup-Jensen, M. Zajdel, H. Claeys, T. E. Petersen, and S. Magnusson, *Proc. Nat. Acad. Sci.* (*U.S.*), **72,** 2577–2581 (1975).
55. J. H. R. Kägi and B. L. Vallee, *J. Biol. Chem.*, **235,** 3460–3465 (1960).
56. C. J. Brandén, personal communication.
57. A. D. Smith and H. Winkler, *Biochem. J.*, **103,** 483–492 (1967).
58. A. Daniels, A. Korda, P. Tanswell, A. Williams, and R. J. P. Williams, *Proc. R. Soc.*, **B187,** 353–363 (1974).

LIST OF INTERVENTIONS

1. Intervention of Ross

What is the connection between a lowering of the free energy and the flexibility of molecular groups of the enzyme, and how is this related to the reason for enzymes being large molecules?

1.1 Intervention of Williams

For a reaction with one elementary step (e.g., electron transfer), conformational mobility at the active site (electron transfer metal atoms such as iron in cytochrome *c*) can be seen as a vibronic coupling; the vibrations are in the metal-ligand bonds. The protein makes the vibration easy by holding iron bond lengths in some slight strain. The vibration could also allow a fluctuation in redox potential so that this potential would match exactly that of a center to which electrons were to be transferred. This would assist tunneling.

For a reaction with many steps, much more complicated internal motions are required, as can be seen in the hydrolysis of a peptide by chymotrypsin. I would say that there the protein provides attacking groups that are in special situations and in specially energized conditions, because of the protein fold, but that the protein also provides mobile surfaces to guide the reactant through the catalytic path. A single rigid structure would not do this job and the importance of a restricted ensemble of conformations could be very great in controlling reaction rates.

I am not greatly impressed by calculations of absolute conformational energies of proteins. I believe that theory of this kind is better applied to the calculation of the energy required to make small changes to excited states of the proteins from known ground-state structures.

2. Intervention of Clementi

I would like to comment on the most interesting report by Professor Williams, who described the static and dynamic aspect of proteins and the reactive aspect (enzyme-substrate) in terms of X-ray and nmr data only. There are new techniques derived from computational and theoretical chemistry that perhaps one could use. Let me summarize the technique as follows:

1. One performs many computations of *ab initio* type, considering different orientations of two molecules (say, two amino acids or one amino acid and one molecule of water).
2. One fits the obtained total energies in form of pair potentials $E = \sum V_{ij}$.
3. One uses such pair potentials as input to statistical mechanics (Monte Carlo, for example) to obtain the structure of water around an amino acid or a protein, or to obtain ΔS and ΔG constants at a given T.

As an example of this technique we can compare the computed and the experimental X-ray diffraction intensity for liquid water at $T = 298°K$ (see, for example, E. Clementi, *Lecture Notes in Chemistry*, Vol. 2, Springer Verlag, 1976). In the same way we can consider water around amino acids. For example, in Fig. 1 we present the isoenergy contour

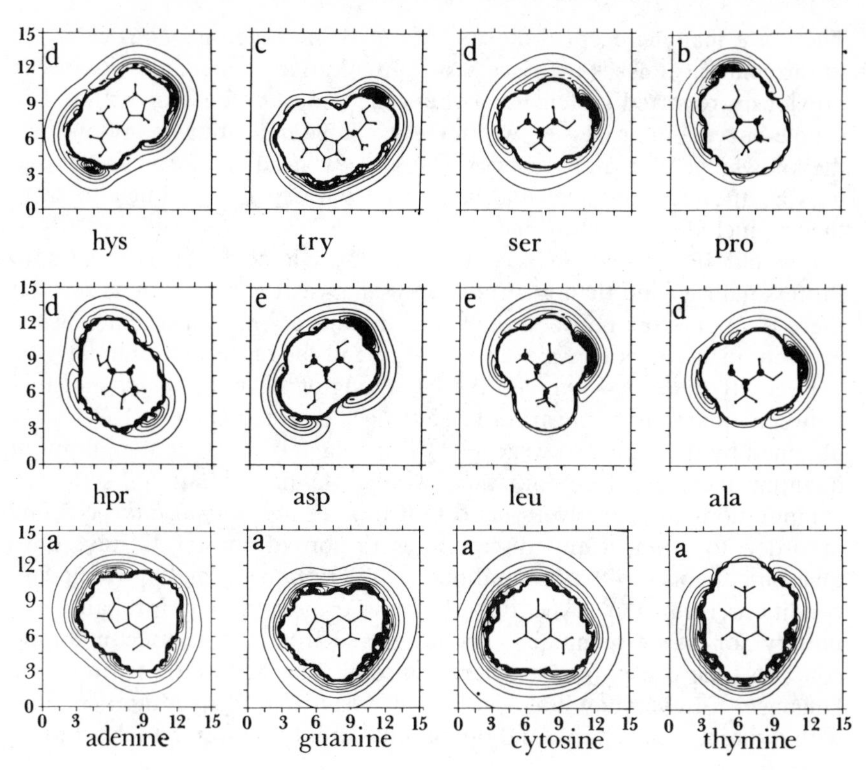

Fig. 1. Isoenergy contours of amino acids in water. Contour interval 1 Kcal/mole.

maps of a number of amino acids and other biomolecules interacting with water. The energy interval (contour to contour) is of 1 kcal/mole: the hydrophobic and hydrophilic regions are very evident. Now *the new aspect of this work is that availability of simple analytical Vij pair potentials allows computation of the interaction of water with a biomolecule several thousand times faster than previously; the interaction of one molecule of water with a protein like lysozyme requires only milliseconds of computer time, and the computer programs are trivially simple.* We are now developing the pair potentials to describe the interaction of any two amino acids, using the preceding three-step technique. It seems to us that the dynamical and the static problem of protein could gain by considering X-ray nmr, and quantum-chemical methods of the type previously outlined.

But there is more. Recently we have quantum-mechanically examined the reaction mechanism of a complex well known in heterogeneous catalyses, namely, C_2H_4 interacting with $Ti—CH_3—(Cl_4)—TiCl_2Al(CH_3)_2$.

There are many concepts one can transfer from heterogeneous catalyses to enzymatic catalyses and vice versa. In addition, the complexity of the previously reported system is of the same order as that encountered in simple enzymatic processes. Again the available indications point out that the knowledge of protein reactions can be advanced at a faster rate if one uses all the methods available not only from experiment but also from theory, including quantum mechanics.

I would like to add a second comment, directed perhaps more to Professor Prigogine than to Professor Williams. It is rather apparent that there is a net break between the two communications, not so much in the problem in itself, but in the terminology and language and foundations.

Recently there has emerged the beginning of a direct, *operational* link between quantum chemistry and statistical thermodynamic. The link is obtained by the ability to write $E = \sum Vij$—namely, to write the output of quantum-mechanical computations as the standard input for statistical computations. It seems very important that an *operational link* be found in order to connect the discrete description of matter (X-ray, nmr, quantum theory) with the continuous description of matter (boundary conditions, diffusion). The link, be it a transformation (probably not unitary) or other technique, should be such that the nonequilibrium concepts, the dissipative structure concepts, can be used not only as a language for everyday biologist, but also as a tool of quantitation value, with a direct, quantitative and operational link to the discrete description of matter.

2.1 Intervention of Magat

I admire very much the X-ray diffraction pattern and the calculated fit you have shown, which are much better than the best agreements of Stillinger and Rahman (J. Chem. Phys. **60,** 1545, 1974). I wonder what was the potential you were using, how was it determined, and what structure of liquid water was assumed?

On the other hand, I agree entirely with your potential graphs, which can explain why the force exerted on a given molecule depends not simply on the dipole moment, but on the charge distribution and approach distances. I tried such calculations in 1932 (*Zeit. Elektrochem. Fasc.* 8*a*), but it was, of course, far too early to achieve such precision.

Finally, I wonder under what conditions you obtain an attractive energy for water by an amino acid that is so much higher than the heat of evaporation of water.

2.2 Intervention of Clementi

The full details of the quantum-mechanical computation to obtain the pair potential are available in O. Matsuoka, M. Yoshimine, and E.

Clementi, *J. Chem. Phys.*, **64,** 1351 (1976); the full details of the computed X-ray diffraction intensity are available in G. C. Lie, M. Yoshimine, and E. Clementi, *J. Chem. Phys.*, **64,** 2314 (1976). The previous computations by Stillinger and Rahman did not use a quantum-mechanically derived potential, but an empirical potential. Present quantum-mechanical techniques, if properly used, can yield remarkably accurate potentials. This fact is not fully appreciated by a large number of chemists, possibly discouraged by the rather large amount of poor theoretical chemistry computations currently in the literature. It is notable that the repulsive part of a potential can be inferred from experiments, in general, with poor accuracy.

3. Intervention of Ubbelohde

A sensitive parameter in the coupling between chemical reaction and diffusion can be the temperature. In many cases the temperature coefficients are markedly different, and a shift of temperature can have a striking effect on systems coupling, compared with temperature effects on simpler molecules.

4. Intervention of Ross

How do you define an "energetic spot" and how does it act in heterogeneous catalysis? Is there any characteristic difference between a protein and a crystal in this respect?

4.1 Intervention of Williams

The best starting point is the heterogeneous catalyst. Take a plane surface of magnesium oxide

Mg O Mg O Mg surface
O Mg O Mg O second layer

Now introduce a defect ⊕

Mg O Mg O Mg
O Mg ⊕ Mg O

The Mg atom above the defect is a better Lewis acid; that is, it has a higher electron affinity than all other Mg atoms in the array, and will be a good catalyst.

In an enzyme the metal atom (or other attacking group) can be put into a geometric or electronic state so as to enhance its acidity (e.g., zinc in carbonic anhydrase). The energization is connected to the fold of the protein and thus fold changes, small in character, can modulate the activity. The metal in the enzyme is linked to the whole protein (or a large part of it) much as is the solid-state defect. It is this feature, together with the requirements for mobility and for specificity, that explains why enzymes are large molecules.

4.2 Intervention of Magat

When you are considering catalysis by amino acids, you speak essentially of what is called the physical adsorption, where the energy of interaction between the "catalyst" and the adsorbed species is of the order of Van der Waals forces that can be sufficient for protein transfer and for other charge transfers that seem to play a very important role in biological reactions but are hardly sufficient for other reactions.

In heterogeneous catalysis we distinguish usually two mechanisms—the acid-base catalysis, which may be of the same type as the amino acid catalysis, and the catalysis by semiconductors and metals. The theory of this last type of catalysis was developed by T. T. Volkenstein in the U.S.S.R., by Germain in France, and by other scientists in Germany and in the United States. This theory is related to what you have indicated for MgO. It is assumed that an electron deficiency or electron excess is introduced as an impurity that creates, ultimately on the surface, a "defect" that can bind "quasi-chemically" electron donors or electron acceptors, respectively.

Finally, concerning irradiations by neutrons and X-rays, the two cases must be distinguished. The neutrons create geometrical defects, displacing nuclei: as you have suggested in the case of MgO, eventually even changing the chemical nature (i.e., realizing the "impurity" situation I mentioned earlier).

The X-rays perturb essentially the electron clouds and the so created defects are cured rather rapidly. All the experiments, particularly those of J. Turkevitch, made in that field led to the conclusion that, whereas very bad catalysts could be slightly improved, the good ones were worsened.

5. Intervention of Magat

There are certain points in the written paper that were not mentioned in the oral presentation on which I would like to comment:

1. It ought perhaps to be underlined that a very strict folding that may be energetically favorable will be extremely unfavorable for the entropy, making an unfavorable contribution to the free enthalpy.
2. Whereas coulombic interactions of ions are spherically symmetrical, the coulomb interactions due to dipoles are, on the contrary, strongly orientation dependent.

3. Finally, the repulsive forces, that as you said play a large role in the definition of "advantageous" foldings, do also play a big role in the definition of crystalline structures of organic compounds and of intermolecular vibration movements. It is very unfortunate that theoretical calculations of repulsive forces are much more difficult than those of attractive forces.

5.1 Intervention of Clementi

In the Schroedinger equation we do not really talk of Van der Waals forces, Coulomb forces, and so on. We simply have a Hamiltonian that includes all those forces and that one attempts to solve for as well as is feasible. Today for systems with 10 to 20 electrons good solutions can be obtained. Few are, however, present in the literature concerning biological systems, where gross oversimplification is still accepted.

6. Intervention of Ubbelohde

An important difference between inorganic heterogeneous catalysts and proteins is that the latter can undergo a whole succession of conformational changes with only a comparatively small energy difference between each "flip." Presumably, quite large activation energies can be built up from such flips. Electron energy levels in the protein molecule can also undergo only moderate changes as a result of conformational flips.

6.1 Intervention of Williams

Conformational flips can be used to assist catalytic activity. Concerted flips could be used to relay information (e.g., in cytochrome *c*). Here there are one tyrosine and one phenylalanine that seem to have concerted onset of motion. Such changes begin to look like second-order phase transitions and are obviously important in membranes.

7. Intervention of Mayer

In your talk you described clearly the power of nmr measurements to give both the static structure and the mean time for many motions in the protein. At the end of your talk you described on the blackboard quite graphically a one-dimensional cut of the catalytic path by a series of hills and valleys on an energy plateau in which, one after another, the valleys deepen, carrying the coordinate position from reactants to products without it ever crossing a high-energy barrier. You gave the impression that you invoked the rapid oscillation of certain subgroups as evidenced by the nmr measurements as reason for this pictorialization.

The usual description that one uses for a very simple reaction

$$Y + AX \Rightarrow YA + X$$

along a line makes a classical analogue with the motion of a billiard ball on a surface with two valleys meeting at an angle at a pass through a high mountain barrier. The solution is that of a purely classical problem fixed surface. *The surface does not pulsate or fluctuate.* The motion of the many atoms involved in enzymatic catalysis is, of course, fantastically more difficult to visualize in the multidimensional space required, much less to compute a rate, but the principle remains the same. The classical motion is in a nonfluctuating potential in coordinate space. The nmr measured speed of "flipping" does not alter this in any direct manner. Only the ensemble average in coordinate space enters the classical description.

7.1 Intervention of Williams

I believe that both potential energy and entropy changes may be required, but I wish to know to what energy does Professor Mayer refer—potential energy or free energy?

There are numerous cases of assisted diffusion but the simplest is that of proton movements in carbonic anhydrase. I do not think that I follow the problem that Professor Mayer has defined.

7.2 Intervention of Mayer

The potential I speak of is usually called the potential of average force. Insofar as it is to be identified to a thermodynamic potential it is a local Helmholtz free energy as a function of the coordinate positions of all the atoms (or radicals) that must change relative positions in the reaction; it may be defined by

$$A_r(\mathbf{r}_1, \mathbf{r}_2, \ldots, \mathbf{r}_n) = -kT \ln \left[\rho_n(\mathbf{r}_1, \ldots, \mathbf{r}_n) \Big/ \prod_{i=1}^{n} \rho_i \right]$$

where $\rho_n(\mathbf{r}_1, \ldots, \mathbf{r}_n)$ is the ensemble probability density that positions $\mathbf{r}_1, \ldots, \mathbf{r}_n$ be occupied by molecules of the specified species and ρ_i the (average) number density of species i.

The *entropy of activation* is determined by a simple integral at right angles to the path of the reaction.

8. Intervention of Caplan

As an attempt to connect the first discussion, which was concerned with diffusion-reaction coupling, with Dr. Williams' presentation of enzymes as dynamic systems, I wanted to direct attention to a number of specific systems. These are the energy-transducing proteins that couple scalar chemical reactions to vectorial flow processes. For example, I am thinking of active transport (Na-K ATPase), muscular contraction (actomyosin ATPase), and the light-driven proton pump of the well-known purple membrane (bacteriorhodopsin). Does the situation described in Dr. Williams' manuscript, where the substrate itself is strained by virtue of binding to the protein, apply to the ATPases mentioned, and if so, how does the strain energy relate to the conversion of chemical free energy to mechanical or osmotic energy?

8.1 Intervention of Williams

I reply to the second question first. I forgot to describe the distortion of substrate on binding to an enzyme surface. To my efforts to describe the protein energies and motions I should add that once the substrate is bound and before the product leaves, there will be parallel (and cooperative) considerations that apply to protein, *E*, and substrate, *S*. However, the reaction path internal to *ES* must leave *E* unchanged and *S* converted to product so that sites on *E* may be initially and finally in unconventional geometric states whereas *S* will only be in such a state after binding and before dissociation. The relationship of strain to the conversion of mechanical energy to chemical energy will occur if the protein does not relax after the reaction but has a period in tension.

The answer to the first question relates to the coupling of proton energies generated by light or oxidation to ATP (adenosine triphosphate) synthesis

$$h\nu + (\mathrm{H}) \rightarrow \leftarrow \mathrm{H}^+ \quad \mathrm{ADP} + \mathrm{P} \rightarrow \mathrm{ATP}$$

My remarks apply equally to Dr. P. Mitchell's views and my own, which were independently formulated [R. J. P. Williams, *J. Theoret. Biol.*, **1,** 1–13 (1961); P. Mitchell, *Nature*, **191,** 144–148 (1961)], for although they have differences, important ones, the differences do not matter here. We both write reactions as follows

$\mathrm{H} \rightarrow \mathrm{H}^+_{(1)}$ (one part of space, creation of protons)

$\mathrm{H}^+_{(1)} \rightarrow \mathrm{H}^+_{(2)}$ (another part of space, diffusion of protons down a gradient)

The drop in energy from one part of space to another is within or across (this a personal choice) a membrane (see Fig. 1) at a phase boundary. I write

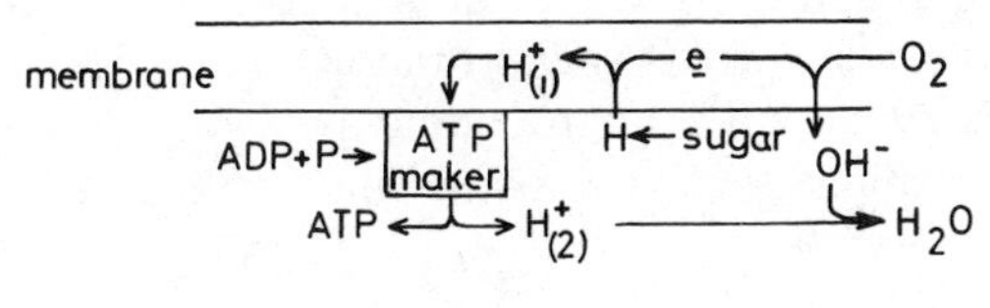

Fig. 1

It is thought that an accumulation of protons, $nH^+_{(i)}$, is needed before ATP is made where n is 2 to 4. Let us compare this with the formulation of Professor Prigogine (see Fig. 2).

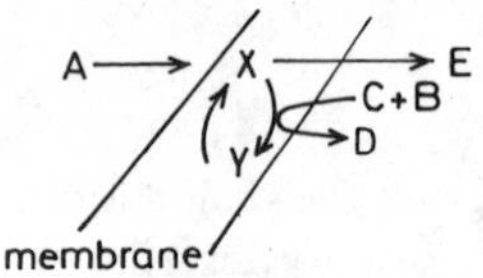

Fig. 2.

Then $A = H$, $X = H^+_{(1)}$, $Y = H^+_{(2)}$, $E = H_2O$, $C + B = ADP + P$, and $D =$ ATP.

The parallel points are obvious but the questions that arise are

1. Do we expect such a system to oscillate?
2. Is this possibly affected by the closeness of the system to equilibrium for we know that oxidative phosphorylation can run backward, and it is not far from equilibrium?
3. Is the diffusion restriction (by the membrane as shown) of any consequence? Note that the enzymes must be ordered in this system.

8.2 Intervention of Hess

In answer to Professor Williams I would like to mention that the overall far-from-equilibrium condition of a process such as oxidative phosphorylation is given as soon as a suitable reductant and oxygen as a source are available for the sink production of CO_2, water, and heat. This overall condition does not define which intermediary steps are near equilibrium and whether some are far off equilibrium. In glycolysis many enzymic reactions are roughly between 8 and 12% near equilibrium. No precise measurements on components of oxidative phosphorylation are available, although it should be mentioned that some components, such as cytochrome b (T) might change their redox mid-potential well over 250 mV, depending on flux conditions as well as the phosphorylation potential (for discussion see B. Hess, "Energy Utilization for Control," in *Energy Transformation in Biological Systems*, ASP, Elsevier-Excerpta Medica, North-Holland, New York, 1975). In addition, it should be recognized that coupling of chemical reactions via large ion diffusion gradients is part of the current chemiosmotic theory. The model Professor Williams has drawn on the board might perfectly describe the overall mechanism and might well fit the oscillatory state of mitochondrial oxidative phosphorylation.

9. Intervention of Caplan

I would like to make a brief remark with reference to the question of efficiency in the phosphorylation step of oxidative phosphorylation according to the models of Mitchell or Williams. Dr. Williams asked whether the system needs to be near equilibrium in order to operate at high efficiency. The point is that two *separate* processes are occurring: proton transport driven by the electrochemical potential gradient of the protons, and phosphorylation against the affinity or negative free energy of the ATP hydrolysis reaction. These two processes need not necessarily be completely coupled to one another; that is, there need not in principle be a fixed or integral stoichiometry between them. Even though each process individually may be relatively far from equilibrium, high efficiencies can nevertheless be achieved if in fact they *are* tightly (if not completely) coupled and the two driving forces are in the appropriate ratio (O. Kedem and S. R. Caplan, *Trans. Faraday Soc.*, **61,** 1897, 1965).

10. Intervention of Goldanskii

One of the main problems mentioned in the talk of Professor Williams was the correlation between the structures and functions of biological compounds, which can be treated as a continuation of the most important chemical problem of correlation between the structures and reactivities. These problems are indeed of extraordinary significance and I would like to mention that besides the nmr (taken as an example of method in Professor Williams' talk), other hyperfine interactions are also very fruitful in the investigation of the previously mentioned correlations.

One variant of ESR method should be mentioned in this connection, namely, the recombination-kinetic method widely used by our laboratory in the Institute of Chemical Physics (Academy of Sciences, Moscow, U.S.S.R). This method opened the possibilities of observation of extremely slow diffusion ($D \gtrsim 10^{-17}$ to 10^{-18} $cm^2 s^{-1}$, linear velocities of paramagnetic centers $\nu \gtrsim 10^{-7}$ $cm\, s^{-1}$) and of the studies of both intraglobular (e.g., caused by segmental motion) and interglobular recombination of paramagnetic labels in proteins and other biopolymers (see Ref. 1 for general description of the method).

Mössbauer spectroscopy should also be mentioned here as a very promising method for combining the structural and dynamic studies of biomolecular systems. The asymmetry of Mössbauer spectra caused by the anisotropy of vibrations of Mössbauer atoms allowed—for example, to find that the mean square amplitude of vibrations of Fe atoms normal to the plane of porphyrin ring (which are responsible for many important biological functions of hemoproteins) is about five times larger than in the ring's plane.[2] The use of time-dependent emission Mössbauer spectroscopy allowed the direct determination of the duration (τ) of intramolecular transfer of electrons—by the example of transformation of Turnbull blue like iron cyanide complexes into Prussian blue type forms of such complexes: $\tau \sim 10^{-8}$ s.[3] This example can be important for the future studies of the rate of electron transfer within various organelles (e.g., mitochondria and chloroplasts). Finally, we should note here the recent use of Mössbauer spectroscopy for getting the evidence of reversible photo-induced conformational change of the whole photoreceptor membranes caused by their illumination and accompanying rhodopsin-prelumirhodopsin transformation at quite low (about 77°K) temperature.[4]

But what deserves particular attention in relation to our symposium is the general question of low-temperature chemical reactivity as a base of completely new approach to the problems of chemical dynamics, of mechanisms of chemical and biological evolution, of avoiding the role of entropy factors in chemical equilibria.

Usual studies of rates of chemical reactions described by Arrhenius law open the possibility of getting only one important dynamic factor—that is, the height of an activation energetical barrier E.

Working at low temperatures: $T < T_t$, where the "tunneling temperature"

$$T_t = \frac{\hbar}{k\pi d\sqrt{2}} \sqrt{\frac{E}{m}}$$

is the "border" temperature below which the quantum tunneling starts to predominate over the classical Arrhenius type above the barrier transitions,[5] one can get all main dynamic characteristics of quantum-chemical reactions, not only the height (E), but also the width of potential barrier (d) and the mass of tunneling particle (m).

Quite recently the studies of radiation-induced solid-state polymerization of formaldehyde over a wide range of temperatures (140 to 4°K) were illuminated by the discovery of quantum-molecular tunneling in chemical reactions, that is, the penetration of the whole molecules or molecular groups through the activation potential barriers of the width of several tenths of angstrom.[6] Such tunnel penetration leads to an appearance of the low-temperature quantum limit of chemical reaction rate,[6,7] which was confirmed recently by the examples of rebinding of ligands to heme proteins[8,9] at very low temperatures.

The existence of chemical reactivity even near absolute zero of temperature when entropy factors play no role in chemical equilibria allows formulation of the idea of "cold prehistory of life," of a cold stage in chemical and biological evolution, to propose the following general principle: Spontaneous formation of even highly organized systems (at any decrease of entropy) can proceed at very low temperatures with only the one condition that such formation is at least slightly exothermal. This process can happen as a very slow but strictly oriented quantum-mechanical tunneling from a zero-vibrational level of initial reactant to various excited levels of final products.

The validity of quantum-chemical kinetics of low-temperature reactions, the existence of the limit of chemical reaction rate,[6] and the applicability of the previously mentioned general principle of cold formation of low-entropy products can be illustrated by recent finding of formaldehyde polymers in interstellar space.[10]

According to J. Lederberg's comment on the previously mentioned findings: "there is no longer any problem in principle about the chemical mechanisms by which the specific monomers and random polymers similar to those in living systems may have been formed." Furthermore, one can think about the significance of low-temperature tunneling not only for the origin of life, but for the general problem of the primitive cosmological condensation taking into account that "condensed cosmic matter belongs almost entirely within the domain of organic chemistry."[11]

References

1. A. I. Mikhailov et al., *Sov. Solid State Phys.* (Fiz. Tverd. Tela, **14,** 1172, 1972 (in Russian).
2. V. I. Goldanskii and R. H. Herber, Eds., *Chemical Applications of Mössbauer Spectroscopy*, Academic, New York, Chaps. 1 and 10.
3. V. P. Alekseev et al., *Sov. JETP Lett.* **16,** 43, 1972 (in English).
4. G. A. Kalamkarov et al., *Dokl.—Biophys.*, **219,** 126, 1974 (in English).
5. V. I. Goldanskii, *Dokl. AN SSSR*, **124,** 1261, 1959 (in Russian).
6. V. I. Goldanskii et al., *Science*, **182,** 1344, 1973.
7. V. I. Goldanskii, *Russ. Chem. Rev.* (Usp. Khim.), **44,** 1019, 1975 (in English).
8. N. Alberding et al., *Science*, **192,** 1002, 1976.
9. V. I. Goldanskii, *Ann. Rev. Phys. Chem.*, **27,** 85, 1976.
10. N. C. Wickramasinghe, *Nature*, **252,** 462, 1974.
11. J. Lederberg, private communication, Sept., 23, 1976.

SYNZYMES: SYNTHETIC POLYMERS WITH ENZYMELIKE CATALYTIC ACTIVITIES

I. M. KLOTZ

Department of Chemistry, and Department of Biochemistry and Molecular Biology, Northwestern University Evanston, Illinois

INTRODUCTION

One of the most remarkable types of homogeneous catalyst is the class of naturally occurring substances called *enzymes.* In the course of a billion years, nature has developed a range of macromolecules with high catalytic efficiency and extraordinary versatility in kinds of reaction whose rates can be accelerated. The history of antivitalism records that in small steps the extraordinary chemistry of a living cell has been slowly unraveled and reproduced externally, de novo, from nonliving materials. It has been a challenge for decades, therefore, to try to reproduce the catalytic properties of enzymes with synthetic macromolecules of nonbiological origin.

An enzyme is a macromolecule. What behavioral features of this biological macromolecule invest it with catalytic powers? It has long been recognized[1-3] that the mode of operation of this macromolecule M involves two sequential processes: (1) binding of the substrate S and (2) provision of a molecular environment conducive to the chemical transformation:

$$M + S \underset{\text{step}}{\overset{\text{binding}}{\rightleftharpoons}} M \cdot S \xrightarrow[\text{step}]{\text{catalytic}} M + \text{products} \tag{1}$$

In terms of transition-state concepts we could also insert the preequilibrium

$$M \cdot S \rightleftharpoons MS^{\ddagger} \rightarrow M + \text{products} \tag{2}$$

into the catalytic step of (1). Particular enzymes show molecular selectivity in the binding step and provide specific functional groups and environments to facilitate the catalytic step, that is, to increase the concentration of $\langle MS \rangle^{\ddagger}$.

To reproduce enzymelike catalytic behavior with a synthetic polymer, we have used, therefore, a two-step approach: (1) fabrication of binding sites on a suitable macromolecular framework; (2) introduction of groups to provide a conducive environment or specific functional side chains to favor formation of the transition state $\langle MS \rangle^{\ddagger}$.

II. FABRICATION OF MOLECULES WITH STRONG BINDING PROPERTIES

Our endeavors toward this goal have been strongly guided by observations in much earlier studies of binding of small molecules by proteins.[4] Serum albumin is extraordinary among nonenzymatic proteins in manifesting strong affinities for small molecules of widely different structure. Molecularly oriented and thermodynamic studies early disclosed the energetic quantities associated with its multiple, stepwise interactions. Table I, for example, lists the free energies, enthalpies, and entropies for the binding of several different anions by serum albumin. It is evident from this list that both apolar and electrostatic interactions are involved in anion binding by albumin.[8] Furthermore, the changes in affinities in successive binding steps of a specific small molecule were interpreted to reflect macromolecular changes induced in the protein.[9] Similarly, the affinities for small molecules with far different structures, apparent, for example, in Table I, were accounted for by attributing "conformational adaptability"[10] to serum albumin macromolecules.

Thus one might expect flexible, water-soluble synthetic polymers with suitable side chains to show strong affinities for small molecules. In the course of 20 years we examined the binding ability of polyvinylpyrrolidone, polyvinylpyridine, polylysine, polyacrylamide, polyisopropylacrylamide, polyvinylimidazole, polyvinylmethyloxazolidinone, poly(vinylmethyloxazolidinone-vinylimidazole), poly(vinylpyrrolidone-vinylimidazole), poly(vinylpyrrolidone-vinylalcohol), poly(vinylpyrrolidone-maleic anhydride), poly(vinylmethyloxazolidinone-maleic anhydride), and poly(2-dimethylaminoethylmethacrylate-methacrylic acid).

TABLE I. Thermodynamics of Binding[a] of Some Anions by Serum Albumin

Ion	$\Delta G°$(cal/mole)	$\Delta H°$(cal/mole)	$\Delta S°$(cal)/(mole)(deg)
Cl	-2220	400	8.7
SCN	−4100	0	13.8
Octyl sulfate	−5010	0	16.7
Decyl sulfate	−6030	−2000	13.3
Dodecyl sulfate	−7220	0	24.0
Methyl orange	−6410	−2100	14.5
Azosulfathiazole	−7150	−2000	17.1

[a] References 5–7.

Other investigators have also studied similar synthetic polymers.[11–17] In our experience no water-soluble polymer binds small molecules with an avidity comparable to serum albumin. A comparison of the latter with two of the best-binding polymers[18] is shown in Fig. 1. Neither polymer binds as strongly as serum albumin. Despite its large cationic charge and many apolar (—CH_2—CH_2—CH_2—CH_2—) side chains, polylysine shows very weak affinity for anions. Polyvinylpyrrolidone is more effective but not impressive.

These polymers have high intrinsic viscosities [e.g., about 22 (ml/g), for polyvinylpyrrolidone], which indicates that the macromolecules are swollen and extended in water. In contrast, serum albumin, with an intrinsic viscosity near 4 (ml/g), must be relatively compact. Promising approaches might be, therefore, to obtain a relatively compact conformation with a water-soluble polymer by introducing cross-linkages or by using a highly branched matrix. The latter has proved to be particularly fruitful.

An interesting polymer constrained to a relatively compact conformation is polyethylenimine (PEI), which can be prepared by suitable polymerization of ethylenimine (Fig. 2) to give a highly branched rather than a linear macromolecule.[19] The structure of a segment of this polymer is shown in Fig. 2. Approximately 25% of its nitrogens are primary amines, 50% secondary, and 25% tertiary.[19] The branching of the polymer may be represented schematically as shown in Fig. 3.

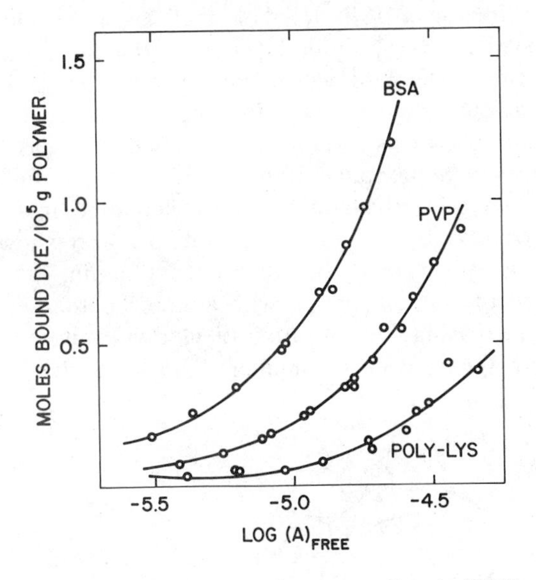

Fig. 1. Comparison of binding affinities of bovine serum albumin (BSA), polyvinylpyrrolidone (PVP), and polylysine (POLYLYS) for methyl orange anions (A), in acetate buffer, pH 5.6, 0.1 ionic strength, and 25°C.

POLYETHYLENEIMINE (PEI)

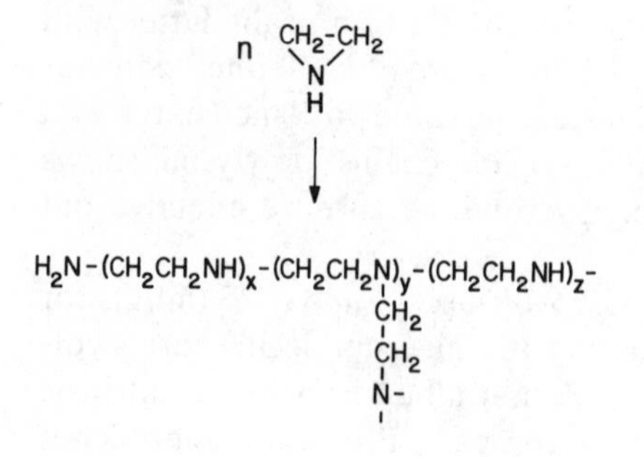

Fig. 2. Structure of a segment of polyethylenimine.

The primary amine groups form a very suitable locus for the attachment of apolar groups to the polymer. We have prepared, therefore, a number of derivatives with different side chains attached to a portion of the primary amine groups. These modified polymers show remarkable binding properties.[20] Figure 4 illustrates the tremendously greater extent of binding by the acylpolyethylenimines as compared with serum albumin. Note that the ordinate axis in this figure extends over the range 0 to 300, whereas that in Fig. 1 covers only the interval 0 to 1.5. At a free methyl orange concentration of 10^{-5} *M*, the lauroyl derivative of polyethylenimine binds over 100 moles of small dye molecule, the hexanoyl about 10, and the butyryl about 1 (Fig. 4), whereas earlier studies with bovine albumin[18,21] lead to values just below 1 (Fig. 1). Admittedly, lauroylpolyethylenimine has a very long apolar side chain. On the other hand, less than 10% of its residues are acylated, whereas serum albumin contains nearly 40% nonpolar amino acid residues. Furthermore, the hexanoyl and butyryl derivatives of polyethylenimine, in which side-chain lengths are comparable to those in proteins, also show substantially greater binding or organic anions than does albumin.

Also remarkable in comparison with albumin is the very steep rise in binding with increasing concentration of methyl orange. In Fig. 4 this is most strikingly apparent in the smaller chain derivatives of the polymer

Fig. 3. Schematic representation of multiple branching in polyethylenimine.

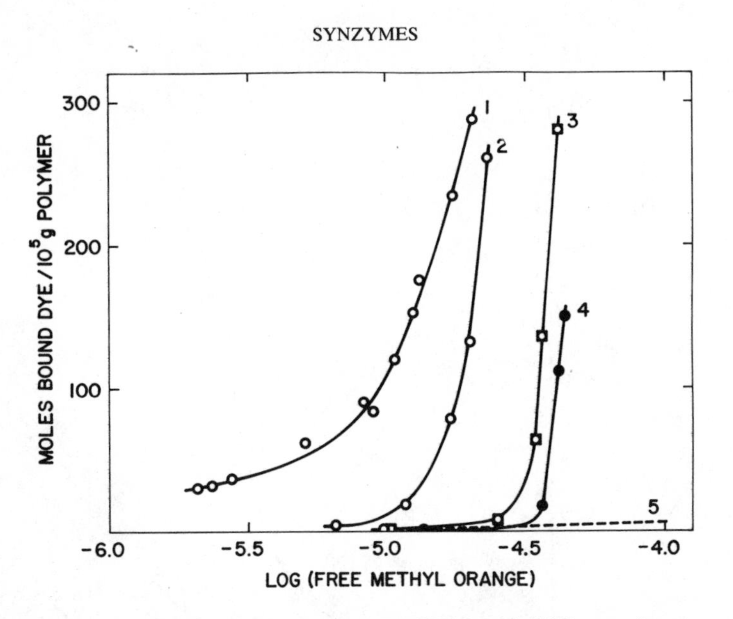

Fig. 4. Extent of binding of methyl orange at pH 7.0 and 25°C as a function of free (nonbound) dye concentration. (*1*) Polyethylenimine with 8.4% of residues acylated by lauroyl groups. (*2*) Polyethylenimine with 11.5% of residues acylated by hexanoyl groups. (*3*) Polyethylenimine with 10% of residues acylated by butyryl (○) or isobutyryl (□) groups. (*4*) Polyethylenimine, PEI-600. (*5*) Bovine serum albumin.

(curves 4, 3, and 2, compared with 5) but, as more sophisticated graphical analyses based on various linear transformations of the fundamental binding equations indicate, it is equally true of the lauroyl derivative.

In connection with widespread interest in the molecular basis of the effect of urea on proteins and their interactions in aqueous solution, it may be pertinent to examine the measurements of binding by substituted polyethylenimines in the presence of this denaturant. These results, together with an extension of the concentration range for binding in the absence of urea, are shown in Fig. 5. Obviously, the presence of urea markedly reduces the binding affinity of lauroylpolyethylenimine. Classically, the effects of urea on proteins have been attributed to the disruption of peptide hydrogen bonds by this solute. In view of the present results with polyethylenimines, however, it seems unlikely that the mechanism of urea action in proteins involves disruption of N—H · · · O═C bonds. Similar conclusions have been reached previously from studies of the effect of urea on binding of anions by polyvinylpyrrolidone[18] and on the acid-base behavior of organic molecules attached to proteins and polymers.[16] Since urea has similar effects on

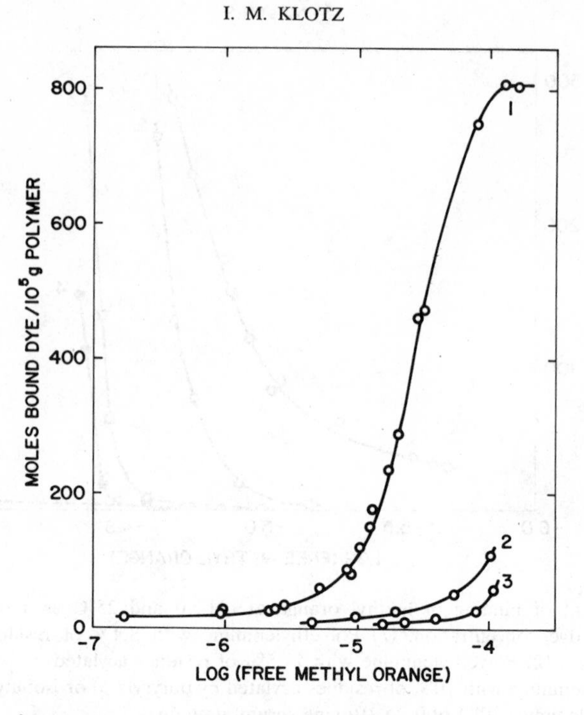

Fig. 5. Effect of urea on extent of binding of methyl orange, at pH 7 and 25°C, by polyethylenimine with 8.4% of residues acylated by lauroyl groups. (*1*) Tris-cacodylate buffer, 0.1*M*. (*2*) Buffer and 6.0 *M* urea. (*3*) Buffer and 9.0 *M* urea.

aqueous solutions of synthetic and biopolymers of such very different molecular structure and conformation, it seems unlikely that it exerts its perturbing effects by direct combination with these different macromolecules. The very fact that large concentrations of urea are necessary for all these effects indicates that its action is due to a change in the character of the solvent environment of the macromolecule.[16,18]

III. ENHANCED REACTIVITY OF NUCLEOPHILES INTRINSICALLY IN THE POLYMER

As was mentioned earlier, polyethylenimine has many amine nitrogens in it, and these have the potential to act as nucleophiles (e.g., in an aminolysis reaction). In addition, they are locally concentrated (Figs. 2 and 3). Furthermore, polymers with attached aliphatic acyl groups provide apolar binding sites in proximity to amine residues of the polymer. One might expect, therefore, to find progressively enhanced rates of aminolysis of substrates with increasingly large apolar substituents. Quantitative measurements of rates of aminolysis indeed reveal enhancements of several orders of magnitude.

For assessment of the aminolytic effectiveness of the polymer amines, we have used[22] a cleavage reaction commonly employed by enzymologists interested in hydrolytic enzymes:

$$O_2N{-}C_6H_4{-}O{-}\overset{\overset{O}{\|}}{C}{-}R \xrightarrow{H_2O} O_2N{-}C_6H_4{-}OH + HO{-}\overset{\overset{O}{\|}}{C}{-}R \qquad (3)$$

If an amine P-NH_2 is used in the aqueous solution, one obtains RCONHP instead of RCOOH. Rates of cleavage of three acyl nitrophenyl esters were followed by the appearance of *p*-nitrophenolate ion as reflected by increased absorbances at 400 nm. The reaction was carried out at pH 9.0, in 0.02 *M* tris(hydroxymethyl)aminomethane buffer, at 25°C. Rate constants were determined from measurements under pseudo-first-order conditions, with the residue molarity of primary amine present in approximately tenfold excess. First-order rate graphs were linear for at least 80% of the reaction. With nitrophenyl acetate and nitrophenyl caproate, the initial ester concentration was 6.66×10^{-5} *M*. With nitrophenyl laurate at this concentration, aminolysis by polymer was too fast to follow and, therefore, both substrate and amine were diluted tenfold for rate measurements.

Table II lists first-order rate constants, corrected for hydrolysis of ester in buffer alone. Propylamine served as a reference amine; in its presence k (in min^{-1}) for aminolysis decreased progressively from 0.98×10^{-2} to 0.51×10^{-2} to 0.05×10^{-2} as the length of the acyl group increased from 2 to 12 carbons (see Table II). The sharp drop for nitrophenyl laurate may be the result of micelle formation[23] even at concentrations of 6×10^{-6} *M*.

With nonacylated polyethylenimines (Table II) the rate constant is increased by a factor of about 4 over that of propylamine. This small enhancement may be due merely to the fact that a greater fraction of primary amine groups in the polymer are in the basic, NH_2 state. With these polyethylenimines, as with propylamine, k drops with increasing length of the hydrocarbon chain of the acyl nitrophenyl ester.

Markedly different trends are seen in the rate constants for aminolysis by lauroylpolyethylenimine (containing 10 residue percent lauroyl groups). For each nitrophenyl ester the rate is substantially greater with lauroylpolyethylenimine than with polymer containing no acyl group. Furthermore, the trend in k is now markedly upward as the acyl group is

TABLE II. First-Order Rate Constants for Amine Acylation by *p*-Nitrophenyl Esters[a]

	$k \times 10^2$ min[b]		
Amine	*p*-Nitrophenyl acetate	*p*-Nitrophenyl caproate	*p*-Nitrophenyl laurate
Propyl	0.98	0.51	0.053
PEI-6[c]	3.60	1.47	0.11
PEI-18[c]	4.38	1.57	0.11
PEI-600[c]	4.60	1.80	0.17
L(10%)-PEI-6[d]	15.2	68.1	698

[a] Measurements made at pH 9.0 in 0.02 *M* tris(hydroxymethyl)-aminomethane buffer, 25°C. Stock solutions of substrate were made in acetonitrile; hence the final buffer also contained 6.7% acetonitrile.

[b] Here $k = k_a - k_0$, where k_a is the measured rate constant in the presence of amine and k_0 is that for the hydrolysis in tris buffer alone; k_0 is 0.94×10^{-2} min^{-1} for the acetyl ester, 0.61×10^{-2} min^{-1} for the caproyl ester, and 0.023×10^{-2} min^{-1} for the lauroyl ester.

[c] The numeral following *PEI* multiplied by 100 is the molecular weight of the polymer sample.

[d] This sample of PEI-6 has 10% of its nitrogens acylated with lauroyl groups.

increased from 2 to 12 carbons (see Table II). Compared to k for propylamine with nitrophenyl laurate, the corresponding k for lauroyl-polyethylenimine is 10^4 times greater. Such a comparison may not be fully appropriate if the low rate with reference amine is due primarily to the micellar state of the lauroyl nitrophenyl ester. If one assumes that in the absence of micelle formation the long-chain ester would show a rate comparable to that of acetyl nitrophenyl ester, the enhancement factor in the presence of lauroylpolyethylenimine still is of the order of 10^3. In any event, it is clear that the introduction of strong binding sites on the polymer leads to marked rate enhancements.

Nitrophenyl esters (3) are activated esters, especially susceptible to nucleophilic groups such as amines. It would be of interest to see if the NH_2 groups of polyethylenimines are also effective in the aminolysis of less-activated esters. As a step in this direction we have examined[24] the kinetics of deacylation of acetylsalicylate (aspirin, I) and of succinyl-disalicylate (diaspirin, II)

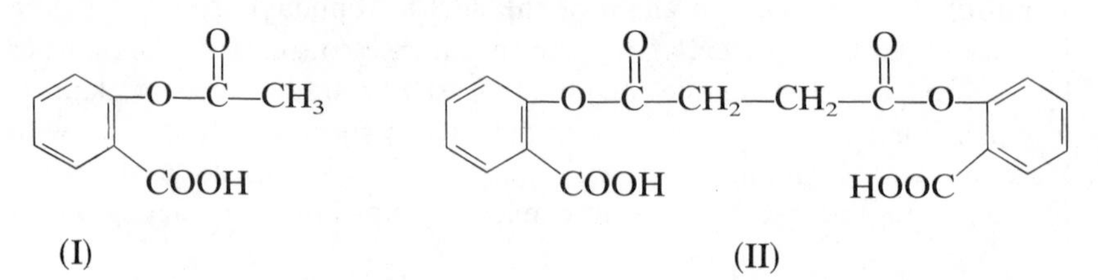

in the presence of these polymers and have compared their effects with that of a simple small-molecule amine.

With these substrates we have made more extensive rate studies and determined more specifically defined kinetic parameters. If each amine-containing binding domain D on the modified polymer binds aspirin A reversibly in a step preceding aminolysis, the kinetic steps may be represented by

$$D + A \underset{k_{-1}}{\overset{k_1}{\rightleftharpoons}} D\cdot A \xrightarrow{k_2} D' + \text{salicylate} \tag{4}$$

For diaspirin S two ester linkages are cleaved in succession and hence the kinetic representation contains an additional intermediate:

$$D + S \underset{k_{-1}}{\overset{k_1}{\rightleftharpoons}} D\cdot S \xrightarrow{k_2} \underset{\text{salicylate}}{\underset{+}{DS'}} \xrightarrow{k_3} \underset{\text{salicylate}}{\underset{+}{D''}} \tag{5}$$

The specific kinetic parameters defined in (4) and (5) have been determined for polyethylenimine and compared with corresponding constants in the presence of CH_3NH_2. These values are summarized in Tables III and IV.

With both substrates, cleavage of salicyl esters is markedly accelerated in the presence of polyethylenimines. The first-order rate constant k_2 in

TABLE III. Kinetic Parameters for Aminolysis of Aspirin

Amine	k_2 (min^{-1})	$\frac{k_2}{\nu K_D}$[a] ($\text{min}^{-1}\ M^{-1}$)	νK_D (M)
$(C_2H_4N)_m$ pH 7.8[b]	0.55	93[c]	0.0059
pH 7.3[b]	0.120		
$(C_2H_4N)_m(C_{12}H_{25})_{0.098m}$ pH 7.8[b]	0.65	78	0.0083
$(C_2H_4N)_m(CH_3CHCH_3)_{0.31m}$ pH 7.3[b]	0.00145		
CH_3NH_2 pH 7.8[b]		0.0255[d]	
None pH 7.8	0.000171[e]		

[a] $\nu = l/n$ = number of primary amines per binding domain, $K_D = (k_{-1} + k_2)/k_1$.

[b] Bis-tris buffer, 25°C.

[c] The average pK_a of the primary amines on $(C_2H_4N)_m$ is near 9.5. If the kinetic parameters are calculated in terms of the concentration of unprotonated amine on the polymer, the following values are obtained: $\nu K_D = 1.15 \times 10^{-4}\ M$; $\nu k_2/K_D = 4.78 \times 10^3\ \text{min}^{-1}\ M^{-1}$.

[d] This value for CH_3NH_2 is actually k_{am}, the second-order rate constant.

[e] This value is the pseudo-first-order rate constant for buffered water with no amine present.

TABLE IV. Kinetic Parameters for Aminolysis[a] of Diaspirin

Amine	Kinetic parameter	Value of parameter
$(C_2H_4N)_m$ pH 7.8[c]	k_2	0.58 min^{-1}
	k_3	8.7 min^{-1}
	νK_D[b]	6.8×10^{-4} M
	k_c[b]	0.53 min^{-1}
	$k_c/\nu K_D$	800 min^{-1} M^{-1}
CH_3NH_2 pH 7.8	k^{α}_{am}[d]	0.0343[e] min^{-1} M^{-1}
	k^{β}_{am}[d]	0.0171[e] min^{-1} M^{-1}
None pH 7.8	k^{α}_{hyd}[d]	0.0003 min^{-1}

[a] Bis-tris buffer, 25°C.

[b] ν = number of primary amines per binding domain, $K_D = k_{-1}/k_1$, $k_c = k_2k_3/(k_2+k_3)$.

[c] The average pK_a of the primary amines on $(C_2H_4N)_m$ is near 9.5. If the kinetic parameters are calculated in terms of the concentration of unprotonated amine on the polymer, the following values are obtained: $\nu K_D = 1.33\times10^{-5}$ M and $\nu k_c/K_D = 4.0\times10^4$ min^{-1} M^{-1}.

[d] k^{α} and k^{β}, respectively, represent the successive rate constants for release of each salicylate of diaspirin.

[e] If the rate constants are calculated in terms of the concentration of unprotonated methylamine, then $k^{\alpha}_{am} = 21.6$ min^{-1} M^{-1} and $k^{\beta}_{am} = 10.8$ min^{-1} M^{-1}.

the polymer environment is 10^3 times greater than in water when acetylsalicylate is cleaved, and corresponding comparisons with succinyldisalicylate show enhancements of the same magnitude. A more appropriate comparison is between polymer and methylamine. With acetylsalicylate, the second-order rate constant $k_2/\nu K_D$ with the polymer is some 10^3 to 10^4 times larger than the second-order constant k_{am} in the presence of small-molecule amine, if the total amine concentrations are used in calculating the rate constants. Similarly for succinyldisalicylate, $k_c/\nu K_D$ is 10^4 to 10^5 times larger than the second-order rate constants for methylamine (see Table IV). If unprotonated amine concentrations are used to calculate rate constants (see Tables III and IV), then $k_2/\nu K_D$ is 10^2 to 10^3 times larger for acetylsalicylate and 10^3 to 10^4 times larger for succinyldisalicylate.

Comparing different polyethylenimines (Table III), we observe that the deacylation rate with the lauryl polymer is approximately equal to that for unmodified polymer. Kinetic analysis reveals that the binding of substrate by the respective polymers, as measured by νK_d, is not appreciably different, nor is the rate constant k_2. With this substrate, then, there is no evidence that added lauryl groups on the polymer increase the effectiveness of the polymer.

It is also of interest to note that the first-order rate constant k_2 for acetylsalicylate in the presence of polyethylenimine is comparable to its counterpart k_c (see Table IV) for succinyldisalicylate. Thus the amine group, whether it is on the polymer or on a small molecule, does not discriminate between acetylsalicylate and succinyldisalicylate. On the other hand, the binding affinity of the polymer for the latter compound is tenfold stronger than for the former. Since the polymer carries a large positive charge it seems reasonable that it should bind dianions more strongly than monoanions.

From previous determinations of the moles of nitrophenyl acylates cleaved by polyethylenimines we presumed that only the primary amine groups of the polymer were functional in aminolysis and hence we assumed that only these would be effective in the deacylation of acetylsalicylate. To test this assumption we have examined the kinetics of aminolysis of aspirin by isopropylated polyethylenimine.

Isopropylation of polyethylenimine by reductive alkylation with acetone and sodium borohydride has been shown to occur selectively with the primary amine residues.[25] Four repeated reductive alkylations gave us preparations with assays for residual primary amines of 0.6 to 0.9%. This could be an artifact reflecting a small extent of reaction of the assay reagent, trinitrobenzenesulfonate, with some secondary amines, or it may be a measure of trace amounts of primary amines in positions on the polymer not conducive to Schiff-base formation with the acetone reagent. Treatment of the isopropylated polymer $(C_2H_4N)_m(CH_3CHCH_3)_{0.31m}$ with acetic anhydride should convert all secondary amines to tertiary amides and any residual primary amines to secondary amides. The proton nmr spectrum of a sample of $(C_2H_4N)_m(CH_3CHCH_3)_{0.31m}(CH_3CO)_{0.67m}$ revealed only one peak in the acetyl region at a frequency corresponding to tertiary amides.[25] Thus no residual primary amines were evident in pmr spectra.

Deacylation rates of acetylsalicylate by isopropylated polymer at pH 7.3 were markedly slower than with unmodified polymer (Table III). With 2.4×10^{-2} total residue molar concentration of isopropylated and unmodified poly(ethylenimine), respectively, k_2, the pseudo-first-order rate for the former was 1.45×10^{-3} min^{-1}, for the latter 1.2×10^{-1} min^{-1}. Clearly, primary amines are the major sites for aminolysis of aspirin.

The most significant insight into the behavior of the polymer with these acylsalicylates is obtained from a comparison of the successive rate constants, k_2 and k_3, for the stepwise cleavage of the salicylate groups in succinyldisalicylate. The second step is faster than the first by a factor of at least 20. Such behavior seems to be a striking manifestation of the kinetic acceleration that is feasible when the substrate is restricted on the binding macromolecule to a region close to a reactive group. In this case the covalent bond holding the succinylmonosalicylate to the polymer [DS' in (5)] most probably restrains translational and rotational motions so that the possibility of interaction with a neighboring primary amine group of the polymer is substantially increased. These kinetic experiments thus suggest that the primary-amine side chains of polyethylenimines are flexible enough and can be close enough that they can act as loci for bifunctional effects in accelerating reaction rates.

These experiments with acyl esters and the intrinsic primary-amine nucleophiles of polyethylenimines thus illustrate that this polymer with binding domains for substrate provides a suitable framework for achieving enhanced reaction rates in aqueous solution. Aminolysis, however, is not a catalytic phenomenon, for the $\mathrm{R\overset{\overset{O}{/\!/}}{C}—}$ adduct is covalently attached to the amines of the polymer and these nucleophiles are thereby removed from action.

IV. CATALYSIS OF ESTER CLEAVAGE BY POLYETHYLENIMINES WITH ATTACHED IMIDAZOLE GROUPS

Having invested the polymers with binding abilities, and having demonstrated their capacity to accelerate rates, we can proceed to graft on to the macromolecular matrix truly catalytic functional groups. We have concentrated largely on the imidazole moiety since it is a well-recognized nucleophile, particularly effective in model systems in the catalysis of hydrolytic reactions.

A derivative of polyethylenimine was prepared containing 15% of its nitrogen residues alkylated with methyleneimidazole

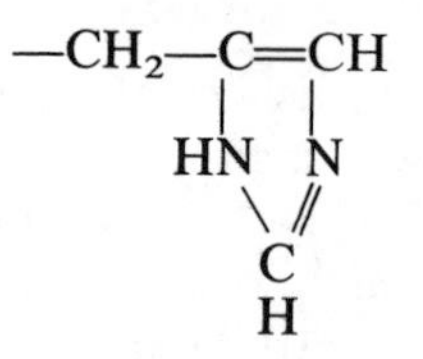

side chains and 10% with dodecyl groups. The former provided nucleophilic catalytic sites; the latter, binding sites. Ninhydrin, used to detect residual primary amines, gave no color with the modified polymer. Hence the alkylation must have occurred preferentially with primary amines of the polymers.

Cleavage of *p*-nitrophenyl caproate [(3)] was followed by the appearance of *p*-nitrophenolate ion, as detected by increased absorbance at

400 nm.[26] The reactions were run in 0.02 *M* tris · HCl buffer (pH 7.3) containing 1.25% acetonitrile, at 25°C.

That the cleavage of nitrophenyl caproate by the polymer is truly hydrolytic was demonstrated in two ways. First, repeated additions were made of 1×10^{-4} *M* substrate, up to five times the concentration of polymer imidazole groups (6×10^{-5} residue molar); nitrophenol was released completely each time. At lower polymer concentrations it was also possible to add a single injection of substrate, and in every case the molar amount of nitrophenol released was substantially in excess of the concentration of imidazole groups on the dodecylpolyethylenimine (Fig. 6).

To compare the catalytic effectiveness of our polymer with that reported for other substances that accentuate nitrophenyl ester cleavage, we[26] have carried out a series of experiments (at pH 7.3) in which the residue molar concentration of polymer imidazole groups was substantially in excess of the concentration of substrate, *p*-nitrophenol acylate. Pseudo-first-order rate constants k_1' were determined at each of a number of polymer concentrations. Under these conditions k_1' was found to be linear with $[\text{P-Im}]_0$, the initial residue concentration of methyleneimidazole groups:

$$k_1' = k[\text{P-Im}]_0 \tag{6}$$

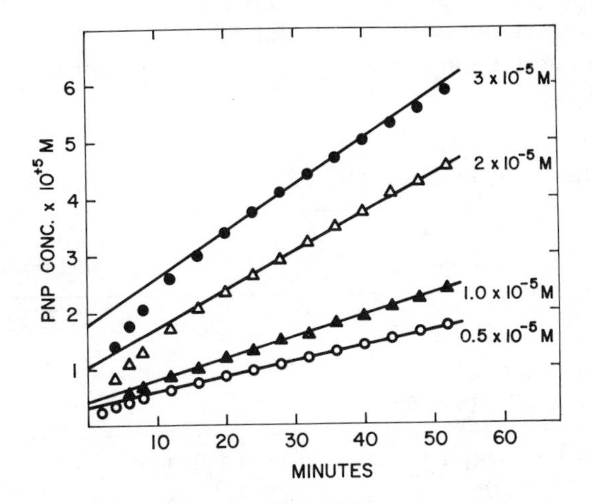

Fig. 6. Rates of esterolysis of *p*-nitrophenyl caproate at pH 7.3 and 25°C in presence of a derivative of polyethylenimine (PEI-600) containing 10% of its residues alkylated with dodecyl groups and 15% alkylated with methyleneimidazole substituents. The numbers shown adjacent to each curve are the residue molar concentrations of imidazole groups in solution. Initial concentration of substrate was 1×10^{-4} *M*.

Values of a "catalytic constant,"[27] k computed from (6) are listed in Table V. It is obvious that the imidazole-substituted dodecylpolyenimines are more than 100 times as effective as the simple imidazole[28,29] molecule itself. The "catalytic constant" for the imidazole-dodecyl polymer in fact approaches that of the enzyme chymotrypsin.

The shapes of the curves in Fig. 6 are consistent with a two-step pathway, analogous to that of a hydrolytic enzyme such as α-chymotrypsin,[30] in which an initial acylation "burst" is followed by a slow deacylation reaction. Following a fast preequilibrium binding, the first kinetic step can be attributed to acylation by substrate of the polymer imidazole residue, accompanied by simultaneous release of nitrophenol(ate). The succeeding kinetic step would then be ascribed to hydrolysis of the acylimidazole leading to carboxylate ion and regenerated imidazole.

It has seemed desirable to try to detect the postulated acylimidazole intermediate by spectroscopic probing. Acetylimidazole has been shown[31] to have an absorbance maximum at 245 nm with an extinction coefficient of 3000. This absorbance should provide a basis for detection of the intermediate. However, in practice the strong absorbances due to the aromatic ester substrate (nitrophenyl caproate) and the product (nitrophenol(ate)), added to the light scattering from the polymer, have made the spectrophotometric observation of the acylimidazole intermediate unfeasible under the reaction conditions previously described.[26]

We have now circumvented these difficulties by using a weakly absorbing substrate, acetic anhydride, in place of the nitrophenyl esters.

TABLE V. Relative Effectiveness of Various Catalysts in Cleavage of Nitrophenyl Esters

Catalyst	Catalytic Constant k (liter/(mol)(min))
Imidazole	10[a]
α-Chymotrypsin	10,000[b]
Polyethylenimine-600 10% dodecyl, 15% imidazole groups[c]	2,700

[a] The substrate used was p-nitrophenylacetate, at a pH near neutrality. Taken from Ref. 29.

[b] Taken from Ref. 27.

[c] This preparation of PEI-600 had 10% of its residues alkylated with dodecyl (i.e., lauryl) substituents and 15% alkylated with methyleneimidazole groups.

When acetic anhydride and laurylimidazole PEI are mixed the following reactions (in addition to spontaneous hydrolysis) are postulated to occur

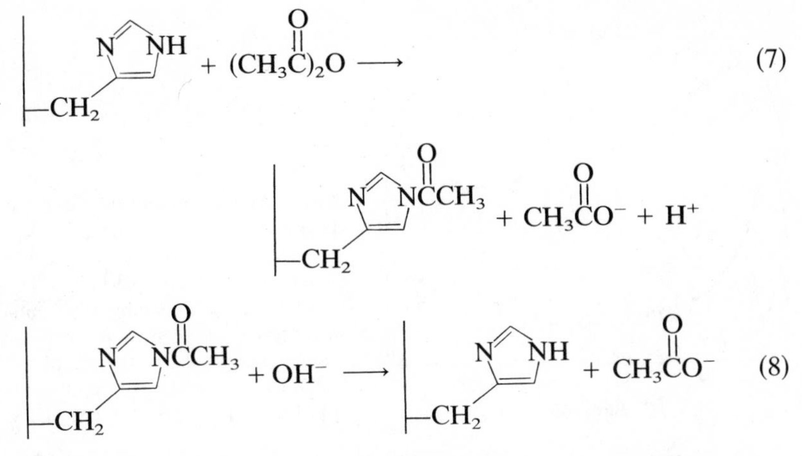

If they do, and if the second reaction is rate limiting, the concentation of laurylacetylimidazole PEI should first rise rapidly, then reach a maximum, and finally decrease. Since the acylimidazole intermediate is the only species that absorbs strongly at 245 nm, the absorbance at this wavelength should follow the same pattern.

When acetic anhydride at concentrations of from 0.5×10^{-3} to 4.0×10^{-3} M is allowed to react with laurylimidazole PEI, with imidazole concentrations of 1.3×10^{-4} to 1.7×10^{-4} M, the absorbance (Fig. 7) follows the predicted bell-shaped profile[32] in accord with the proposed two-step mechanism. The initial rates at pH 4.5 (and also those at pH 8.6) indicate a greater than zero-order dependence on the acetic anhydride concentration. The maximum absorbances at pH 4.5 (and at pH 8.6) increase approximately twofold when initial anhydride concentration is increased from 0.5×10^{-3} to 1.0×10^{-3} M. The maximum absorbance change possible in these experiments, assuming complete conversion of the imidazole to acetylimidazole is 0.5. The observed absorbance changes of 0.045 to 0.30 imply that conversions of about 10 to 60% are actually attained. Shortly after the time of maximum absorbance in each experiment, the decay of the intermediate becomes exponential and remains so over at least three half-lives. The rate of deacylation on the polymer increases with pH over the range studied, in contrast with simple acetylimidazole, which displays a rate minimum at about pH 7. The

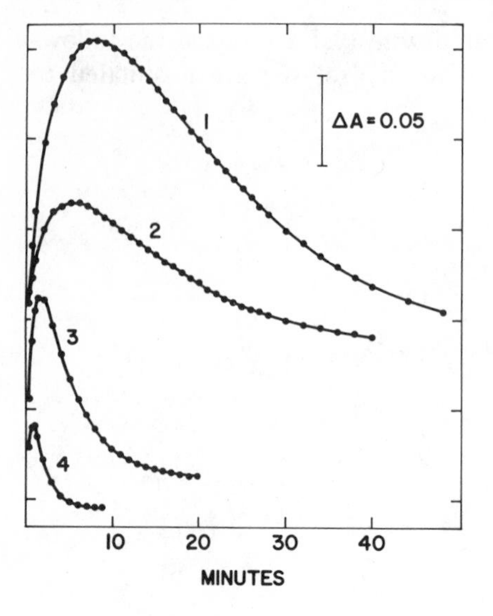

Fig. 7. Formation and subsequent decomposition of an acetylimidazole intermediate in aqueous solutions of laurylimidazole PEI ([imidazole] = 1.7×10^{-4} *M*) and acetic anhydride. Initial concentrations of acetic anhydride and pH values are (*1*) 1.0×10^{-3} *M*, pH 4.5; (*2*) 0.5×10^{-3} *M*, pH 4.5; (*3*) 0.5×10^{-3} *M*, pH 7.8; and (*4*) 0.5×10^{-3} *M*, pH 8.6.

behavior of the polymer is not surprising, however, since the charge on the polymer suppresses protonation of the acetylimidazole attached to it. An acylimidazolium ion, the species responsible for the increase in rates at low pH,[33] is thus formed to a lesser extent on the polymer than in the small model compound. At higher pH the cationic charge on the polymer increases the concentration of hydroxyl ion in the vicinity of the attached acetylimidazole groups and hence rates of deacylation are enhanced relative to those of small molecules.

In the earlier kinetic study[26] the rate constant for hydrolysis of laurylcaproylimidazole PEI to caproate ion and laurylimidazole PEI was estimated to be 0.06 min^{-1} at pH 7.3. In the present study we have found the rate constant for hydrolysis of laurylacetylimidazole PEI to acetate ion and laurylimidazole PEI under similar conditions to be 0.1 to 0.2 min^{-1}. In the earlier study the hydrolysis rate was inferred by an indirect method from the turnover rate in a steady-state situation. In view of the uncertainties in the indirect method and the difference in size of the acyl group in the two cases, the approximate equality of the deacylation rate constants is gratifying.

To resolve any uncertainty caused by the difference between acetate and caproate in the active substrate an experiment was attempted using caproic anhydride as the substrate. The reaction could not be followed, however, presumably because of the low solubility of caproic anhydride.

On the other hand, with propionic anhydride as substrate, in which case there is no solubility problem, the course of the reaction was very much the same as in the experiments with acetic anhydride. The rate constants for decomposition of propionylimidazole are strikingly similar to those for acetylimidazole.

The polyethylenimines are also effective in the cleavage of nitrophenylsulfate esters and nitrophenylphosphate esters. These have not yet been studied as extensively as the acyl esters, but interesting kinetic accelerations are already apparent. Nitrocatechol sulfate, for example, is very stable in aqueous solution at ambient temperature. In fact, even in the presence of 2 *M* imidazole no hydrolysis can be detected at room temperature. At 95°C in the presence of 2 *M* imidazole cleavage is barely perceptible. In contrast, a modified polyethylenimine with attached imidazole groups cleaves the sulfate ester at 20°C.[34] Some kinetic parameters are compared in Table VI. It is obvious that accelerations of many orders of magnitude are effected by the polymer.

Some exploratory experiments have also been carried out with some phosphate esters. Rate constants for one of these, dinitrophenylphosphate, are listed in Table VII. With this substrate, surprisingly, the intrinsic primary amines of the polymer seem to be acting in a turnover pathway, that is, are phosphorylated and then dephosphorylated, but this mechanism needs to be better substantiated. In any event it, it is clear that the rates in the presence of polymer are accelerated more than 10^3-fold with dinitrophenylphosphate. Even larger accelerations have been observed with other phosphate esters.

TABLE VI. Comparison of Rates of Hydrolysis of 4-Nitrocatechol Sulfate

	Polyethylenimine 600 10% dodecyl, 15% imidazole groups, (20°C)	Spontaneous hydroylsis (20°C)	Imidazole catalyzed (20°C)
$k_{\text{acylation}}$	$5.2\times10^{-2}\ s^{-1}$	–	–
$k_{\text{deacylation}}$	$1.1\times10^{-3}\ s^{-1}$	$<3\times10^{-12}\ s^{-1}$ [a]	–
$K_{\text{prequilibrium binding}}$	$3\times10^{-3}\ M$	–	–
$\frac{k_{\text{acylation}}}{K_{\text{binding}}}$	$18\ s^{-1}\ M^{-1}$	–	$5\times10^{-12}\ M^{-1}$ [b]

[a] In 0.02 *M* tris, pH 9.2. The entry is the pseudo-first-order rate constant.

[b] In 0.02 *M* tris-2 *M* imidazole. The entry is the second-order rate constant at 20°C calculated from measurements at 95°C.

TABLE VII. Comparison of Rates of Hydrolysis of 2,4-Dinitrophenylphosphate

Catalyst	Active group	Rate of hydrolysis (M^{-1} s^{-1})
$PEI_{10\%\ dodecyl}$	Free primary amines	0.9
$PEI_{10\%\ dodecyl,\ isopropylated}$	Secondary amines only	Not detectable
$PEI_{10\%\ dodecyl,\ 12\%\ imidazole,\ isopropylated}$	Free imidazole Secondary amines only	Not detectable
N-Isopropylethylenediamine	Free primary amines	0.0006

Thus polyethylenimine has been found to cleave catalytically esters of widely differing molecular nature.

V. PROBE OF CONFORMATION OF POLYETHYLENIMINES WITH ^{19}F NUCLEAR MAGNETIC RESONANCE

In view of the marked kinetic effects described so far, it has seemed wise to carry out some examination of the three-dimensional disposition of the groups on the polyethylenimines. It has been our hope that even a rudimentary insight into the conformation of the polymer would provide a guide to the search for other types of chemical reaction that might be facilitated in the presence of these polymers. The results of investigations with two conformational probes, ^{19}F nuclear magnetic resonance and excimer fluorescence, are described.

Fluorine magnetic resonance chemical shifts of fluorine-labeled analogs of organic compounds have been demonstrated to give information on the local environment and thus on the location of the ^{19}F label.[35,36] Our strategy, therefore, has been to use 10,10,10-trifluorodecanoyl substituents in place of dodecanoyl, that is, to introduce an $^{19}F_3C$-terminal group on the long-chain hydrophobic moiety attached to the polymer. In this way we[37,38] have taken advantage of the high nmr sensitivity of ^{19}F to local environment.

Figure 8 displays the nmr spectra of a polyethylenimine derivative in which 5.5% of the amine nitrogens had been acylated [designated $(C_2H_4N)_m(CF_3(CH_2)_8CO)_{0.055m}$]. To a good approximation the spectrum is well represented in each case as either one or two broad, symmetrical resonances, due, respectively, to one or two environments for the fluorine probes. (The resonances are sufficiently broad that the spin-spin multiplet structure is entirely obscured.) Clearly, the presence of two resonances is consistent either with slow exchange of fluorines between one environment and the other or with rapid exchange in a situation in which one

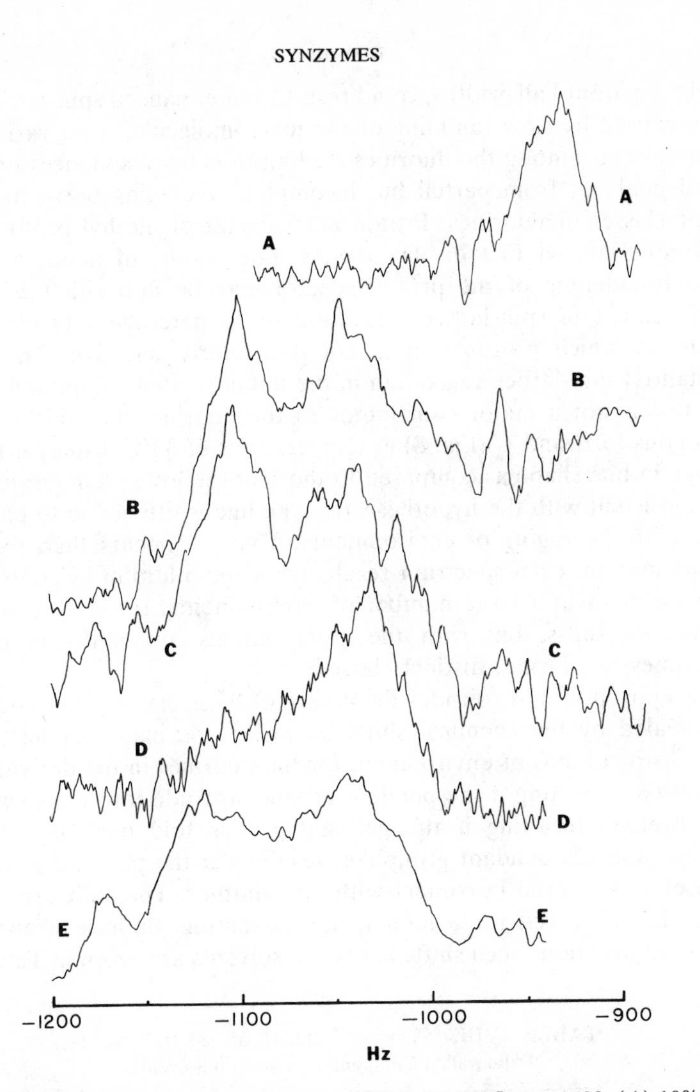

Fig. 8. Fluorine nmr spectra of $[(C_2H_4)_m(CF_3(CH_2)_8CO)_{0.055m}]$, $m = 1400$. (*A*) 10% solution in absolute ethanol; (*B*) 4% solution in water, pH 7.0; (*C*) 2% solution in water, pH 6.7; (*D*) 2% solution in water, pH 8.9; and (*E*) 1.5% solution in 6 *M* urea, pH 9.

fraction of the fluorines experiences a high probability of being in one environment and the remainder have a high probability of being in the other environment. The thermodynamic distinguishability of residues implicit in the second possibility is consistent with the known heterogeneity of PEI and its derivatives.

The substantial line width of the individual resonances, from 20 to

60 Hz apparent half-widths, could result from enhanced spin-lattice relaxation caused by slow tumbling of the macromolecule, from variability of environments among the fluorines leading to many resonances under one broad peak, or from partial but incomplete averaging between the two major classes of fluorines. Proton nmr spectra of methyl protons in the analogous lauroyl PEI display modest line widths of about 5 Hz. The slight broadening of the proton spectra may be due either to a small enhancement in spin-lattice relaxation or to heterogeneity of environments, to which proton nmr is not particularly sensitive. The lack of substantial spin-lattice relaxation in the unfluorinated compound suggests that this is not a major contributor to the fluorine line widths. Spectra analogous to *B* and *C* (Fig. 8) at temperatures of 33°C display little or no change in line shape as compared to those at the lower temperature. This is inconsistent with the hypothesis that the line width is due to partial but incomplete averaging of environments. Thus it appears that, to a good approximation, each spectrum results from the addition of contributions from fluorines in a large number of environments, resonating at a large number of shifts, but with the environments and shifts in one and sometimes two broad, distinct classes.

It is appropriate to consider the nature of these classes of environments as revealed by the chemical shifts. A priori one may visualize at least three distinct kinds of environment for the fluorines in the derivative: (*1*) solventlike, resulting if the pendant groups protrude into the solvent; (*2*) PEI-corelike, resulting if the pendant groups fold back into the PEI interior; and (*3*) pendant grouplike, resulting if the pendant groups find themselves clustered in contact with one another. The shift expected for (*1*) is that observed for a model, nonassociating, fluorine probe in the solvent in question. Such shifts in several solvents are given in Table VIII.

TABLE VIII. Chemical Shifts of Model 1,1,1-Trifluoralkyl Compounds in Several Solvents

Solvent	Chemical shift[a] (Hz at 84.67 MHz)
Ethanol	−905
N-Methylacetamide	−983
In micellar state	−1000 to −1030
Water	−1091
6 *M* Urea	−1118
PEI-6 (extrapolated)	−1175
50% Aqueous PEI-6	−1200

[a] External trifluoroacetic acid was used as reference.

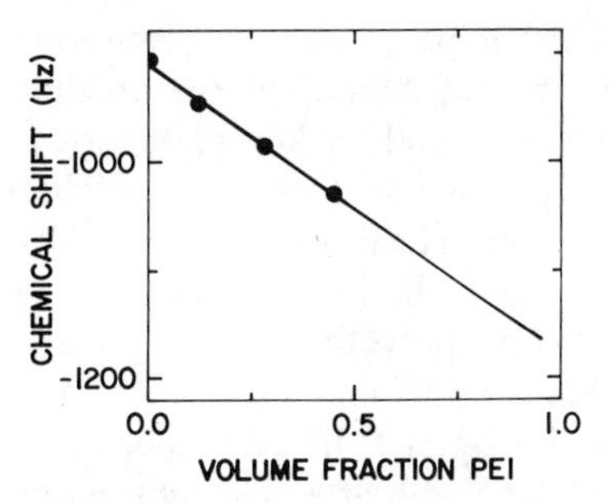

Fig. 9. Fluorine chemical shifts of 10,10,10-trifluorodecanoic acid in PEI-6-ethanol mixtures.

The shift expected for (*2*) depends somewhat on the model taken for the PEI core. If the PEI core is water-free, an estimate of the shift of a model compound is obtained from the data in Fig. 9 as −1175 Hz. In the more likely event that the PEI core is wet, for example, containing 50% water, the shift is estimated as −1200 Hz (Table VIII). In either case the shift is to very low field, lower than for any other environment examined. The shift expected for (*3*) is that for a micellelike environment, which has been shown in several studies[35,36,39] to be −1000 to −1030 Hz, depending on the nature of the aggregating species.

Figure 8*A* gives the spectrum of the polymer derivative in absolute ethanol. The chemical shift is the same, within a few hertz, as expected for a fluorine probe free in the solution in question. Spectra 8*B* and 8*C* display two resonances in each case, one centered at −1045 Hz and the other at −1105 Hz. These shifts are close to being micellelike and waterlike, respectively. Each of the shifts is slightly to low field from what would have been expected for this assignment, positions suggesting that the fluorines in the waterlike environment have some limited exposure to the PEI core and that the fluorines in the micellelike environment have some limited exposure to the PEI core or are exposed to water somewhat more than similar fluorines in detergent micelles. Spectrum *D* displays one dominant resonance at −1030 Hz and a hint of a small resonance at about −1105 Hz. The assignments in this case also are micellelike and waterlike. Spectrum *E* exhibits two resonances, at −1040 and −1120 Hz. Again the shifts can be ascribed to a micellelike and a waterlike (or 6*M* urea-like) environment. In no case is a very low chemical shift found. This indicates that in all the samples examined, the fluorine probe does not spend an appreciable fraction of the time in the PEI core.

Spectra *B* and *C* in Fig. 8 represent two different concentrations of the same polymer under similar conditions. The absence of a noticeable concentration effect on the relative areas indicates that the micellelike structure is not an intermolecular one, which could be further dissociated by a reduction in concentration. The effect of pH on the distribution (compare *C* with *D*) is as expected, since the substantial charge on the polymer at pH 7 should expand the polymer and make interactions among the side chains less likely. Similarly, the effect of urea on the distribution (compare *D* with *E*) is as expected, since urea is known to destabilize micelles [see Ref. 36 and references contained therein).

The fluorine nmr spectra of modified polyethylenimine with trifluorodecanoyl groups attached to 9.5% of the amines, designated $(C_2H_4N)_m(CF_3(CH_2)_8CO)_{0.095m}$ are shown in Fig. 10. In each spectrum there is only one broad resonance (40 to 60 Hz width at half-height) with chemical shifts from 0.8 ppm at low pH to 1.0 ppm at high pH. In this polymer all the trifluorodecanoyl side chains spend most of the time in apolar clusters. Furthermore, the clusters, as far as can be determined

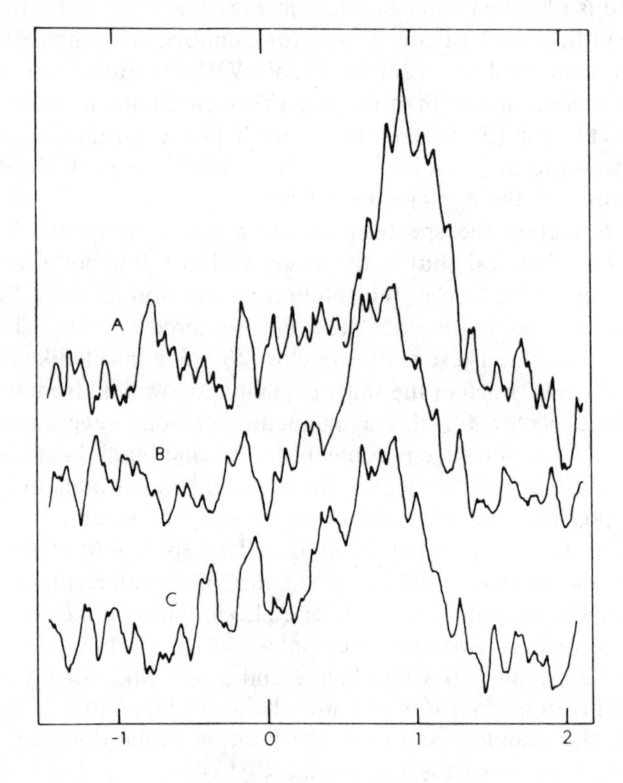

Fig. 10. Fluorine nmr spectra of $(10,10,10\text{-trifluorodecanoyl})_{0.095}$ PEI-600 in water (2% solution) at pH values (*A*) 9.2, (*B*) 6.7, and (*C*) 3.8.

from positions of the chemical shifts, are similar to those formed by detergent ions. The small pH dependence in the chemical shift suggests that, at least at low pH, there is exchange of some of the apolar side chains between the aqueous and the cluster phases. If one assumes, however, that the chemical shift of the label in the cluster on this PEI derivative is about the same as that in typical detergent micelles, one must conclude that even at low pH nearly all the apolar residues are within the clusters. In view of the results for $(\text{trifluorodecanoyl})_{0.095}$PEI-600, the shift of the high-field resonance of $(\text{trifluorodecanoyl})_{0.055}$PEI-600 of only about 0.7 ppm upfield from the monomer shift implies either that the residues giving that resonance spend a substantial part of the time in an aqueous environment or that the cluster in which they are incorporated is substantially different from the conventional detergent micelle.

The trifluorodecanoyl PEI derivatives have binding properties similar to those of the long-chain hydrocarbon derivatives.[20] Thus, in view of the structural and functional similarities between the fluorine containing PEI derivatives and the other hydrophobic PEI derivatives, we may assume that apolar clusters are also present in the latter. Nevertheless, there may be small differences in their properties. First, the substitution of three terminal fluorines for three terminal protons in long-chain aliphatic hydrocarbons reduces the hydrophobic tendencies of these chains.[35] Thus it is to be expected that $(\text{decanoyl})_{0.095}$PEI would have (at any instant) a greater fraction of its apolar side chains in micellelike structures than would $(\text{trifluorodecanoyl})_{0.095}$PEI. Second, the basic PEI derivative of primary interest in binding and catalytic studies is $(\text{dodecanoyl})_{0.10}$PEI, which has two additional methylene groups per aliphatic chain to increase its hydrophobic tendencies. It is to be expected, therefore, that $(\text{dodecanoyl})_{0.10}$PEI will have a greater fraction of its apolar groups in clusters than will $(\text{decanoyl})_{0.10}$PEI. Thus both changes (hydrogen for fluorine, addition of two methylenes) contribute in the same direction, in favor of the cluster type of environment. Since even in the ^{19}F-labeled polymer [$(\text{trifluorodecanoyl})_{0.095}$PEI], examined by nmr spectroscopy, most of the apolar residues reside in micellelike structures, in $(\text{dodecanoyl})_{0.10}$PEI the apolar residues must be so incorporated to an even greater extent, approaching 100%.

The clustering of the apolar pendant groups, which has been clearly revealed by nmr spectroscopy, places rather strong constraints on the type of conformation that the hydrophobic PEI derivatives can assume in aqueous solution. We have built a three-dimensional model that illustrates one type of conformation consonant with these constraints. A photograph of this model giving a full view of a large segment of polymer is displayed in Fig. 11.

Fig. 11. Molecular model of the conformation of a large segment of laurylimidazole polyethylenimine in aqueous solution, with some molecules of bound nitrophenylcaproate added. One of the bound molecules is pointed to by a white arrow; the nitrophenyl group

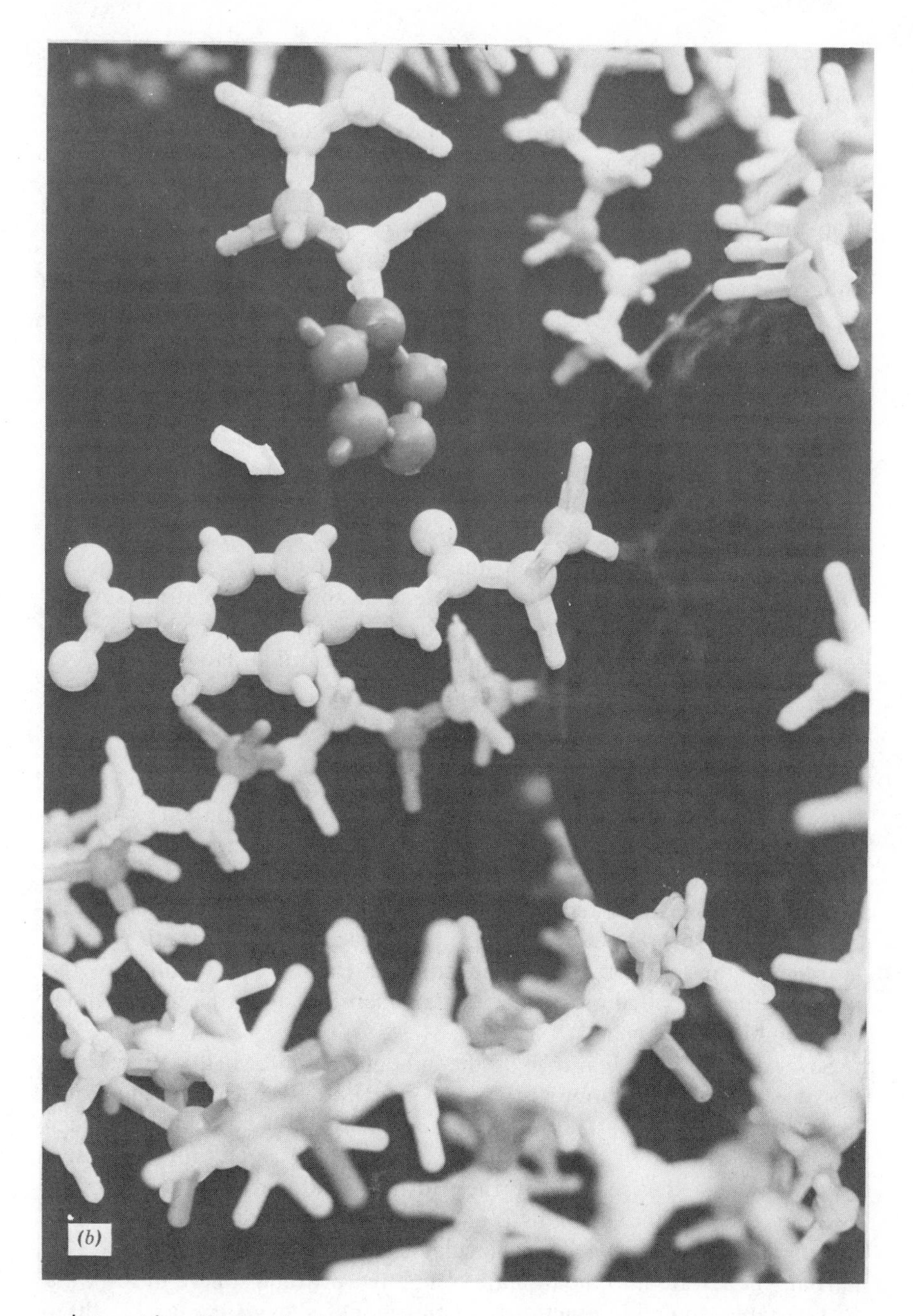

projects out from the domain of the apolar cluster; the caproyl chain is buried in the cluster and hence is not explicitly indicated. In (*a*) the width corresponds to a scale of about 30 Å; in (*b*), to a scale of about 15 Å.

In this model the large central sphere represents a cluster of apolar groups. The individual chains in the cluster are not depicted separately, since by analogy with associated apolar chains of detergent micelles they are presumed to be closely packed, with a density about that of a liquid, long-chain hydrocarbon. Thus the cluster is represented as a homogeneous spherical domain, since individual chains would not stand out in a uniform region. The number of chains within the cluster is determined by the size of the sphere, which could be chosen over a rather wide range and still be consonant with the constraints of the polymer framework. For construction of this model, clusters were constituted of 20 apolar groups, and would be about 20 Å in diameter.

The choice of 20 as the number of apolar groups per cluster is somewhat arbitrary. Having built the present model and another using eight apolar groups in a smaller cluster, we conclude that the features of importance are not qualitatively changed by a change in cluster size. In the limit of very large size the clusters would have to take on a nonspherical shape.[40] Coincident with this, the area per apolar group at the surface of the cluster would be somewhat reduced, eventually to such an extent that the polyamine framework would become crowded. However, before such a condition could be attained, electrostatic effects and solvation effects would presumably stop the further growth of the cluster.

At the surface of an apolar spherical cluster, each of the constituent 20 chains is attached to a point of the PEI framework, this depictment corresponding to the covalent linkage of the hydrocarbon side chain to the polymer macromolecular backbone. The PEI framework (Fig. 11) is represented explicitly in terms of its constituent —CH_2CH_2N— residues, illustrated by the white and gray balls and connectors. Since the sample of PEI used has substantial branching in the macromolecule, there is much latitude in the manner of building polymer chains. The features of the model being described are not particularly sensitive to the precise manner of assembly of the chains.

Several additional features of the model are noteworthy. First, it is possible to build it without straining chemical bonds or causing unfavorable steric interactions. The polyamine chain is sufficiently long to reach around a cluster, but it is not so long or so bulky as to cause excessive crowding near the surface of the cluster. The spaces at this surface between the polyamine chains (Fig. 11) are likely binding sites for small apolar molecules, since such molecules can be bound at the interface or partially penetrate into the domain of the hydrocarbon sphere in response to favorable apolar interactions. In the model shown in Fig. 11, three bound *p*-nitrophenyl caproate molecules have also been included to illustrate possible modes of binding. An arrow points to one of these small molecules.

The regions of the polymer domain that are not occupied by either the framework or bound molecules will contain water and counterions to the charged residues. It is evident from the model that the polymer forms a very open, porous matrix into which water molecules have easy access.

In addition to the lauryl groups and the polyamine framework, attached imidazole residues are also represented in the model. These may be recognized as the five-membered rings. One of them is particularly noticeable at the upper right of the white arrow. Since in practice they are linked primarily to primary amines, the imidazole groups in this model are attached at the termini of flexible segments of the polymer, as is apparent in Fig. 11. It is readily possible for the imidazole residue to approach a bound substrate molecule, as would be required for its observed catalytic effect.

Thus this model provides a very serviceable conformational representation for the interpretation of the unusual behavior of modified polyethylenimines.

VI. FLUORESCENCE SPECTROSCOPY AS A PROBE OF CONFORMATION OF POLYETHYLENIMINES

Observations from fluorescence spectroscopy are also in concordance with this model and provide some additional insight into the microscopic features of the conformational behavior of the primary-amine side chains of the polymer.

Excimer formation can serve as a sensitive probe of group proximities Excimers make evident the interaction of an excited molecule M^*, (typically an aromatic hydrocarbon), with a molecule in the ground state M producing an excited dimer M_2^* (or D^*). The dimer must be formed within the lifetime of the excited species (e.g., for pyrene derivatives, about 100 nsec). For molecules such as pyrene, excimer formation and fluorescence are contingent on attainment of a well-defined steric arrangement in the dimer.[41]

We have investigated the ability of covalently bound pyrenyl derivatives of polyethylenimine to form excimers within the macromolecular matrix.[42] For covalently bound probes, the increase in average local concentration and chain flexibility should facilitate excimer formation.

To keep the probe specifically at the primary amine side chains, we have used pyrenecarboxaldehyde and attached it covalently to polyethylenimine by reduction with sodium borohydride of the Schiff base formed between the probe aldehyde and the polymer primary amine. Such reductive alkylation has been used widely with primary amines of enzymes.[43,44] For three different adducts to the polymer, the extent of coupling with pyrene, expressed relative to units of monomer residues, was 0.1%, 1%, and 5%, respectively. The sample with 0.1% of pyrene, that is, with an average of 1.4 pyrene groups per 1400 monomer units on the macromolecule (of 60,000 molecular weight), provides a basis for determining the fluorescence of an individual pyrene group when it is attached to the polymer. With this sample we can also observe the influence of the polymer amines and environment on monomeric pyrene fluorescence.

Relative fluorescence intensities and spectra at a series of pH values are shown in Fig. 12, and some pertinent parameters are summarized in Table IX. At a specific pH the fluorescence intensity as a function of wavelength has been expressed as relative to the intensity at the maximum at the same pH. The maximum intensities were different at different pH values. Nevertheless, the ratio of intensities of dimer and monomer, I_D/I_M, being independent of polymer concentration, may be compared for all the samples and pH values.

A progressive decrease in the intensity ratio I_D/I_M for pyrenylmethyl polyethylenimine is observed as the pH is decreased (Table IX). The values of I_D/I_M at pH 6.15 are about 30% of those at 9.6 and at pH$<$1 they drop to about 2.5%. At high pH, where the molecule is uncharged and relatively compact, excimer formation is facilitated because of a high local concentration of pyrene and the ability of these moieties to form intramolecular clusters. In contrast, at low pH, charge repulsion between chains causes polymer expansion and disruption of clusters, with a resultant decline in excimer formation. The same trend for almost all the derivatives indicates that the effects are averaged out over the entire polymer. At pH$<$1 the 5% derivative still manifests substantial excimer formation, evidently because the clusters are not fully dispersed. Fluorine nmr probes with hydrocarbon moieties on the polymer revealed similar behavior at high degrees of protonation.[37]

The 0.1% polymer derivative contains about one molecule of pyrene per macromolecule, whereas the 5% one has 50 to 60. The distribution of ligand on the polymer is at best binominal.[42] Thus a significant fraction of the former derivative contains two to five molecules of pyrene per macromolecule. It is surprising to find excimer formation in the 0.1% derivative since so few molecules are on a single polymer chain. It seems apparent, therefore, that in aqueous solution these moieties on the polymer are clustered in the ground state so that excimer formation is facilitated.

To circumvent any problems caused by cluster quenching by amines on the polymer, a preparation of pyrenylmethyl polyethylenimine was made in which all the primary and almost all the secondary amines were acetylated. Steric factors in the polymer should block quenching interactions with the tertiary amines. In addition, these samples were examined

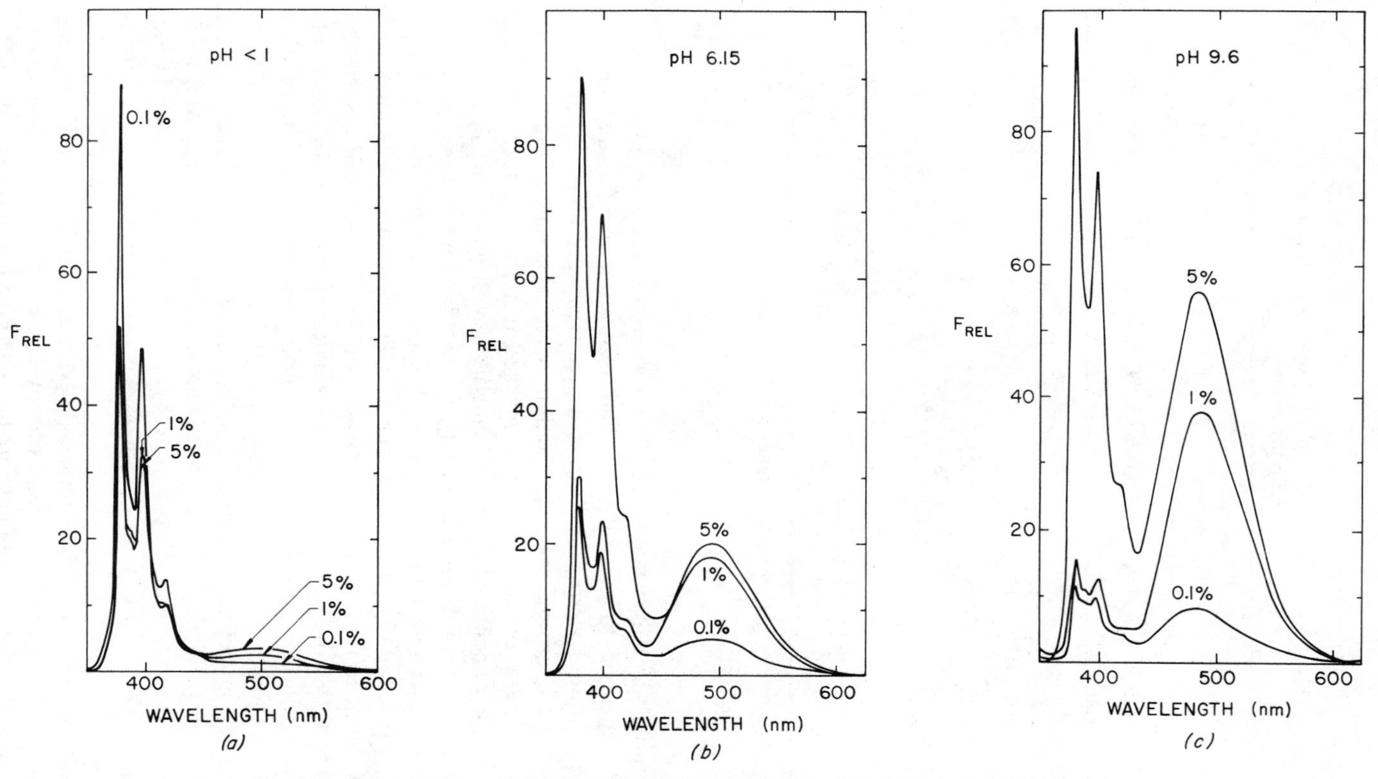

Fig. 12. Fluorescence emission spectra at different pH values [(*a*) pH<1; (*b*) pH 6.15; (*c*) pH 9.6)] for pyrene covalently attached to polyethylenimine. Percentages designate the fractional number of monomer residues with attached pyrene moieties.

TABLE IX. Fluorescence Characteristics of Covalently Bound Pyrene Adducts to Polyethylenimine

Pyrene content (%)	pH	Fluorescence intensity[a] 400 nm	Fluorescence intensity[a] 500 nm	Ratio of intensities $R = \frac{\text{intensity at 500}}{\text{intensity at 400}}$	$R_{pH=x}/R_{pH=9.6}$ (×100)
5	9.6	100	75.4	0.754	100
1	9.6	17.1	50.4	2.95	100
0.1	9.6	12.7	10.7	0.84	100
5	6.15	100	28.2	0.28	37
1	6.15	33.4	25.8	0.77	26
0.1	6.15	27.0	7.7	0.284	33
5	<1	64.4	7.1	0.110	14.6
1	<1	65.3	4.9	0.075	2.53
0.1	<1	100	2.2	0.022	2.6

[a] Within a particular pH set the maximum was normalized to 100.

in ethanol, which has been found to disperse clusters of hydrophobic residues attached to the polymer. The absorption spectra, and fluorescence excitation spectra, of all these derivatives were identical, as would be expected if ground-state interactions were eliminated. Fluorescence spectra of these derivatives show a marked increase of excimer fluorescence, at the expense of monomer emission, with increasing ligation.

The temperature dependence of I_D/I_M was determined for a peracetylated 5.3% pyrene-polymer derivative. For pyrene in ethanol, log (I_D/I_M) versus $1/T$ follows a bell-shaped curve. The high- and low-temperature side of the peak represent regions where the dissociation and association constants, respectively, of the monomer-dimer equilibrium are rate limiting. The position of the peak is qualitatively an indicator of the binding energy of the excimer. For pyrene the peak occurs at 30 to 40°C in ethanol. For molecules such as naphthalene or benzene with lower binding energies, the peak falls below room temperature. The temperature dependence of peracetylated pyrenylmethyl polyethylenimine is seen in Fig. 13. The peak at 40°C indicates that true pyrene excimer formation is occurring.

A peracetylated polyethylenimine derivative with only 0.043% pyrene contains on the average less than one pyrene group per macromolecule. The emission from this appears to be that of all monomers. A model

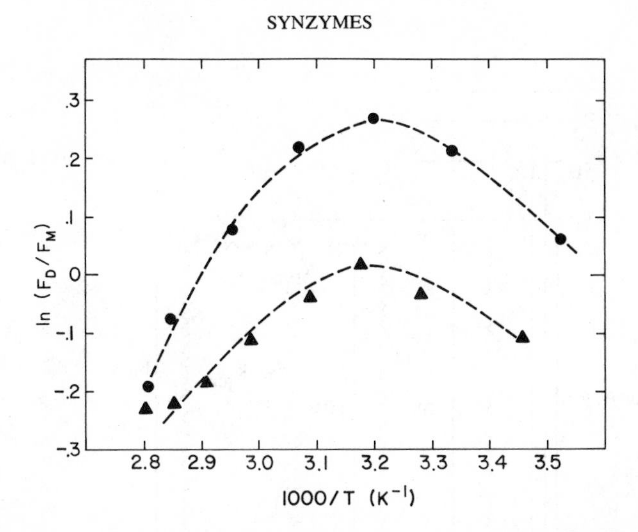

Fig. 13. Temperature dependence of excimer formation for the peracetylated 5.3% pyrenyl methyl polyethylenimine. The two curves represent results following raising (●) and lowering (▲) the temperature. The difference between them is due to photochemical degradation.

small molecule, pyrenylmethyl *N*-acetyl hexylamine, had a relative quantum yield about 40 times that of the 0.043% peracetylated pyrenylmethyl polyethylenimine and an identical emission spectrum, assuming the lifetime for free pyrene in ethanol to be about 300 nsec. The lifetime of monomeric pyrene in the polymer matrix is about 7 to 9 nsec. From the quenching of the fluorescence of a 5.3% peracetylated pyrenylmethyl polyethylenimine and the concentration of dissolved oxygen in ethanol a fluorescence lifetime of about 8 nsec can be estimated according to the method of Berlman.[45] This is in reasonable agreement with the lifetime estimated above for monomer fluorescence.

A fourfold decrease in the I_D/I_M ratio was observed for the 5.3% peracetylated pyrenylmethyl polyethylenimine derivative in glycerol compared to methanol. The higher viscosity of the glycerol limits the mobility of the attached pyrene group necessary to form excimer, decreases the association rate, and hence lowers I_D/I_M. These samples at 77°C showed essentially no excimer emission. Clearly, diffusion of the pyrene moieties attached to the polymer side chains is necessary for excimer formation.

Fluorescence measurements should ultimately be related to quantum yields. To define the parameters required, we adopt the general scheme shown in Fig. 14 for excited state transitions. The ground states and excited states are designated by S_0 and $S_1, S_2, \ldots, S_1'$, respectively. The rates of transitions from higher to lower energy levels are designated by

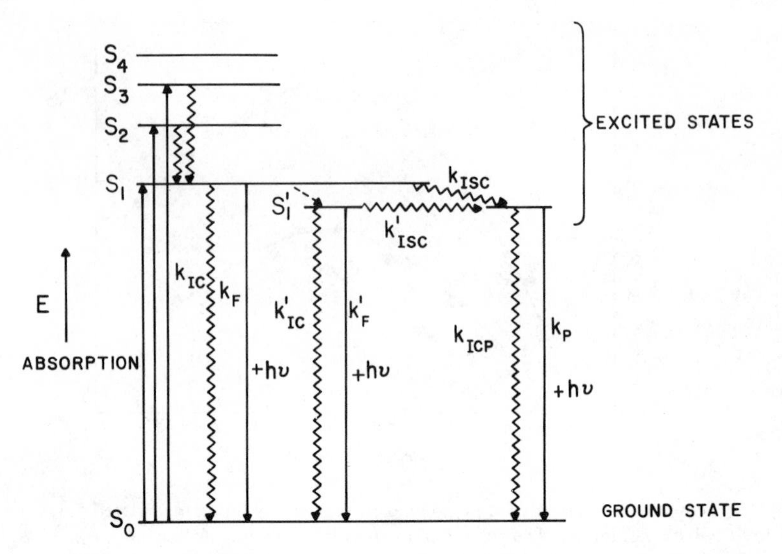

Fig. 14. Schematic representation of energy levels and transitions for fluorescence and related processes: k_{IC}, rate constant for interval conversion; k_F, rate constant for fluorescence; k_{ISC}, rate constant for intersystems crossing; k_{ICP}, rate constant for internal conversion from triplet state; k_P, rate constant for phosphorescence; S_1', energy level for the first excited singlet state after solvent rearrangement for a polarity probe in a polar solvent.

appropriate k's (see legend to Fig. 14). In addition we define the association and dissociation rate constants in the monomer-dimer equilibrium, k_{DM} and k_{MD}, respectively:

$$M + M^* \underset{k_{MD}}{\overset{k_{DM}}{\rightleftharpoons}} D^* \tag{9}$$

The parameter k_{DM} should have values typical of diffusion-controlled processes, and its magnitude would depend on viscosity and steric factors. On the other hand, the value of k_{MD} should depend on the energy of binding in the excimer.

A more quantitative expression of the pyrenyl side chain interactions can then be generated by the following analysis. Excimer formation on a polymer surface is dependent on three factors: (*1*) the fraction of ligation, (*2*) the macromolecular constraints on the number of ligands any ligand could reach and interact with, and (*3*) the probability of interaction, that is, the dependency on the time course of excimer formation and the efficiency of the reaction.

The first two variables can be examined by distribution analysis. The fraction of molecules with "n" neighbors can be expressed as a function of ligation and the interaction number. By *interaction number* we mean the number of ligands accessible to any specific ligand at 100% ligation. For example, in linear polyvinylnaphthalene, any specific naphthalene group can only interact with the naphthalenes on either side of it in the chain; therefore the interaction number is two. For more compact and globular structures, interaction numbers can be quite high and would be so in intramacromolecular clusters. The distribution expression can be obtained by analogy to N selections of black and white balls from a box. Thus the fraction of groups with i-neighbors and an interaction number of N is given by

$$f_i = \binom{N}{i} f_L^i (1 - f_L)^{(N-i)} \tag{10}$$

where f_L is the fraction of ligation.

For excimeric systems in which the fluorescing molecules are not attached to a polymer chain, the observed quantum yield Φ_M for monomer follows[41] the equation

$$\Phi_M = \frac{q_{FM}}{1+[M]A} \tag{11}$$

where q_{FM} is the fluorescence quantum efficiency, $[M]$ is the concentration of the molecule, and A is defined by

$$A = \frac{k_D}{k_M}\left(\frac{k_{DM}}{k_{MD}+k_D}\right) \tag{12}$$

where k_D and k_M are the decay constants for the excimer and monomer, respectively. Similarly, for dimer

$$\Phi_D = \frac{q_{FD}}{(1+1/[M]A)} \tag{13}$$

Thus for simple systems

$$\frac{\Phi_D}{\Phi_M} = \frac{q_{FD}}{q_{FM}} A[M] \tag{14}$$

A linear plot should be obtained for Φ_D/Φ_M plotted versus $[M]$.

For polymeric systems it is difficult to utilize such a plot since $[M]$ is not known at the polymer surface. In addition, the polymer-bound ligands are limited to a fairly discrete number of ligand neighbors with which to form an excimer. To circumvent these difficulties we can redefine $k_{DM}[M]$ in terms of $k'_{DM}N$, where N is the number of ligands that are neighbors of any specific ligand and k'_{DM} represents an average first-order rate constant for the association of an excimer from two polymer-bound monomers, each with only one neighbor. For (13) we would then have

$$\Phi_D = \sum_{N=1}^{\infty} f_N \frac{q_{FD}}{1+(1/NA')} \tag{15}$$

where f_N is the fraction of ligands with N neighbors and A' is defined by an equation like (12) but with k'_{DM}. For a ligand with three neighbors the first-order rate constant is $3k'_{DM}$. Thus (14) becomes

$$\frac{\Phi_D}{\Phi_M} = \frac{\sum_{N=1}^{\infty} f_N[q_{FD}/(1+1/NA')]}{\sum_{N=1}^{\infty} f_N[q_{FM}/(1+NA')]+f_0 q_{FM}} \tag{16}$$

$$\frac{\Phi_D}{\Phi_M} = \frac{q_{FD}}{q_{FM}} \left[\frac{\sum_{N=1}^{\infty} f_N[1/(1+1/NA')]}{\sum_{N=1}^{\infty} f_N[1/(1+NA')]+f_0} \right] \tag{17}$$

In (16) and (17), f_0 is the fraction of ligands with no neighbors.

From (10) we can obtain values for f_N assuming different interaction numbers. The parameter A is difficult to determine, particularly for a polymeric system. Nevertheless, we can make calculations assuming values for A, notably $A \ll 1$, $A = 1$, $A \gg 1$. Each of these cases has significance in terms of the physical processes. For $A \ll 1$ the formation of the excimer is a relatively slow process and the excimer fluorescence is dependent on the number of neighbors it has. Equation 17 reduces to

$$\frac{\Phi_D}{\Phi_M} = \frac{q_{FD}}{q_{FM}} (A') \sum_{N=1}^{\infty} N f_N \tag{18}$$

In this instance a plot of Φ_D/Φ_M versus the percent ligation of the polymer should be linear. The polymer experimental data are not in accord with this prediction (Fig. 15*a*).

In turn, for $A \gg 1$,

$$\frac{\Phi_D}{\Phi_M} = \frac{q_{FD}(A')}{q_{FM}} \frac{\sum_{N=1}^{\infty} f_N}{\sum_{N=1}^{\infty} (f_N/N)+Af_0} \tag{19}$$

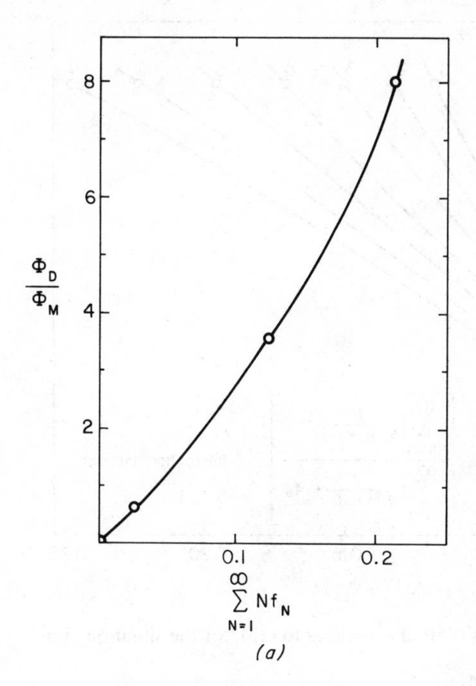

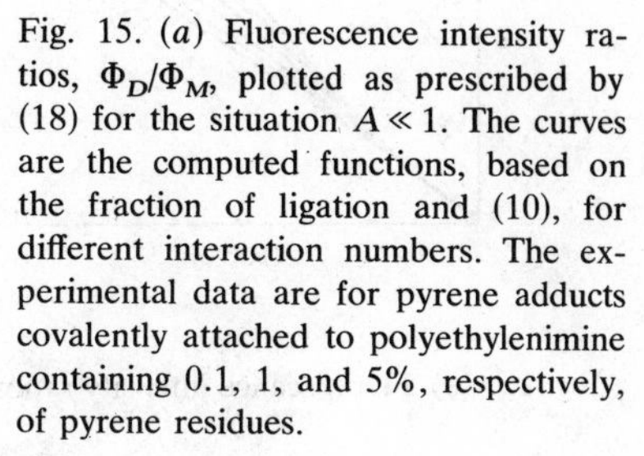
Fig. 15. (*a*) Fluorescence intensity ratios, Φ_D/Φ_M, plotted as prescribed by (18) for the situation $A \ll 1$. The curves are the computed functions, based on the fraction of ligation and (10), for different interaction numbers. The experimental data are for pyrene adducts covalently attached to polyethylenimine containing 0.1, 1, and 5%, respectively, of pyrene residues.

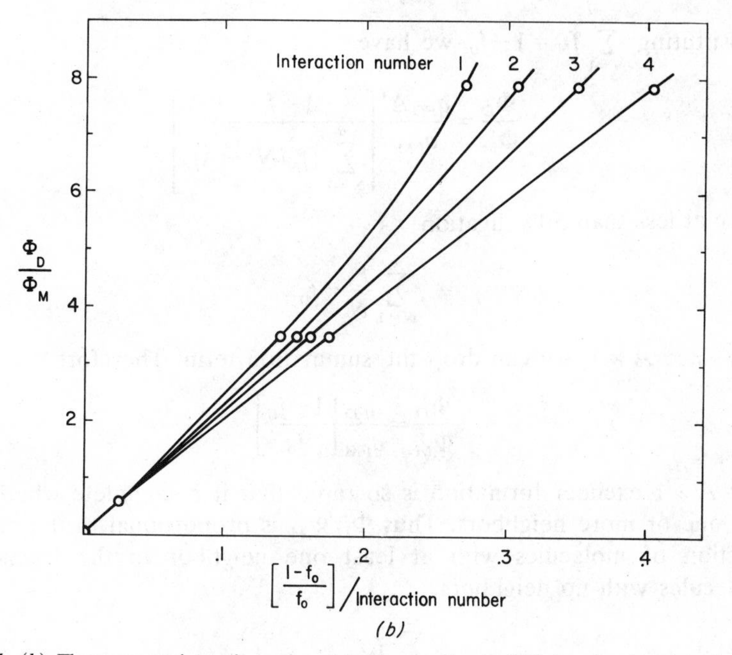

Fig. 15. (*b*) Fluorescence intensity ratios plotted according to (22), for the situation $A \gg 1$.

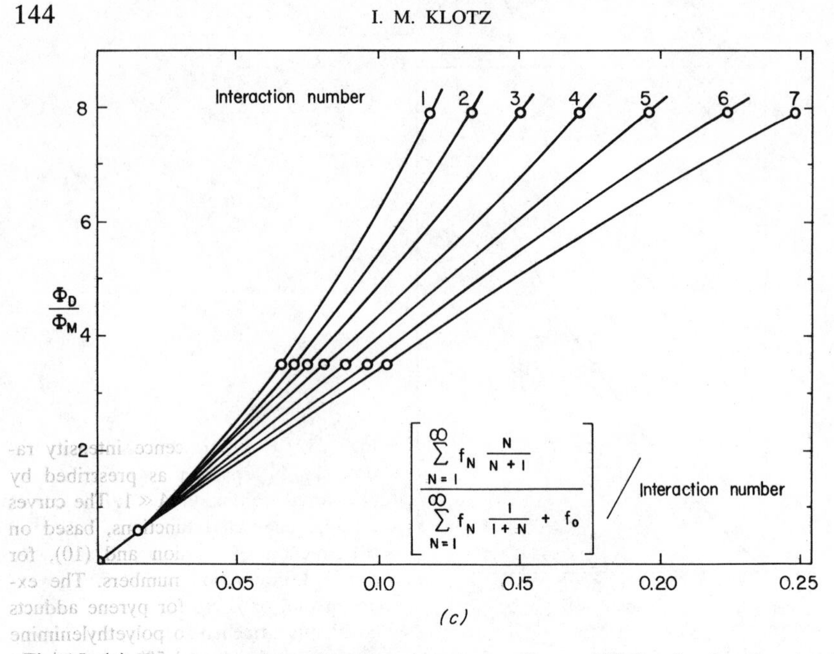

Fig. 15. (*c*) Fluorescence intensity ratios plotted according to (23), for the situation $A = 1$.

Substituting $\sum_{N=1}^{\infty} f_N = 1 - f_0$, we have

$$\frac{\Phi_D}{\Phi_M} = \frac{q_{FD}A'}{q_{FM}}\left[\frac{1-f_0}{\sum_{N=1}^{\infty}(f_N/N) + Af_0}\right] \tag{20}$$

Since at less than 50% ligation

$$\sum_{N=1}^{\infty} \frac{f_N}{N} = f_0$$

and since $A \gg 1$, we can drop the summation term. Therefore

$$\frac{\Phi_D}{\Phi_M} = \frac{q_{FD}}{q_{FM}}\left[\frac{1-f_0}{f_0}\right] \tag{21}$$

For $A \gg 1$ excimer formation is so rapid that it is complete whether M has one or more neighbors. Thus Φ_D/Φ_M is proportional to the ratio of fraction of molecules with at least one neighbor to the fraction of molecules with no neighbors,

$$\frac{\sum_{N=1}^{\infty} f_N}{f_0} = \frac{1-f_0}{f_0} \tag{22}$$

Figure 15b illustrates the dependence of fluorescence intensity ratios on the f_0 function prescribed by (21), with different assumed values for the interaction number.

For $A = 1$, a result between the extremes is produced, with (17) becoming

$$\frac{\Phi_D}{\Phi_M} = \frac{q_{FD}}{q_{FM}} \frac{\sum_{N=1}^{\infty} f_N[N/(N+1)]}{\sum_{N=1}^{\infty} f_N[1/(1+N)] + f_0} \tag{23}$$

Figure 15c gives the graphical representation of the fluorescence intensity ratios as prescribed by (23).

For $A \ll 1$, the theoretical curve (Fig. 15a) is the same irrespective of the assumed interaction number. Clearly, the observations cannot be fitted to a straight line, as is required by (18). Thus $A \ll 1$ is not an appropriate parameter for the pyrenylpolyethylenimine adducts.

As Fig. 15b illustrates, the graphical relation appears to be linear for an interaction number of 3 to 4, if $A \gg 1$. Alternatively, for $A = 1$, linearity is evident (Fig. 15c) when the interaction number is 5 to 6. Thus a large value of A is compatible with the smallest interaction number. Excimer formation occurs within the fluorescence lifetime, about 8 nsec. Within that time the pyrene-labeled amine side chains must approach within about 4 Å of each other. For the 5.3% pyrenylpolyethylenimine derivative in ethanol, where no ground-state association occurs, the effective local concentration of pyrene on the polymer matrix is about 10^{-2} M, as calculated from excimer fluorescence. In aqueous solution, where clusters form within the polymer matrix, the effective local concentration of pyrene adduct must be even greater. The quantitative assessment of fluorescence intensities (Fig. 15) points to a minimum interaction number of 3 to 4 pyrenyl-labeled amine side chains, within the 8 nsec lifetime. Since $A \gg 1$, it appears from (12) that $k_{DM}(A) \gg k_{MD} + k_D$. Thus excimer formation must be very rapid in the polymer environment. We can conclude, therefore, that the primary-amine side chains of polyethylenimine are very flexible and mobile.

It is of interest to note in passing that thermodynamic responses, acid-base titrations, of these polymer primary amines also reflect their behaviorial insularity from the macromolecular matrix. Gross titrations of polyethylenimines with acid (or base) manifest overlapping uptake (or loss) of protons by the primary, secondary, and tertiary nitrogens. Nevertheless, it has proved possible to follow the primary amines without obstruction from the others by converting the former, through specific reductive alkylation of the Schiff base with acetone, to $[CH_3]_2CH$—NH— groups on the polymer (T. Johnson and I. M. Klotz, unpublished investigations). The CH_3 protons are readily resolved in the proton magnetic

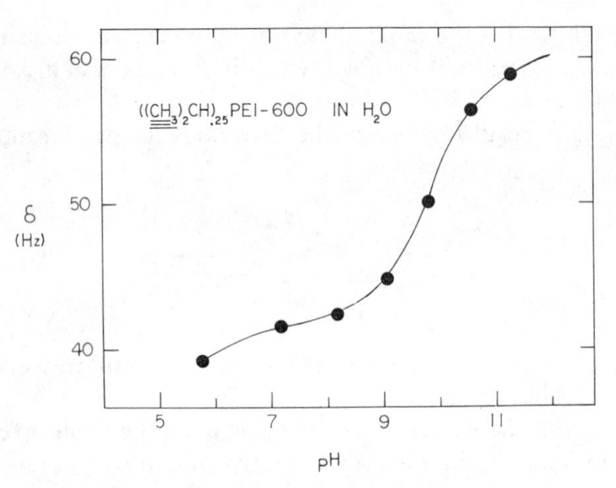

Fig. 16. Chemical shift of isopropyl (methyl) protons of isopropylated polyethylenimine (600) in water with 0.1 M carbonate, 0.1 M tris. The pH was adjusted with HCl. Scale reference point provided by CH_3CN.

resonance spectroscopy of the modified polymer. The pH-titration curve of these amines (Fig. 16) leads to a pK_a on the polymer near 10, not much below that for an isolated small-molecule amine. Thus a thermodynamic measure of interaction with H^+ ions also is in concordance with a model for the polymer in which the primary amines are relatively free from motional constraints imposed in the more interior regions of the macromolecular matrix.

These characteristics presumably would also be imparted to the apolar binding groups and nucleophilic catalytic moieties covalently bonded to the polymer. A structure with such features has the potential of being an effective catalyst in a variety of reactions with a range of substrates of widely differing structure and chemical nature.

VII. CATALYSIS OF DECARBOXYLATION REACTIONS BY POLYETHYLENIMINES

The model of the polymer derived from magnetic resonance and fluorescence spectroscopy studies has proved very useful in suggesting other types or reactions besides hydrolytic ones in which catalytic effects might be achieved. For example, it has been shown by Kemp and Paul[46,47] that the decarboxylation of certain benzisoxazole carboxylic acids is very markedly accelerated in apolar, aprotic solvents, in contrast to water. Such an apolar solvent apparently lowers the energy of the charge-delocalized transition state in this decarboxylation reaction. Since our modified polyethylenimines contain apolar domains (the hydrophobic clusters shown in Fig. 11), it seemed to us that they ought to provide an environment favorable to the decarboxylation transition state and hence should be effective catalysts *in a fully aqueous solvent.* Furthermore, since these polymers carry a large positive charge, they should bind the anionic carboxylate substrate and hence accentuate decarboxylations in a Michaelis-Menten manner, that is, in a pathway involving binding to the macromolecular matrix followed by environmental catalysis. We have examined, therefore, the kinetics of catalysis in these systems[48] and analyzed them in mechanistic terms analogous to those used in enzymic catalyses.

To minimize any ambiguity in interpretation, we have prepared modified polyethylenimines in which all the nitrogens have been quaternized, with various apolar groups, so that they cannot function in a nucleophilic mechanism. Some of the derivatives prepared may be represented by the stoichiometric formulas

$$[(C_2H_4N)_m(C_{12}H_{25})_{0.25m}(CH_3)_{1.75m}]Cl, \qquad m = 1400$$
$$[(C_2H_4N)_m(C_{12}H_{25})_{0.25m}(C_2H_5)_{1.75m}]Cl, \qquad m = 1400$$
$$[(C_2H_4N)_m(C_{12}H_{25})_{0.1m}(CH_3)_{1.9m}]Cl, \qquad m = 1400$$
$$[(C_2H_4N)_m(C_{12}H_{25})_{0.25m}]Cl, \qquad m = 1400$$

Any catalytic effects of these modified polymers must reflect, therefore, only the influence of their apolar domains and their cationic charges.

The reaction studied with the benzisoxazole(III) is

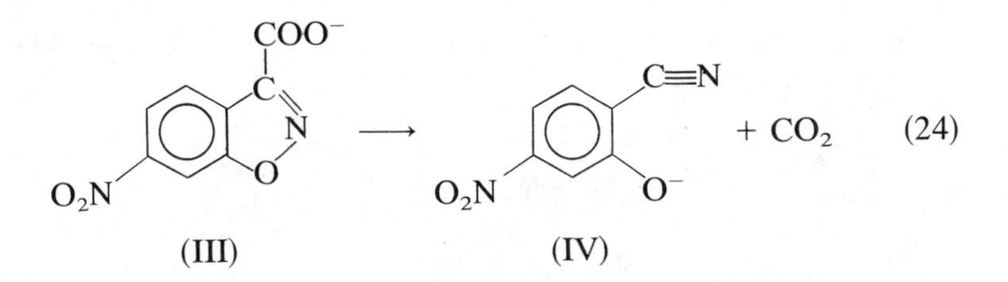

(24)

Its rate can be followed by measurement of the increase in absorbance due to the release of 2-cyano-5-nitrophenol (IV).

Rates of decarboxylation were used to analyze the kinetics following the general format established by Michaelis and Menten.[3] If C represents one catalytic site on the polymer and S the carboxylate substrate, the following scheme may be formulated:

$$S + C \underset{k_{-1}}{\overset{k_1}{\rightleftharpoons}} SC \xrightarrow{k_2} \text{products} + C \qquad (25)$$

For certain sets of the experiments the initial concentration of catalytic sites C_0 was kept in great excess over that of substrate S_0. Under these conditions, $C_0 \gg S_0$, the decarboxylation follows pseudo-first-order kinetics, and a corresponding rate constant k_{obs} may be extracted from the observed rates. Also under these conditions, the steady-state rate expression for the mechanism of (25) becomes

$$k_{obs} = \frac{k_2 C_0}{K_M + C_0} \tag{26}$$

where $K_M = (k_{-1} + k_2)/k_1$. If it is assumed that the n catalytic sites on one molecule of polymer P behave independently of each other, then

$$k_{obs} = \frac{nk_2 P_0}{K_M + nP_0} \tag{27}$$

where P_0 is the initial concentration of polymer. A linear transform of (27) is

$$\frac{1}{k_{obs}} = \frac{K_M/n}{k_2} \frac{1}{P_0} + \frac{1}{k_2} \tag{28}$$

Alternatively, experiments were also carried out under conditions of excess of substrate, that is, $S_0 \gg C_0$. Under these circumstances initial velocities V_0 are measured and the steady-state expression for the scheme in (25) is

$$k_0 = \frac{k_2 C_0}{K_M + S_0} \tag{29}$$

where k_0 is the initial rate constant, V_0/S_0. Again if the n catalytic sites on a polymer molecule behave independently, then

$$k_0 = \frac{nk_2 P_0}{K_M + S_0} \tag{30}$$

and

$$\frac{P_0}{k_0} = \frac{K_M}{nk_2} + \frac{1}{nk_2} S_0 \tag{31}$$

Fitting experimental data at $P_0 \gg S_0$ to (28), one can evaluate k_2 and K_M/n. A corresponding treatment of observations for $S_0 \gg P_0$ by means of (31) gives nk_2 and K_M. Combination of the information provides an explicit value for n, as well as of k_2 and K_M.

Figure 17 illustrates a typical decarboxylation experiment with nitrobenzisoxazole carboxylate under conditions of $C_0 \gg S_0$ (curve A). With

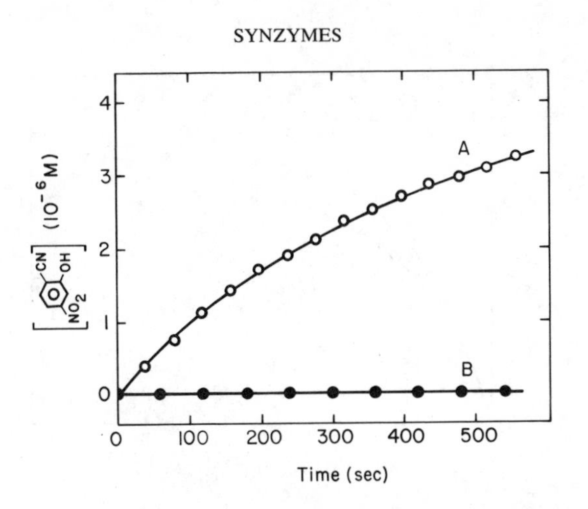

Fig. 17. Rate of release of 5-nitro-2-cyanophenolate during decarboxylation of 6-nitrobenzisoxazole-3-carboxylate at 25°C, pH 7.4, $\mu = 0.05$. Initial concentration of substrate, $S_0 = 4.18 \times 10^{-6}$ *M*. (*A*) In presence of quaternized polyethylenimine $(C_2H_4N)_m(C_{12}H_{25})_{0.25m}(C_2H_5)_{1.75m}$, $P_0 = 7.47 \times 10^{-6}$ *M*. Molecular weight of modified polymer is 240,000 with chloride ions. (*B*) Spontaneous decarboxylation in absence of polymer.

low concentrations of substrate, spontaneous decomposition (curve *B*) cannot be detected. A series of experiments similar to those in Fig. 17 but with $S_0 \gg C_0$ is summarized in Fig. 18. This illustrates the saturation behavior required by the scheme of (27). Correspondingly a series of experiments with $C_0 \gg S_0$ also exhibits a plateau in a graph of observed rate versus polymer concentration. Since S and C appear symmetrically in the rate steps of (25), each must exhibit saturation behavior, and each does in this decarboxylation reaction. The kinetics of decarboxylation in the presence of polymer thus show concentration dependencies analogous to those found in the natural macromolecular catalysts—the enzymes.

Table X summarizes the kinetic parameters derived from the decarboxylation rates of nitrobenzisoxazole carboxylate catalyzed by each of four different lauryl polyethylenimines.

Comparing catalytic effects of different modified polyethylenimines on the decarboxylation of nitrobenzisoxazole carboxylate, we can discern several interesting features. Comparison of the 25% laurylated polymer in the quaternized and nonquaternized forms, *A* and *D*, respectively, in Table X, shows that the former is about three times more effective as a catalyst than the latter. For the quaternized polymer the first-order catalytic constant k_2 and the second-order rate constant nk_2/K_M are greater and the binding of substrate (measured by K_M^{-1}) is stronger.

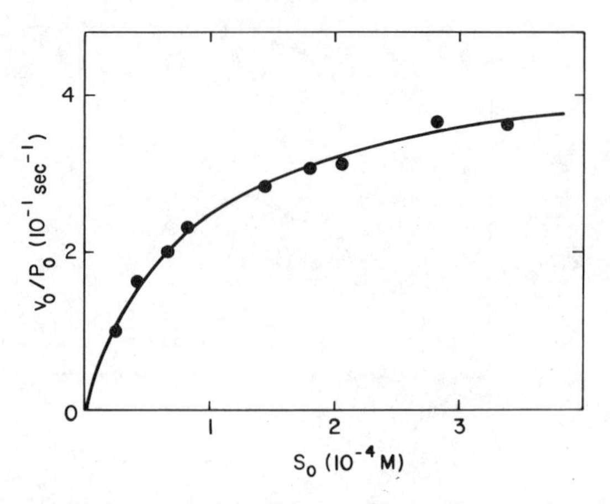

Fig. 18. Variation of initial rate constant V_0/P_0 for decarboxylation of 6-nitrobenzisoxazole-3-carboxylate as a function of substrate concentration under conditions of excess substrate. Initial velocities were corrected for the spontaneous hydrolysis of substrate in absence of polymer. Polymer is $(C_2H_4N)_m$ $(C_{12}H_{25})_{0.25m}(C_2H_5)_{1.75m}$. The curve was drawn according to (30), with $nk_2 = 0.458$ sec^{-1}, and $K_M = 8.59 \times 10^{-5}$ M.

Stronger binding of substrate probably reflects the greater charge on the quaternized polymer at pH 7.4, since, in contrast to the nonquaternized polyethylenimine, it does not dissociate H^+ ions with increasing pH. However, stronger binding alone is not responsible for the increased catalytic effectiveness of the quaternized polymer, for k_2 in its environment is also twofold greater. Evidently the cationic environment it provides is more favorable for the anionic transition state[46,47] in the mechanistic pathway of the decarboxylation reaction.

The ethyl-quaternized polymer is more effective as a catalyst than the comparable methyl-quaternized polymer (*A* and *B*, respectively, in Table X). This is due in large part to the threefold increase in k_2, which can be interpreted as a reflection of the contribution of the more apolar environment provided by ethyl as contrasted to methyl groups.

The very strong role played by the apolar environment is also evident in a comparison of 25% and 10% laurylated polethylenimines (*A* and *C*, respectively, in Table X). The catalytic constant k_2, reflecting the effect of polymer environment on the transition state, drops by a factor of 5 in the less apolar environment.

All the polymers show marked catalytic effects on the decarboxylation in the water solvent. In the polymer environment the intrinsic first-order rate constant k_2 can be 10^3-fold greater than the pseudo-first-order rate constant in the aqueous solvent alone (*B* and *E*, respectively, in Table X).

TABLE X. Kinetic Parameters for Catalyzed Decarboxylation of Nitrobenzisoxazole Carboxylate by Modified Polyethylenimines[a]

Polymer derivative	(nk_2/K_M) $10^3\ s^{-1}\ M^{-1}$	k_2 $10^{-3}\ s^{-1}$	nk_2 $10^{-2}\ s^{-1}$	K_M $10^{-5}\ M$	K_M/n $10^{-7}\ M$	n^b	Average number[b] ethylenimine units per catalytic site	Average number[b] lauryl residues per catalytic site
A. $(C_2H_4N)_m(C_{12}H_{25})_{0.25m}(CH_3)_{1.75m}$								
$C_0 \gg S_0$	1.44	1.30			9.0	56; 72	25; 19	6.2; 4.9
$C_0 \ll S_0$	1.12		7.3	6.5				
B. $(C_2H_4N)_m(C_{12}H_{25})_{0.25m}(C_2H_5)_{1.75m}$								
$C_0 \gg S_0$	4.63	3.92			8.5	117; 107	12; 13	3.0; 3.2
$C_0 \ll S_0$	5.33		45.8	8.6				
C. $(C_2H_4N)_m(C_{12}H_{25})_{0.1m}(CH_3)_{1.9m}$								
$C_0 \gg S_0$	0.11	0.27			25			
D. $(C_2H_4N)_m(C_{12}H_{25})_{0.25m}$								
$C_0 \gg S_0$	0.50	0.65			13	86; 78	16; 18	4.1; 4.5
$C_0 \ll S_0$	0.55		5.6	10.2				
E. None		0.003[c]						

[a] Reactions followed at 25°C, pH 7.4, ionic strength 0.05.

[b] The first entry has been calculated from nk_2 and k_2, the second from K_M/n and K_M.

[c] Spontaneous first-order rate constant in absence of polymer; the value listed is from Ref. 46 and our less precise measurements gave similar results.

Furthermore, the polymer is an effective catalyst at concentrations of the order of 10^{-6} M, corresponding to catalytic site concentrations of about 10^{-4} M. This means that the decarboxylation rate can be enhanced more than 1000 times by addition of 5×10^{-6} M polymer to the aqueous solution. In contrast, for micellar systems the maximum enhancement reported is at best 100-fold.[49]

Thus it is evident that the modified polyethylenimines provide a matrix for achieving homogeneous catalysis of decarboxylation of activated anionic substrates in an aqueous environment. Clearly, the modified polyethylenimines provide solvation features that stabilize the anionic transition structure in the state with particularly sensitive bonds. Large solvent effects have been observed in kinetic studies of many reactions involving anions.[46,47,50,51] Suitable derivatives of polyethylenimine should manifest interesting effects in many of these reactions also.

Although the activated benzisoxazole carboxylates serve admirably as substrates to reveal certain mechanistic details of importance in designing a macromolecular catalyst, they are not of biological interest per se. We have also been anxious to achieve catalyses of reactions that occur in a cellular environment. With respect to decarboxylation, we have been focusing on oxalacetate as a suitable substrate.

In enzymic decarboxylations the mechanistic pathway seems to involve Schiff base formation between an —NH_2 from a lysine residue and a C=O of the keto acid.[52] Likewise, with small-molecule primary amines, catalysis of decarboxylation of β-ketoacids[53–58] has been ascribed to a Schiff base intermediate. The overall reaction for oxalacetate is

$$HO_2C{-}\underset{\underset{O}{\|}}{C}{-}CH_2{-}CO_2^- + RNH_2 + H^+ \rightarrow HO_2C{-}\underset{\underset{O}{\|}}{C}{-}CH_3 + CO_2 + RNH_2 \qquad (32)$$

We have prepared, therefore, a modified polyethylenimine with about 10% of its residues containing free primary —NH_2 groups, 20% of its residues containing long-chain apolar hydrocarbon groups, and the remaining nitrogens quaternized to block secondary and tertiary amines and to maintain a positive charge on the polymer. This polymer, $[(C_2H_4N)_{0.90m}(C_2H_4NH_2)_{0.10m}(C_{12}H_{25})_{0.20m}(CH_3)_m]Cl$, catalyzes markedly the rate of decarboxylation of oxalacetate.[59]

Rates have been monitored by ultraviolet absorption,[60] following the decay of oxalacetate, and by enzymatic assays[61,62] responding to the production of pyruvate [see (32)]. Figure 19 illustrates the rapid drop in absorbance with time for oxalacetate in the presence of the primary-amine-containing polymer. Even a sixtyfold excess of substrate over primary-amine concentration was completely decarboxylated. This and

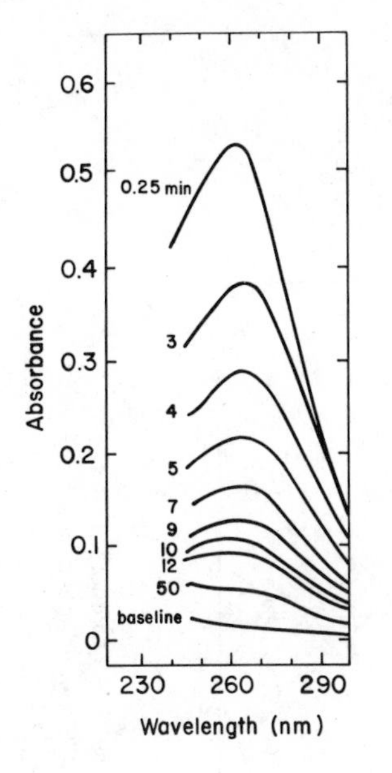

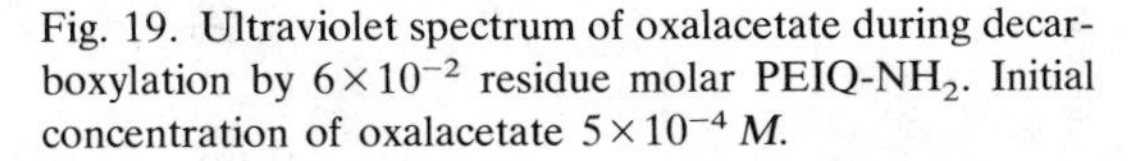
Fig. 19. Ultraviolet spectrum of oxalacetate during decarboxylation by 6×10^{-2} residue molar PEIQ-NH_2. Initial concentration of oxalacetate 5×10^{-4} *M*.

other experiments established that catalytic turnover is effected by the modified polymer.

A kinetic analysis of the observed rates was made following the mechanistic scheme described by (25) to (31). The rate parameters at two pH values, 4.5 and 7.0, are shown in Table XI.

In this system k_2 values were easily measured in solutions with large excesses (about 150:1) of catalytic sites over substrate concentrations. Under these conditions the observed first-order rate constant was unaffected by increasing polymer concentration. The variation of k_2 over the pH range 3 to 7 is displayed in Fig. 20. It is bell shaped and exhibits a maximum at pH 4.5. This bell-shaped pH-rate profile is similar to that of other model primary amines[57,58,63] as well as the enzyme acetoacetate decarboxylase.[52]

As is evident from Table XI, the decarboxylation of oxalacetate is markedly faster in the presence of amine-containing modified polyethylenimine (PEIQ—NH_2). The first-order rate constant k_2 in the polymer environment is 10^2 to 10^3 times greater than that in aqueous solvent alone. Furthermore, at pH 4.5 the amine groups in the polymer matrix are 10^5 times more effective than those of the small molecule $C_2H_5NH_2$, as measured by the second-order rate constants. Alternatively, one can make the comparison by calculating from its second-order rate constant, the concentration of $C_2H_5NH_2$ that would give a first-order constant

TABLE XI. Kinetic Parameters for Catalyzed Decarboxylation of Oxalacetate by Modified Polyethylenimine

Catalyst	pH	k_2 (min^{-1})	K_M (M)	n[a]	k_2/K_M ($M^{-1}\ min^{-1}$)
PEIQ—NH_2; $[(C_2H_4N)_{0.90m}(C_2H_4NH_2)_{0.10m}(C_{12}H_{25})_{0.20m}(CH_3)_m]Cl$	4.5	2.1	3.5×10^{-4}	90	6.0×10^{3}
	7.0	0.8	2.3×10^{-3}	150	3.4×10^{2}
$C_2H_5NH_2$[b]	4.5				4.5×10^{-2}
	7.0				4.8×10^{-2}
PEIQ;[c] $[(C_2H_4N)_m(C_{12}H_{25})_{0.25m}(CH_3)_{1.75m}]Cl$	5.0	3.6×10^{-2}			
	7.0	7.9×10^{-3}			
None[d]	4.5	4.3×10^{-3}			
	7.0	1.0×10^{-3}			

[a] The number of catalytic sites per macromolecule.
[b] The second-order rate constants were calculated from the data of Pedersen, Ref. 63.
[c] Spontaneous decarboxylation in the presence of 6.7×10^{-2} residue molar polymer.
[d] Spontaneous decarboxylation reported in Ref. 64.

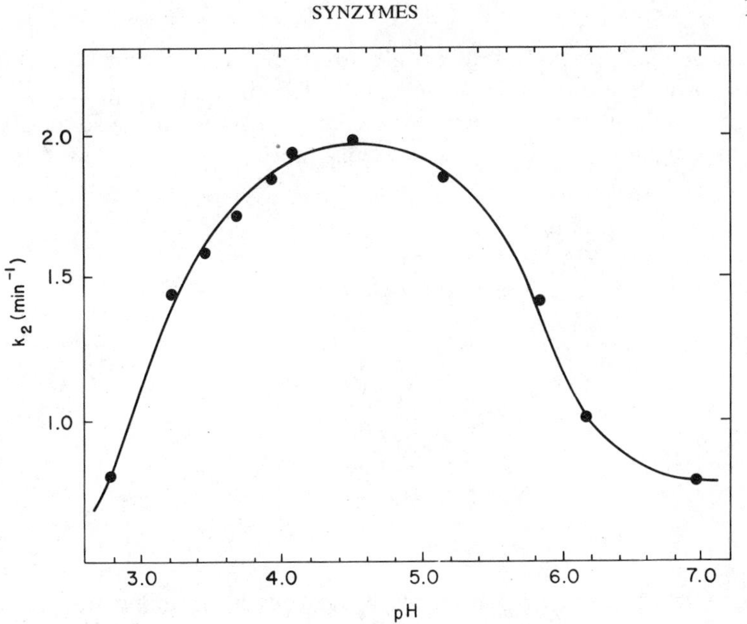

Fig. 20. Rate-constant-pH profile for PEIQ-NH_2 catalyzed decarboxylation of oxalacetate.

equal to k_2 for the polymer; the small-molecule amine would have to be 50 M in concentration.

The effectiveness of the polymer reflects several of its chemical and structural properties. Its cationic charge and apolar groups endow it with a very strong affinity for small molecules, particularly anions, as has been known for some time.[20] For oxalacetate specifically, binding by polymer is also directly evident in the ultraviolet and infrared spectroscopic changes accompanying the shift in tautomeric equilibrium in the environment of the macromolecule.

A further manifestation of an initial binding step is saturation kinetics. For example, in the presence of excess catalyst, (26) predicts a hyperbolic relation between k_{obs} and concentration of nucleophilic sites $C_0(=nP_0)$. Such behavior is actually what is observed.

The number of nucleophilic sites per macromolecule n can be established from kinetic measurements in the presence of excess substrate and complementary ones in the presence of excess polymer.[48] In the decarboxylation of oxalacetate, n is 90 at pH 4.5 and 150 at pH 7.0. The modified polyethylenimine polymer used contains 140 primary amine groups per macromolecule. Thus 65 to 100% of these act as nucleophilic sites for the decarboxylation reaction.

The similarity in pH-rate profile for the polymer (Fig. 20) and for simple primary amines or acetoacetate decarboxylase suggests that the mechanistic pathways of the decarboxylation reaction may also be alike. In the model amine and enzyme systems there is good evidence for the pathway shown in (33):

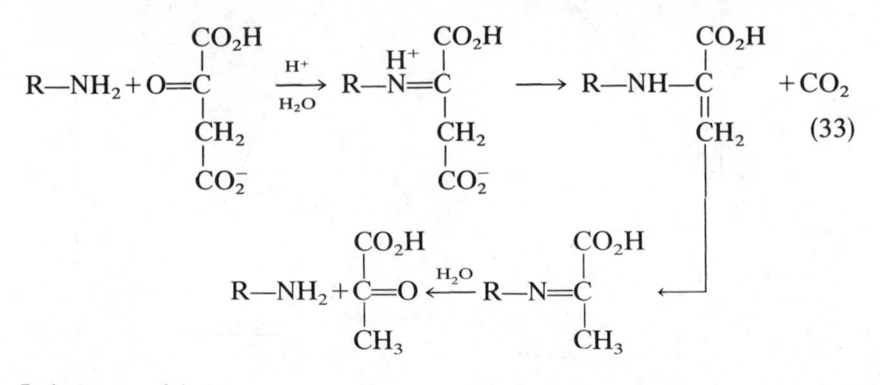

It became of interest to see if we could obtain any indication of Schiff base formation with the polymer. Since spectroscopic probes would be obscured with the actual substrate, oxalacetate, because of the progress of the decarboxylation reaction (32), we have examined instead the spectra of oxalacetate-4-ethyl ester in solutions of the same modified poly-(ethylenimine) PEIQ—NH_2. Such solutions develop a new absorption band at 290 nm. Furthermore, this band is essentially abolished if $NaBH_4$ is added to the solution (Fig. 21). As is well known, $NaBH_4$ reduces Schiff base linkages to amine groups.[43,44]

It has also been possible to confirm the presence of the reduction product of a Schiff base on the polymer by proton magnetic resonance. For this purpose we have used unmodified poly(ethylenimine), since it too catalyzes the decarboxylation of oxalacetate to its product, pyruvate. Unmodified polyethylenimine was mixed with oxalacetate-4-ethyl ester. One-half of this solution was treated with $NaBH_4$; the second half was exposed to a similar environment but no $NaBH_4$ was added. The borohydride-treated polymer exhibited a strong triplet in the nmr spectrum centered at 3.4 ppm upfield from the HOD resonance. This new feature would be expected from the terminal methyl protons of the oxalacetate ester attached to the polymer. Only a very weak triplet was found in the control sample not treated with borohydride. These observations are strong evidence for the formation of Schiff bases with the polymer primary amine groups. Evidently the mechanistic pathway for decarboxylation by the polymer catalyst is similar to that used enzymatically.

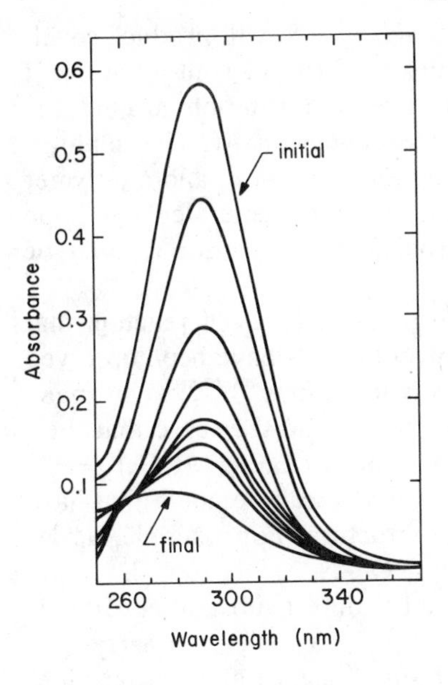

Fig. 21. Changes in ultraviolet spectra on addition of $NaBH_4$ to a mixture of oxalacetate-4-ethyl ester and PEIQ-NH_2.

VIII. FUTURE PROSPECTS

This description of the behavior of polyethylenimines demonstrates that it is possible to construct synthetic polymers with traits analogous to those of enzymes. The essential structural features in the effective polymers are high local density of functional groups and hydrophobic apolar domains of submicroscopic size, all embedded in a gossamerlike macromolecular framework readily permeated by the aqueous solvent. Such structures can be catalytically effective under ambient conditions, that is, at room temperature and pressure in aqueous environments near physiological pH 7.

As modified so far the polyethylenimines, in contrast to enzymes, are weak in structural specificity toward substrates. This need not be a defect, however, for these macromolecular catalysts do not have to operate in a cellular environment and hence need not be subject to constraints designed to maintain the stability of a very complex, integrated biochemical network. Nevertheless, circumstances may arise where substrate specificity may be an essential requirement. We have some ideas on how this might be achieved with these relatively elastic macromolecular frameworks. For example, preliminary experiments show that we can attach —SH groups covalently to the polymer. It should be possible thereafter to add to the polymer solution an "inhibitor" with a structure analogous to the potential substrate and then to expose the solution to air or to pure O_2. Oxidation of sulfhydryls, $2SH \rightarrow S—S$, will produce covalent cross-links that will freeze the polymer into a rigid conformation. If the inhibitor is bound to the polymer, the conformation around the binding site should also become fixed three-dimensionally. The inhibitor could then be removed, by dialysis or column separation, and a polymer with some specificity in choice of substrate should be generated. We hope in this way to achieve some measure of specificity in addition to catalytic effectiveness.

It should also be feasible to extend further the types of reaction that can be accelerated. For example, large solvent effects have been observed in kinetic studies of many reactions involving anions.[46,47,50,51] In many cases the solvents are aprotic but not truly apolar, in the sense that their molecules have large dipole moments, for example, $(CH_3)_2S{=}O$, $CH_3CON(CH_3)_2$. Derivatives of polyethylenimine can be made that have substituents mimicking these in chemical structure. For example, acylation of the polymer will produce $CH_3CO—N{<}$ loci on the macromolecule. Such modified polymers should manifest substantial catalytic effects.

It should also be possible to introduce bifunctional catalytic groups into the polymer. Also very enticing is an exploration of the potential of other nucleophilic groups besides the imidazole that we have attached to the polymer or the primary amines that are intrinsically present in it. Attractive possibilities at the moment are —SH, $>C{=}N—OH$, and

$$—\overset{\overset{\displaystyle O}{\|}}{C}—\overset{\overset{\displaystyle CH_3}{|}}{N}—OH,$$

the mercaptan because it appears in the cysteine proteases (e.g., papain) and the others because they are effective in small-molecule model systems.

The modified polyethylenimines described so far are only a few of many possibilities. It is obvious that this polymer provides a remarkably versatile macromolecular matrix for the attachment of a wide variety of different types of functional groups. Furthermore, the polymer framework makes it possible to juxtapose a binding site, a catalytic group, and an apolar-aqueous interface in a locally compact array. Thus a wide range of local environments can be created on this macromolecular water-soluble catalyst. We hope to be able to exploit these to obtain a series of synthetic macromolecules with tailor-made catalytic properties.

References

1. A. J. Brown, *J. Chem. Soc.*, **81,** 373 (1902).
2. V. Henri, *Lois Générales de l'action des Diastases*, Hermann, Paris, 1903.
3. L. Michaelis and M. L. Menten, *Biochem. Z.*, **49,** 333 (1913).

4. I. M. Klotz, in *The Proteins*, H. Neurath and K. Bailey, Eds., Vol. I, Part B, Academic, New York, 1953, Chap. 8.
5. I. M. Klotz and J. M. Urquhart, *J. Am. Chem. Soc.*, **71,** 847 (1949).
6. F. Karush and M. Sonenberg, *J. Am. Chem. Soc.*, **71,** 1369 (1949).
7. G. Scatchard, I. H. Scheinberg, and S. H. Armstrong, Jr., *J. Am. Chem. Soc.*, **72,** 535 (1950).
8. I. M. Klotz, *J. Am. Chem. Soc.*, **68,** 2299 (1946).
9. I. M. Klotz, *Cold Spring Harbor Symp. Quant. Biol.*, **14,** 97 (1949).
10. F. Karush, *J. Am. Chem. Soc.*, **72,** 2705 (1950).
11. H. Bennhold and R. Schubert, *Z. Ges. Exp. Med.*, **113,** 722 (1943).
12. C. Wunderly, *Arzneim.-Forsch.*, **4,** I, 29 (1950).
13. U. P. Strauss and E. G. Jackson, *J. Polym. Sci.*, **6,** 649 (1951).
14. W. Scholtan, *Makromol. Chem.*, **11,** 131 (1953).
15. S. Saito, *Kolloid Z.*, **154,** 19 (1957).
16. I. M. Klotz and V. H. Stryker, *J. Am. Chem. Soc.*, **82,** 5169 (1960).
17. P. Molyneux and H. P. Frank, *J. Am. Chem. Soc.*, **83,** 3169 (1961).
18. I. M. Klotz and K. Shikama, *Arch. Biochem. Biophys.*, **123,** 551 (1968).
19. L. E. Davis, in *Water Soluble Resins*, R. L. Davidson and M. Sittig, Eds., Reinhold, New York, p. 216.
20. I. M. Klotz, G. P. Royer, and A. R. Sloniewsky, *Biochemistry*, **8,** 4752 (1969).
21. I. M. Klotz, F. M. Walker, and R. B. Pivan, *J. Am. Chem. Soc.*, **68,** 1486 (1946).
22. G. P. Royer and I. M. Klotz, *J. Am. Chem. Soc.*, **91,** 5885 (1969).
23. F. M. Menger and C. E. Portnoy, *J. Am., Chem. Soc.*, **90,** 1875 (1968).
24. R. S. Johnson and I. M. Klotz, in preparation.
25. T. W. Johnson and I. M. Klotz, *Macromolecules*, **7,** 149 (1974).
26. I. M. Klotz, G. P. Royer, and I. S. Scarpa, *Proc. Nat. Acad. Sci.* (*U.S.*), **68,** 263 (1971).
27. E. Katchalski, G. D. Fasman, E. Simons, E. R. Blout, F. R. N. Gurd, and W. L. Koltun, *Arch. Biochem. Biophys.*, **88,** 361 (1960).
28. M. L. Bender and B. W. Turnquest, *J. Am. Chem. Soc.*, **79,** 1652 (1951).
29. T. C. Bruice and G. L. Schmir, *J. Am. Chem. Soc.*, **79,** 1663 (1957).
30. F. J. Kezdy and M. L. Bender, *Biochemistry*, **1,** 1097 (1962), and references therein.
31. E. R. Stadtman, in *The Mechanism of Enzyme Action*, W. D. McElroy and B. Glass, Eds., Johns Hopkins, Baltimore, 1954, p. 581.
32. T. W. Johnson and I. M. Klotz, *Macromolecules*, **6,** 788 (1973).
33. W. P. Jencks and J. Carriuolo, *J. Biol. Chem.*, **234,** 1272 (1959).
34. H. C. Kiefer, W. I. Congdon, I. S. Scarpa, and I. M. Klotz, *Proc. Nat. Acad. Sci.* (*U.S.*), **69,** 2155 (1972).
35. N. Muller and R. H. Birkhahn, *J. Phys. Chem.*, **71,** 957 (1967).
36. N. Muller and T. W. Johnson, *J. Phys. Chem.*, **73,** 2042 (1969).
37. T. W. Johnson and I. M. Klotz, *J. Phys. Chem.*, **75,** 4061 (1971).
38. T. W. Johnson and I. M. Klotz, *Macromolecules*, **7,** 618 (1974).
39. T. W. Johnson and N. Muller, *Biochemistry*, **9,** 1943 (1970).
40. C. Tanford, *J. Phys. Chem.*, **76,** 3020 (1972).
41. J. B. Birks, *Photophysics of Aromatic Molecules*, Wiley-Interscience, London, 1970, Chap. 7.
42. R. A. Pranis and I. M. Klotz, *Biopolymers*, **16,** 299 (1977).
43. E. H. Fischer, A. B. Kent, E. R. Snyder, and E. G. Krebs, *J. Am. Chem. Soc.*, **80,** 2906 (1958).
44. B. L. Horecker, S. Pontremoli, C. Ricci, and T. Cheng, *Proc. Nat. Acad. Sci.* (*U.S.*), **47,** 1949 (1961).

45. I. B. Berlman, *Handbook of Fluorescence Spectra of Aromatic Molecules*, Academic, New York, 1965, p. 35.
46. D. S. Kemp and K. Paul, *J. Am. Chem. Soc.*, **92,** 2553 (1970).
47. D. S. Kemp and K. Paul, *J. Am. Chem. Soc.*, **97,** 7305 (1975).
48. J. Suh, I. S. Scarpa, and I. M. Klotz, *J. Am., Chem. Soc.*, **98,** 7060 (1976).
49. C. A. Bunton and M. Minch, *Tetrahedron Lett.*, 3881 (1970).
50. A. J. Parker, *Adv. Phys. Org. Chem.*, **5,** 173 (1967).
51. A. J. Parker, *Chem. Rev.*, **69**, 1 (1969).
52. F. H. Westheimer, *Proc. Chem. Soc.*, 253 (1963).
53. K. J. Pedersen, *J. Am. Chem. Soc.*, **51,** 2098 (1929).
54. K. J. Pedersen, *J. Phys. Chem.*, **38,** 559 (1932).
55. S. Kaneoko, *J. Biochem.* (Japan), **28,** 1 (1938).
56. R. Steinberger and F. H. Westheimer, *J. Am. Chem. Soc.*, **71,** 4158 (1949).
57. R. W. Hay, *Aust. J. Chem.*, **18,** 337 (1965).
58. L. K. M. Lam and D. E. Schmidt, Jr., *Can. J. Chem.*, **51,** 1959 (1973).
59. W. J. Spetnagel and I. M. Klotz, *J. Am. Chem. Soc.*, **98,** 8199 (1976).
60. G. W. Kosichi and S. N. Lipovac, *Can. J. Chem.*, **42,** 403 (1964).
61. A Kornberg and W. E. Pricer, Jr., *J. Biol. Chem.*, **193,** 481 (1951).
62. R. W. Von Korff, in *Methods in Enzymology*, J. M. Lowenstein, Ed., Vol. XIII, Academic, New York, 1969, p. 521.
63. K. J. Pedersen, *Acta Chem. Scand.*, **8,** 710 (1954).
64. C. S. Tasi, *Can. J. Chem.*, **45,** 873 (1967).

LIST OF INTERVENTIONS

1. Somorjai
   1.1 Klotz
2. Ubbelohde
   2.1 Williams
   2.2 Klotz
   2.3 Claesson
   2.4 Clementi
   2.5 Claesson
3. Mandel
4. Klotz
   4.1 Mandel
   4.2 Klotz
5. Claesson
   5.1 Klotz
6. Simon
   6.1 Klotz
7. Caplan
   7.1 Klotz
8. Thomas
   8.1 Klotz
9. Thomas
   9.1 Klotz
10. Williams
   10.1 Klotz

1. Intervention of Somorjai

1. Have you investigated the pH, temperature, and so on, dependence of the catalytic activity of synzymes?

2. What, if any, specificity does PEI have? Can one control specificity, and, if so, what are the important parameters?

1.1 Intervention of Klotz

Yes, we have made some pH activity studies. One example is shown in the article (Fig. 20) for the decarboxylation of oxalacetate. Interestingly, the bell shape is similar to that found by others in the catalysis of this reaction by small-molecule amines.

2. Intervention of Ubbelohde

When complex chemical processes are being investigated, this usually implies quite large entropy changes, as well as the usual enthalpy changes. In such cases the volume change on passing to the activated state can be even more informative than the more often studied activation energy. Studies of the effects of pressure on chemical reaction routes can then be particularly rewarding. To extend such measurements up to useful pressures no longer calls for a very specialized high-pressure laboratory.

Accordingly, the pressure variable should be much more widely studied. The use of pressure is not reserved to specialists.

2.1 Intervention of Williams

The work of Drickamer should be remembered. He has shown that spin-state transitions of metal ions in proteins such as myoglobin can be driven by pressure. He has followed many properties of proteins under pressure [see, for example, J. Chem. Phys. **55,** 1633 (1971)]

2.2 Intervention of Klotz

Professor Ubbelohde's suggestions are certainly worthwhile. But there are many directions that we are trying to pursue: more detailed studies of the conformation of the polymers, introduction of new functional groups, macromolecular characterization by hydrodynamic methods, and so on. If I were a professor at Imperial College, I could assign all these problems, and those you suggest, to members of the staff; but I am constrained by a smaller research group.

2.3 Intervention of Claesson

I have been asked to make a few remarks about the fact that it is nowadays quite easy to make measurements under high pressure. Consequently, pressure can be used as a variable almost as conveniently as temperature in chemical research. How large is the effect of pressure? Of course, no general answer can be given—not even the direction of the effect, which depends on the signs of ΔV (which is quite large for macromolecules). However, as a rule of thumb we can say that a change in pressure of 10 atm has about the same effect as a change of temperature of 1°C. Therefore a moderate pressure of a few thousand atmospheres is sufficient to give large effects (e.g., dissociating or denaturing proteins). We have developed cells with quartz windows for optical measurements (spectrophotometry, light scattering, circular dichroism) that are small enough to be carried in the pocket, with a pressure of 3000 atm inside. [S. Claesson, S. Malmrud, and B. Lundgren, *Trans. Faraday Soc.*, **66,** 3048 (1970), (large pressure cell), S. Claesson and L. D. Hayward, *Chem. Scripta,* **9,** 18 (1976) (small cell).] This pressure can be reached by using the small type of hydraulic press commonly available for making pellets for infrared work. The cells fit into most standard commercial instruments for optical measurements. A typical example is the effect of pressure on the second virial coefficient of polystyrene in good solvents, which increases about 50% at about 1000 atm and then remains constant at further pressure increase. This behavior can be explained qualitatively by the theory based on corresponding states as developed by Prigogine and co-workers. [C. J. McDonald and S. Claesson, *Chem. Scripta*, **4,** 115 (1973).]

2.4 Intervention of Clementi

On one hand, keeping in mind Mr. Solvay's invitation to remember that chemical sciences must attempt to be relevant to the needs of society and, on the other hand, noticing some enthusiasm for chemical processes at 5000 atm, I would like to ask the following question: accepted for the sake of argument the equivalence (from the viewpoint of kinetics) between $\Delta T = 1°C$ and $\Delta P = 10$ atm, and given a reasonable threshold of pressure (say, 500 atm) and temperature (say, 200°C), what is the increment in cost for an increment of 1°C relative to the increment of 10 atm (clearly, not in a laboratory experiment, but in a chemical plant)?

2.5 Intervention of Claesson

Though it is difficult to generalize, increases of pressure are often economically wholly feasible.

3. Intervention of Mandel

In connection with the observations of Professor Klotz on the effect of structural modifications on the binding affinities of synthetic macromolecules, I would like to make the following comments. It is well known to polyelectrolyte chemists that the binding of small molecules can be strongly affected by changing the structure and the conformation of macromolecules. Surprisingly enough, this has been studied much more extensively for negatively charged macromolecules than for positively charged ones such as these studied by Professor Klotz. One of the most striking examples is that of PMA, poly-(methacrylic acid) $\left[CH_2\!-\!\underset{COOH}{\overset{CH_3}{C}}\!- \right]_n$ as compared to PAA, poly-(acrylic acid) $\left[CH_2\!-\!\underset{COOH}{CH}\!- \right]_n$. The former binds small molecules in a much more effective way than the latter, the binding constant being at least two orders magnitude larger (e.g., in the case of positively charged dye molecules). PMA at low degrees of dissociation can even bind polycyclic aromatic hydrocarbons to a nonnegligible extent. Clearly, no electrostatic forces are involved here. In the case of positively charged dyes, the amount of dye molecules that can be bound per macromolecule is determined by the charge of the macromolecule. In analogy to what was pointed out by Professor Klotz, at low degrees of dissociation α, PMA has a very tightly coiled conformation, as revealed, for example, by a very small intrinsic viscosity, in contrast to what is observed with PAA. This tightly coiled conformation unfolds on neutralization in a small region around $\alpha = 0.2$, in what can be called a conformational transition, not observed for PAA.

The following may be of interest as it shows some analogy to what is observed with proteins. The positively charged dye auramine-O binds more strongly to PMA in its tightly coiled conformation than to the extended coil one (after the conformational transition). At the same time the fluorescence of that dye (practically nonfluorescent when free in solution) is found to be enhanced by binding to PMA. However, the quantum yield per bound molecule is much larger for the dye bound to the tightly coiled conformation than to the extended one. (The difference is several orders of magnitude.) Apparently the properties of the molecule bound to the macromolecule are not only strongly influenced by the macromolecule but also by its conformational properties. On the other hand, when an appreciable amount of dye is bound to PMA in the extended conformational form, a shift in the conformation of PMA toward the tightly coiled form is observed. Thus the binding of the small molecule in its turn may also have a nonnegligible influence on the conformational properties of the macromolecule. In the area of the catalytic effects of synthetic polymers, the work of Morawetz who started more than 10 years ago to investigate such effects, should be acknowledged. Others, such as Ise, also have probed this field. In general, catalytic activity was found to be due to the accumulation effect of the same group along the macromolecular chain, that is, to the locally increased concentration of these groups.

4. Intervention of Klotz

I do *not* consider trivial the change from

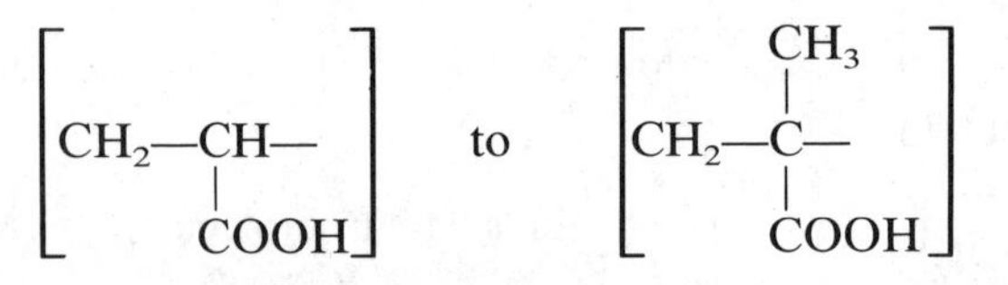

A more proper representation of the polymer character of the structure,

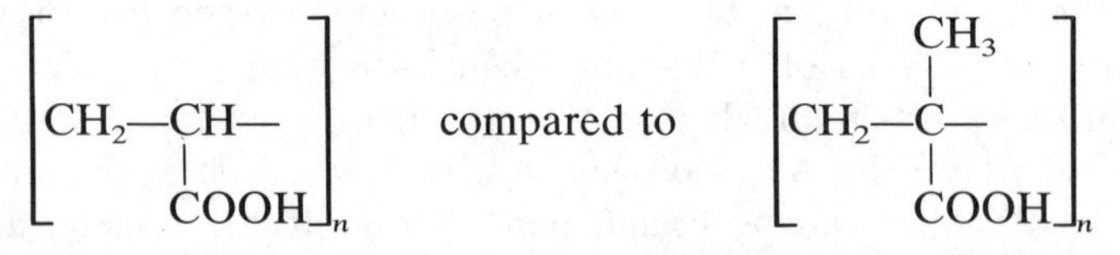

will make my point clear. Thus there are n CH_3 groups present in a chain of polymethacrylate. That is a very substantial amount of apolar material.

The addition of 1 CH_3 to three-carbon monomer is a change comparable to adding 1 $C_{12}H_{25}$ to 10% of the C_2H_4N residues in polyethylenimine. In the latter case the 10% laurylpolyethylenimine shows much increased binding compared to unmodified polymer, as is evident in Fig. 4 of my chapter. With respect to fluorescence, many fluorophores in solution showing negligible fluorescence become strongly fluorescent when bound to proteins or to polymers with apolar domains. The change in conformation of a polymer on adding a component to its solution is really no great surprise. Gibbsian thermodynamics requires that

$$N_1\, d\mu_1 + N_2\, d\mu_2 + \cdots = 0$$

Thus the chemical potentials (μ_i), or the free energies, of the components are inextricably interlinked. If you change μ of the solvent by adding something to it [changing the mole fraction (N_i)], you must change μ of the solute. If the chemical potential is changed, it is a reasonable expectation that the molecular structure is also perturbed. Of course, if a ligand is bound directly to the polymer, a mutual accommodation of conformational structures is also likely to occur.

4.1 Intervention of Mandel

In reply to what Professor Klotz has said, I think that I cannot agree with his statement that the increased binding affinity of small molecules by poly-(methacrylic) as compared to poly-(acrylic acid) is just related to the large number of CH_3 groups borne by the polymer chain. I do not feel that the effect is so trivial, particularly as it seems to be connected to the special conformational properties of PMA. At this stage it seems that it is still quite difficult to understand why the small CH_3 group by itself can bring on the tightly coiled conformation as observed for PMA. In fact, in other polyelectrolytes, such as the copolymers of maleic acid and alkenes or ethylene-alkyl-ethers,

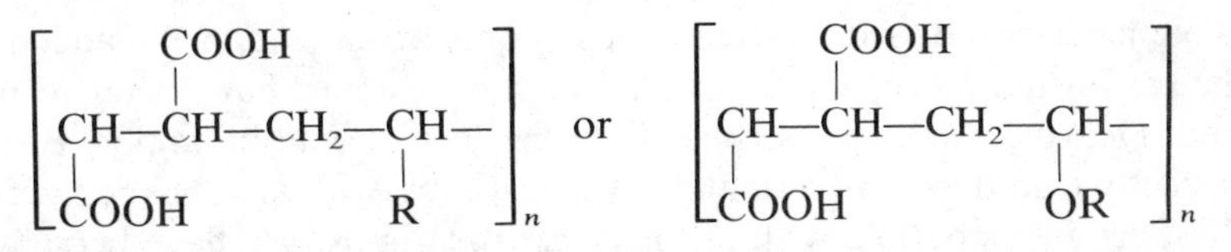

the alkyl chain R must contain three, four, or even more C atoms before a conformational form analogous to that of PMA at low α is obtained.

I cannot but agree with Professor Klotz that the influence of the binding of small molecules on the properties of these molecules and the conformational properties of the polymer must follow from thermodynamics. Of course such effects, when observed, must follow the rules of thermodynamics. But these thermodynamic rules do not tell us what is happening on the molecular level.

4.2 Intervention of Klotz

I have no substantive disagreement with Dr. Mandel. It was my impression from his previous comment that he viewed the addition of a CH_3 group to the acrylate residue as trivial. For reasons that I just described I consider the addition of 1 CH_3 to a three-carbon monomer as a substantial change in the context of a polymer (comparable to adding 1 $C_{12}H_{25}$ group to 10% of the monomer residues in polyethylenimine).

5. Intervention of Claesson

Have you made measurements of fluorescence depolarization in order to get information about internal molecular motion in your catalysts?

5.1 Intervention of Klotz

We have not measured fluorescence depolarization with fluorophors and our polymers, but such measurements have been made by others, particularly with proteins; and, as you indicate, it is possible to determine rotational relaxation times for the macromolecule and thus to obtain some insight into its behavior in solution.

6. Intervention of Simon

Fluorinated hydrocarbons are known to interact strongly even with rather hydrophilic anions. By introducing trifluoromethyl groups into the polymers you mentioned, I would expect a substantial increase in the binding of anions. Did you observe such an effect when introducing the trifluoromethyl labels?

6.1 Intervention of Klotz

We have introduced fluorine (as CF_3) only at the end of the added side chains. Furthermore, less than 10% of the residues have been so modified. Hence I think the perturbation due to F should be minor. But it is perfectly true that *perfluor*inated hydrocarbons, with all CH's replaced by CF's, are incompatible with ordinary hydrocarbons. We have been aware of this incompatibility, and hence have kept the fraction of CF bonds in our polymer very low.

7. Intervention of Caplan

In view of the potential importance of the systems he has described in chemical engineering, I should like to know whether Dr. Klotz has attempted to insolubilize his synzymes in the form of particles or membranes, either by cross-linking them or otherwise immobilizing them in an appropriate matrix. In this way their performance as heterogeneous catalysts could be studied.

7.1 Intervention of Klotz

We have not joined our polymer to insoluble supports, but it is obvious that with so many primary-NH_2 groups in the polymer, coupling to a support should be straightforward.

We have also made some experiments aiming to make locally dense polymers by means of disulfide cross-links. We start with a polymer, such as polylysine, introduce —SH groups to give

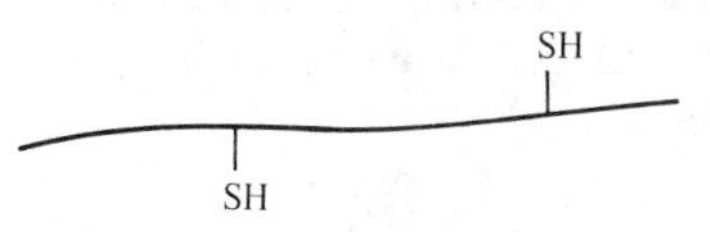

bring the polymer to high pH so that it becomes compact

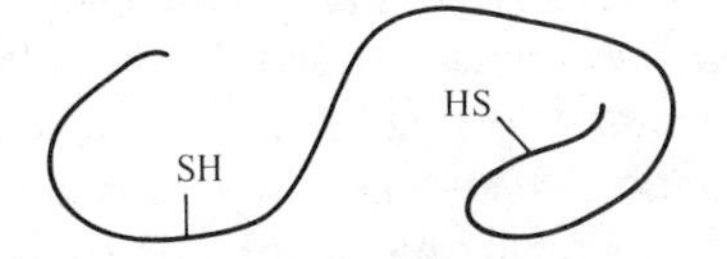

and then allow air to come in to produce S—S cross-linkages

We have not pursued this in detail but it offers a good possibility for developing catalytic polymers and for producing insoluble forms.

8. Intervention of Thomas

What is the stability of the catalytic activity of your polymers, both as a function of time and as a function of temperature, in comparison with enzymes?

8.1 Intervention of Klotz

We have made no detailed or systematic studies of change of activity with temperature or time. There are some differences in behavior of different polymers stored below 0°C over long periods of time. In some cases activity is lost. This seems to be due to some aggregation phenomenon, for the polymer also loses its solubility in water. If it is dissolved in an organic solvent, dialyzed against water, and lyophilized, solubility and activity are regained.

9. Intervention of Thomas

What is the evaluation of the cost of your polymers?

9.1 Intervention of Klotz

I have no feeling for economic aspects. Industry does find it possible to use solid support enzymes. The unmodified polymer is relatively cheap. We have received 5 gal samples (of a 30% aqueous solution) at zero cost from industrial sources. Of course, our chemical modifications would add to the cost. Nevertheless, the initial material should present no economic problems, since it is available to us because it is being manufactured for some large-scale uses.

10. Intervention of Williams

A comparison of peroxidase and cytochrome P-450 illustrates the problems of comparing enzymes and their related catalysts such as synzymes. Peroxidase has low substrate specificity and a simple free-radical oxidation reaction. The substrate site is 10 Å from the iron (H_2O_2 site) and is probably just an oily "droplet" region of the protein. This proteins has parallels with Professor Klotz's systems. Proximity is perhaps sufficient to explain the activation of the organic substrate (but not for that of H_2O_2).

Cytochrome P-450 can oxidize such substrates as long-chain fatty acids at ω-groups. This introduces two problems—specificity and a *large* change in activation energy. I do not think that proximity is enough here to explain insertion of O-atom into a hydrocarbon chain.

Can Professor Klotz comment on the possibility of (*a*) an increase in specificity, (*b*) an increase in the ability to do really difficult reactions with synzymes?

10.1 Intervention of Klotz

It is quite true that we have no specificity as yet. However, I refuse to become a prisoner of enzymologists. We *are not* trying to make enzymes; we are making *catalysts*. We are not aiming toward enzyme substitutes for insertion in cellular systems. There specificity is crucial, for cells would not have evolved if enzymes were nonspecific (and hence lethal in a cell). With our broad catalysts, we have choice of the process to be accelerated by selection of the substrate added.

Nevertheless, we have made some exploratory studies toward investing specificity into the polymer. This too is being done by the introduction of S—S cross-links:

—SH HS— $\xrightarrow{\text{air}}$ —S—S—

Before oxidation, a ligand (methyl orange) is introduced and is bound by polymer. This oxidation may "freeze" the binding-site conformation. There are indications that some specificity, at least in the dye binding, may be achieved. Compared to nature, we have not been very successful yet. But after all, nature has had 10^9 years to carry out her experiments, whereas we have had only 10^1.

11. Intervention of Hess

Did you study the influence of the ionic strength on the catalytic functions of the synzymes?

11.1 Intervention of Klotz

We have made no systematic studies of ionic strength effects. Ionic strength does not seem important in the catalytic hydrolysis of the acylnitrophenolates, which carry no charge. There are substantial salt effects in the decarboxylation reaction with nitrobenzisoxazole, which is an anionic substrate. These may reflect general ion cloud effects or anionic competition for binding to the polymer.

11.2 Intervention of Ubbelohde

Is the effect of ionic strength caused by a localized influence on catalytic centers, or does it point to the role of overall electrostriction on catalytic activity? Comparative experiments on the effect of pressure on catalytic activity might help to elucidate this.

12. Intervention of Hess

What is the minimum size of synzymes? And is there any control effect observed that is analogous to the allosteric control of enzyme activities?

12.1 Intervention of Klotz

We have compared three different polymer samples in the molecular weight range of 600 to 60,000 but only in regard to the nucleophilic reaction with nitrophenyl esters. The differences are small, less than a factor of 10.

We have made no detailed or systematic comparisons of molecular weight effects.

As I described in our excimer experiments, the polymer shows much flexibility. For allosteric effects it is not essential to have subunit systems. Thus the simple polymer macromolecule has the intrinsic ability to undergo conformational changes with accompanying changes in activity. But this too is an area in which we have made no experiments. In all experiments the kinetics has been Michaelian, with no evidence of interaction between sites.

13. Intervention of Hess

The cleft size of phosphotransferases might be on the order of a diameter of 10 Å. However, it should be recognized that the size is varied under the influence of controlling ligands such as observed in allosteric enzymes.

I would like to remind the audience that Mildred Cohn had shown that in some phosphotransferases the binding of ligands to the active site of the enzyme leads to an exclusion of water molecules in the neighborhood of manganese as part of the active site function.

14. Intervention of Prelog

I would like to mention a geometrical feature of many enzymes that Professor Klotz's synzymes do not show. The active sites of most enzymes are located in clefts or channels, that is, in concave parts of the molecule. Synzymes are globular polymers and therefore do not show the concavity that is especially important for stereospecificity.

14.1 Intervention of Klotz

I agree with Professor Prelog, but only about 95%. Not all active centers are "clefts." In chymotrypsin, as nearly as I can recall, the active site is more like a flat surface left on a sphere after a slice has been cut off from the top. Perhaps even more complementary to a cleft is the first "male protein" described last year in PNAS by one of the Swedish crystallography groups. In any event, clefts are not all-pervasive.

14.2 Intervention of Prelog

The dimensions of cavities in enzymes differ considerably, depending on their physiological function. In many cases the clefts are occupied by clusters of organized water molecules. Such clusters can be seen in certain X-ray structures of enzymes (e.g., the structure of carboxypeptidase A determined by Lipscomb). If the clefts are deep, as in horse liver alcohol dehydrogenase, a channel for removal of water is present (Brändén).

14.3 Intervention of Klotz

Another example that I now recall of poor fit between shape of a binding site and that of ligand is the immunoglobulin, or Bence-Jones protein, that binds the hapten phosphorylcholine. The X-ray structure of the antibody has been worked out and it is evident that the binding area is much larger than that of the hapten, and hardly complementary in shape.

15. Intervention of Lehn

Up to now we have mainly considered macromolecules, but suitably designed smaller molecules may also show features that are usually associated with enzyme-type systems; for instance, they may contain a cleft, an intramolecular cavity and display very stable and selective binding.

Thus the spherical macrotricyclic molecule represented below gives a tetraprotonated species that has very unusual binding properties. It binds very strongly the spherical chloride anion, holding it in an ideal tetrahedral arrangement of N^{+}—H · · · Cl^{-} hydrogen bonds. This complex is much more stable and the Cl^{-}/Br^{-} selectivity is much higher than any other known to date. These properties may be attributed to the presence

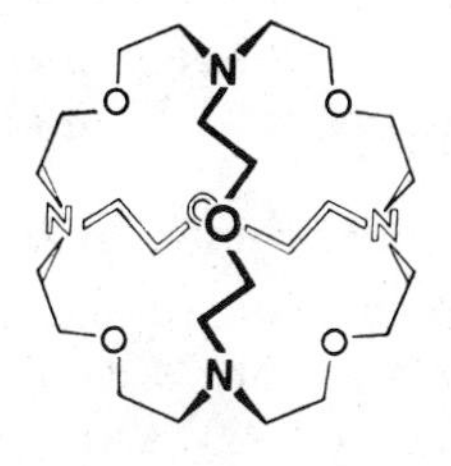

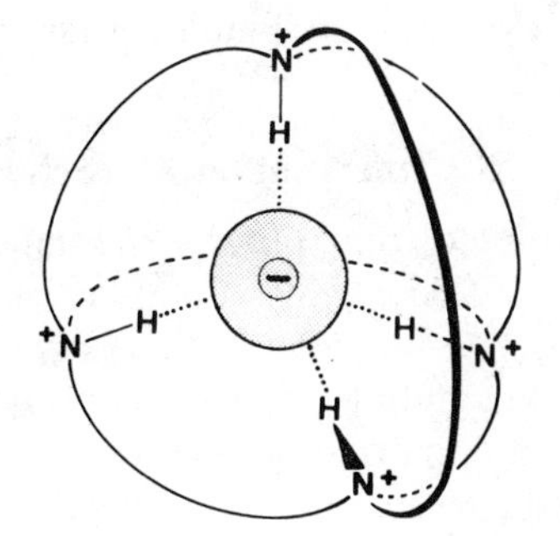

of a closed intramolecular cavity of suitable size, which hinders solvation and substantially resists deformation because of its high connectivity [see *J. Am. Chem. Soc.*, **98,** 6403 (1976)].

One may stress that in proteins clefts are built by coiling back and forth a linear thread maintained by intermolecular binding forces (and eventually a few disulfide bridges). In the case of smaller synthetic molecules, organic synthesis provides means for building the molecular architecture using more highly connected cornerstones.

16. Intervention of Somorjai

I am not convinced of the importance of distinguishing betwen convexity and concavity of the enzyme surface: Entropic factors might mitigate against the geometric advantage of having concavity. Geometric factors are necessary but not sufficient in explaining the catalytic power and specificity of enzymes. It is essential to understand the dynamical and energetic aspects of the structure, in particular the mechanism of energy transduction to the active site-substrate complex. The active site cannot be considered in isolation and thus experiments that mimic the active site by means of a rigid, excised analogue can provide rate enhancements that are due only to geometric-entropic factors. The analogue is in a lower energy state than the true active site is in its three-dimensional enzymic environment; thus the triggering effect of the right substrate is less explosive in the analogue. It would be interesting if the analogue-substrate kinetic experiments could be carried out while maintaining the analogue in an excited state by some form of energy pumping (laser?).

17. Intervention of Williams

Lysozyme is a hinge-protein. Phillips and his coworkers describing the solid state [see *The Enzymes*, 3rd Ed., P. D. Boyer ed., Vol VII, Academic, New York, p. 665 (1972)] and our own solution studies both indicate that the two sides of the protein can close together when substrate is bound. The whole protein is involved in activity. It may be that other enzymes close up on binding substrates (e.g., kinases and P–450 cytochromes).

18. Intervention of Prelog

By my remarks I wanted to emphasize the importance of concavity of enzymes as a feature that determines the specificity and stereospecificity. I didn't want to say that it is a feature that is necessary for all enzymes. However, starting with a globular convex substrate, a concave active site allows a strategic distribution of a number of small and large interactions that make the enzyme specific and stereospecific.

19. Intervention of Ubbelohde

It is important to remember that water molecules in a cavity may assume a different packing and even a different crystallinity from water in bulk.

20. Intervention of Clementi

Concerning the number of molecules in a cleft and around an enzyme, I feel that there is some confusion both experimentally and theoretically. If we consider *experiments* (e.g., lysozyme), the quoted number for the molecules of water per unit cell change from 150 to 250 according to the authors; the number of bound (and therefore seen by X-ray) molecules vary from 5 to 10 (and not the same one); in addition, we recall that H_2O can easily be confused with other inorganic small molecules. The theoretical side is equally unclear, *if one uses old concepts* as the Van der Waals radii, which are not sufficient to differentiate one atom from another during interaction with a given atom.

However, the use of the *new interaction potentials obtained from quantum mechanics* (those used to compute the contour maps; see discussion of the report of Dr. Williams) allows a rather accurate determination of how many molecules are placed in a given region (e.g., the cleft) and where each molecule is placed, with a precision far in excess of present-day X-ray resolution for proteins. Specifically, the count of molecules of water is accurate to 2 or 3% and the position is accurate to about 0.2 to 0.3 Å.

21. Intervention of Williams

Let me put it to Professor Lehn that biology can do better than he can with small molecules. From uroporphyrin biology makes (*1*) *copper* porphyrins (turacin feathers), (*2*) *heme* (iron) porphyrins, (*3*) *magnesium* chlorins, (*4*) *cobalt* corrins (B_{12}). These compounds are formed with very little confusion between metals and partners. I cannot give selectivity factors, but they are very big. How is this managed? Is it not better than man can do? Maximum size selectivity for anions is not yet understood as few anion-binding sites in biology are known.

22. Intervention of Lehn

There are examples where small structural changes in small molecules cause very large changes in selectivity of complexation. Thus whereas *A* binds Cd^{2+} about 10^6 times stronger than Zn^{2+}, the ligand *B*, which contains a smaller intramolecular cavity, binds Cd^{2+} stronger than Zn^{2+} by only a factor of about 10. A number of other examples could be cited.

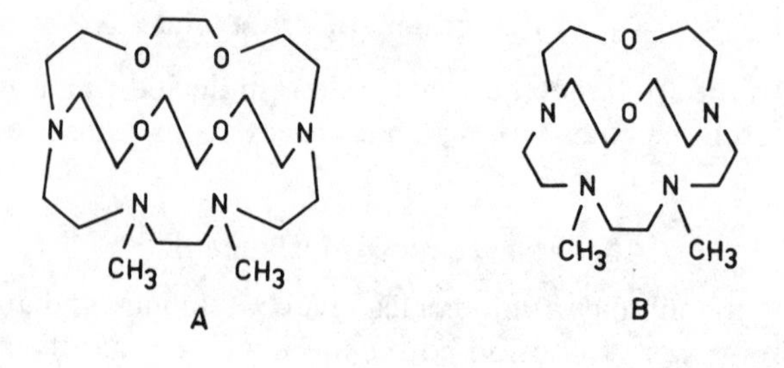

23. Intervention of Hammes

I believe too much emphasis is being put on the binding function of the clefts (hydrophobic pockets) commonly found in enzymes. Although the binding specificity of enzymes is important, the primary function of enzymes is catalysis. During catalysis, the cleft almost certainly must be coupled to the rest of the enzyme molecule. This is why it is important to use weak noncovalent interactions rather than rigid covalent interactions to obtain substrate specificity. These noncovalent interactions can be used to couple the cleft to the rest of the enzyme molecule through cooperative conformational changes. Cases are well known where the properties of an amino acid residue quite far from the binding site are altered by the binding of a substrate or inhibitor to the enzyme. The coordination of weak but specific interactions throughout the macromolecule (i.e., cooperativity) is undoubtedly essential for efficient catalysis.

24. Intervention of Ross

We have to be concerned with two separate problems: First, we face the problem of the dynamics of a many-body mechanical system such as an enzyme. It may well be impossible, probably undesirable as well, to think of a molecular dynamics calculation for such a complicated system. A separate problem is the one of a statistical (thermodynamic) analysis of a membrane; that has to do with the number of molecules involved, and the magnitude of the fluctuations expected.

25. Intervention of Mandel

Coming back to the remark of Professor Prelog about the role of the cleft in the catalytic activity of the enzyme, I think that this should be stated in a more general manner. What is needed is a macromolecule that has a very specific conformation, both from the static and dynamic point of view. Such a highly specific conformation can only be brought about by a very *specific distribution* on the molecule, of well-defined groups of different nature. It is in fact the interaction between these groups and eventually some of these with the water that determines the conformation of the protein. Now in a synthetic polymer in general no such specific distribution of interacting groups exists and no specificity of the average conformation results. In order to bring about such specificity it would be necessary to introduce into the polymer chain chemical modifications in a nonrandom way, but not knowing beforehand what distribution exactly is needed for a given specific conformation and what specific conformation may be needed for optimal catalytic activity. I therefore feel that from that point of view the situation looks quite hopeless and that the chances to obtain synthetic analogues of specific enzymes are very low.

26. Intervention of Williams

The great difficulty in understanding the evolution of a protein can be illustrated by cytochromes *c*, which are known to have a methionine and a histidine as heme ligands. The position of the methionine, especially, in the sequence of the proteins is very variable. Again cytochromes *c* have very different molecular weights. If this is true what is the implication for the overall fold? The immediate neighbors of the heme, not bound to iron, include a tryptophan, almost invariable in structure but not in sequence position. How has this been evolved? There is but one cytochrome *c* that has no tryptophan. I do not believe that we could follow evolution *with confidence* from sequences of single enzymes. Cytochrome c_3 sequences show this very clearly, for these proteins, all from sulfur bacteria, show an enormous sequence variation [R. P. Ambler, Syst. Zool. **22,** 554 (1974)]. There is no evidence that this is related to their evolution.

Peroxidases are even more difficult, for one biological species (i.e., a given plant) may contain 10 isoenzymes that may vary among themselves both with regard to molecular weight and sequence, although all have a similar active site region. Why does the turnip have so many peroxidases and why are they so different? They can use O_2 or H_2O_2; they oxidize all kinds of substrates, including growth hormones; they are protective devices. There is an obvious difficulty in understanding evolution if function is so confusing. Peroxidase is of course common to plants and animals and must be a very old enzyme. But have these enzymes evolved by divergent or convergent evolution using heme as a basic coenzyme?

27. Intervention of Kirschner

I would like to return to the original theme of synthetic enzymes. I think that Dr. Klotz is being too modest in calling his synzymes mere catalysts. Surely his work is aimed at producing a catalyst as similar as possible to a real enzyme. Considering the heterogeneous nature of polyethylenimine and the rather random chemical modifications converting these to catalysts (with astonishing properties), the average properties are bound to be contributed to by a small fraction of extremely efficient catalysts. One would suppose that imposition of a selective process might be able to single out such enzymes. Dr. Klotz, have you tried to fractionate your synzyme mixture (e.g., by affinity chromatography)? And what results have you obtained?

27.1 Intervention of Klotz

Dr. Kirschner is quite right, of course, that a synthetic polymer consists of a mixture of macromolecules of different size and a distribution of catalytic activities. The kinetic parameters I have listed are some kind of average for such a mixture. It makes good sense to try to fractionate such systems. We have decided, however, to postpone such attempts until we have obtained more active polymers by further chemical modification.

It is true that our chemical modifications for introducing functional groups on the polymer involve a stochastic approach. The implication of the comment is that nature does things differently. If my understanding of evolution is correct, evolution is the example *par excellence* of a stochastic approach. Nature tries everything; the environment makes the selections. It is my hope that our conscious selectivity process may be more effective.

Let me repeat once again that we are *not* trying to synthesize enzymes. Contrary to Dr. Kirschner's assertion, the term *synzyme* was not coined to imply that we are trying to do so. True, we have wanted to attract the attention of enzymologists. In my long experience in the scientific world, I have learned that to attract attention it is more important to invent a catchy term than to discover a phenomenon.

CONTROL OF THE CATALYTIC ACTIVITY OF ENZYMES BY THE NEAR AND REMOTE ENVIRONMENT OF A POLYATOMIC FRAMEWORK

GORDON G. HAMMES

Department of Chemistry, Cornell University, Ithaca, New York

Abstract

The elementary steps involved in enzyme-substrate interactions and in interactions within the polypeptide chain of an enzyme primarily involve hydrogen bonding, electrostatic interactions, and hydrophobic interactions. The thermodynamic stabilities of individual interactions of these types are low in water, but a large number of these interactions acting cooperatively in a macromolecule permits a stable catalytic structure to be formed. The kinetics of the elementary steps are rapid and are typically characterized by rate constants $\geqslant 10^8\ s^{-1}$. However, in enzymes consisting of single polypeptide chains or weakly interacting polypeptide chains, the cooperative nature of the noncovalent interactions results in conformational changes with typical rate constants of 10^3 to $10^4\ s^{-1}$. The mechanism of enzyme catalysis can be pictured as the binding of substrate to a sterically specific site on the enzyme which triggers a series of conformational changes that optimize the active site environment for catalysis. This optimization involves proper orientation of functional groups on the enzyme, restriction of possible substrate structures, and formation of a hydrophobic pocket for the substrate that strengthens specific noncovalent interactions and permits efficient acid-base catalysis. Thus an enzymatic reaction can be thought of as having a great entropic advantage over a reaction catalyzed by small molecules in solution. Cooperative transitions of a flexible macromolecule can assist in catalysis not only by structural optimization, but also by participating dynamically in the catalysis and by the formation of covalent intermediates. The mechanism of action of the enzyme ribonuclease A is discussed as an example illustrating the general principles developed. Enzymes containing multiple polypeptide chains can regulate enzyme catalysis through conformational changes that alter subunit (interpolypeptide) interactions. The interactions promoted by the binding of ligands at different sites (allosterism) are considered in terms of two limiting models: concerted subunit conformational changes (Monod–Wyman–Changeux model) and sequential subunit conformational changes (Adair–Koshland–Nemethy–Filmer model). Polymerization depolymerization of enzyme subunits provides another mechanism for regulation through changes in subunit interactions. The regulation is manifested as changes in the shape of the normal hyperbolic binding isotherms for substrates and regulatory effectors and/or as changes in the maximum catalytic efficiency of the enzyme. The rates of regulatory conformational changes are generally somewhat slower than those associated with catalysis and in some cases are extremely slow (min and hr). Aspartate transcarbamylase (from

Escherichia coli) and phosphofructokinase (from rabbit muscle) are considered as examples of regulatory enzymes. Multienzyme complexes contain several different enzymes catalyzing a sequence of reactions. The efficiency of catalysis is increased by restricting the intermediates in the sequence of enzyme reactions to within the complex. In the case of tryptophan synthetase (*E. coli*), the two different enzymes also interact to stabilize particular polypeptide conformations and binding sites. The pyruvate dehydrogenase complex from *E. coli* may utilize a covalently bound lipoic acid to transfer molecules between the three different catalytic sites by forming covalent compounds. Enzymic activity also can be modulated by interactions with a membrane. This may involve the requirement of a hydrophobic environment or may be due to specific interactions between protein and phospholipids, as with β hydroxybutyrate dehydrogenase. A membrane can localize an enzyme to where it is physiologically needed. In some cases the function of the enzyme can be coupled with another membrane function, such as transport. For example, the ATP synthesizing enzyme used in oxidative phosphorylation and photophosphorylation requires a proton gradient across the membrane in order to be functional. Enzymic activity also can be modulated by interaction of the enzyme with other membrane proteins. In the case of adenylate cyclase, the catalytic activity of the enzyme bound to the inside of a membrane is triggered by the binding of a hormone to a receptor on the outside of the membrane. In both enzymic catalysis and regulation a common thread appears to be the modulation of macromolecular conformation.

I. INTRODUCTION

The control of enzyme activity by the environment of a polyatomic framework is a vast topic, which I shall not attempt to cover fully in this report. Instead I will concentrate on some selected interactions between and within polypeptide chains that influence enzymatic activity. First, elementary steps involved in ligand-protein, intraprotein, and interprotein interactions are considered. Then enzymes consisting of a single polypeptide chain are discussed, followed by enzymes consisting of multiple polypeptide chains. The concluding sections are concerned with multienzyme complexes and enzymes associated with membranes.

II. ELEMENTARY STEPS IN PROTEIN INTERACTIONS

The interactions of ligands, such as substrates, with proteins and the interactions within proteins are dominated by the influence of water. Although covalent bonds are involved in some cases, noncovalent interactions are more prevalent. As a starting point, the thermodynamics and dynamics of hydrogen bonding, electrostatic interactions, and hydrophobic interactions are considered, since these interactions are of fundamental importance in proteins. Hydrogen bonding has been studied extensively in simple systems. Some of the thermodynamic and kinetic parameters characterizing the formation of a hydrogen-bonded dimer of two pyridone molecules in several different solvents are presented in

TABLE I. Thermodynamic and Kinetic Parameters for the Reaction[a]

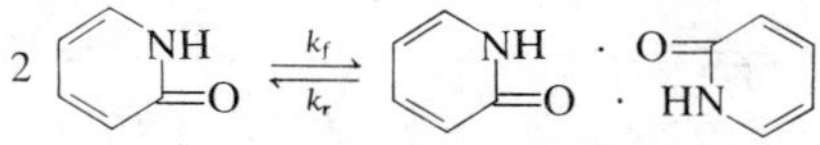

$$2\ \text{(NH, =O)} \underset{k_r}{\overset{k_f}{\rightleftharpoons}} \text{(NH, =O)} \cdot \text{(O=, HN)}$$

Solvent	$\Delta G°$ (kcal/mole)	$\Delta H°$ (kcal/mole)	$10^{-9}\,k_f$ ($M^{-1}\,s^{-1}$)	$10^{-7}\,k_r$ (s^{-1})	Reference
$CHCl_3$	−3.0	−5.9	3.3	2.2	1
50 wt.% dioxane-CCl_4	−2.5	−4.6	2.1	2.9	2
Dioxane	−1.6	−1.7	2.1	13.0	3
1% water-dioxane	−1.3	—	1.7	17.0	3
CCl_4-dimethyl sulfoxide (1.1 m)	−0.4	—	0.26	14.8	2
CCl_4-dimethyl sulfoxide (5.5 m)	0.9	—	0.069	2.7	2

[a] At 25°C, except for 1% water-dioxane where the temperature was 13°C.

Table I. The entries in the table are arranged in order of the increasing hydrogen bonding capabilities of the solvent. As expected, the standard free energy and enthalpy changes become increasingly more positive as the hydrogen bonding capability of the solvent increases, reflecting the fact that the strength of a hydrogen bond within the dimer becomes very similar to the strength of a hydrogen bond between a solvent molecule and 2-pyridone. For the last three entries $\Delta H°$ is unknown, but is probably very close to zero. The dimerization reaction cannot be studied in water, but the standard free energy change (1 M standard state) at 25°C is certainly greater than zero, and the enthalpy change is probably close to zero. Thus with water as a solvent, hydrogen bonding interactions not involving water are not likely to provide strong structural stabilization. However, regions can be created within a macromolecule where water is excluded (or its effective concentration is very low), and in such cases significant structural stabilization of the macromolecular structure or ligand-macromolecule interaction can occur through hydrogen bonding. Hydrogen bonding also can provide great specificity because of the different possible types of hydrogen bonds and the strong preference for linear bonding.

Since enzyme catalysis and regulation are dynamic processes, the dynamics of hydrogen bonding is also an important consideration. In relatively weakly hydrogen bonding solvents, such as the first four entries in Table I, the association rate is essentially diffusion controlled, whereas the association rate constants for the last two entries are considerably less than expected for a diffusion-controlled process. This can be understood in terms of the mechanism

$$2\,P \underset{k_{-1}}{\overset{k_1}{\rightleftharpoons}} P\cdots P \underset{k_{-2}}{\overset{k_2}{\rightleftharpoons}} P\text{—}P \underset{k_{-3}}{\overset{k_3}{\rightleftharpoons}} P\text{=}P \tag{1}$$

where P represents pyridone, $P\cdots P$ is a nonhydrogen-bonded dimer that forms and dissociates by diffusion-controlled rates, P—P is the dimer with one hydrogen bond formed, and P=P is the complex with two hydrogen bonds. If the intermediates are assumed to be in a steady state,

$$\begin{aligned} k_f &= \frac{k_1}{1+(k_{-1}/k_2)(1+k_{-2}/k_3)} \\ k_r &= \frac{k_{-3}}{1+(k_3/k_{-2})(1+k_2/k_{-1})} \end{aligned} \tag{2}$$

If the reaction is diffusion controlled, $k_f \approx k_1$; this is true when $k_2 > k_{-1}$ or when desolvation of the solutes and formation of the first hydrogen bond is faster than diffusion apart of the reactants. The value of k_{-1} is about $10^{10}\,s^{-1}$; therefore k_2 must have a value of 10^{11} to $10^{12}\,s^{-1}$, which is only 10 to 100 molecular vibrations. In solvents where the solvent itself can form strong hydrogen bonds, such as the last two entries in Table I where the concentration of dimethylsulfoxide is large, the association rate is no longer diffusion controlled. In these cases desolvation of the solute becomes rate determining with a specific rate constant of about $10^8\,s^{-1}$. This rate constant is characteristic of solvation-desolvation processes involving hydrogen bonds. Note that when dimer formation is diffusion controlled, $k_r = k_{-1}(k_{-2}/k_2)(k_{-3}/k_3)$. Since k_{-1} is about the same in all cases, the reverse rate constant is a measure of the thermodynamic stability of the two pyridone-pyridone hydrogen bonds formed. The conclusion to be derived from these results is that the rate constant characterizing the making and breaking of single hydrogen bonds in water is very probably $\geqslant 10^8\,s^{-1}$.

Electrostatic interactions, such as ion pairing, are well known and undoubtedly play an important role in biological processes. The high dielectric constant of water considerably weakens electrostatic interactions. For example, the potential energy of interaction of a positive and negative charge a few angstroms apart is only 1 kcal/mole in water; however, in a hydrophobic (organic solventlike) environment this energy can be 20 to 30 times larger. In addition, clusters of charges on a macromolecule interacting with a charged ligand or other charges on the marcromolecule can give rise to large energies of interaction. Ion-dipole interactions also may be of great significance because water has a strong dipole; the potential energy of interaction between ions and dipoles is orientation dependent: It is considerably less than 1 kcal/mole in a medium with a high dielectric constant such as water but can approach a few kcal/mole in a medium of organic solventlike dielectric constant. The dynamics of simple electrostatic interactions is not well documented. Ion pairing appears to be a diffusion-controlled process in both formation and dissociation. A useful model for ion-dipole interactions is the interaction of metal ions with the first hydration shell of metal ions.[4] The hydration shell is highly oriented because of the field of the ion. The water molecules dissociate readily from univalent cations such as alkali metals, the specific rate constants being approximately $10^9\ s^{-1}$. For more highly charged metal ions, the rate of dissociation of water depends on the ionic radii and on specific electronic properties of the metals, and the rate constants range from 1 to $10^9\ s^{-1}$. Thus charged species can have large effects on the energetics and dynamics of water interactions.

Hydrophobic interactions represent the fact that hydrocarbon structures and water do not like to associate. Thus if a hydrocarbon is put into water (and is soluble), water will tend to form a sheath around the hydrocarbon in which the water dipoles are strongly oriented through hydrogen bonding. The free energy associated with such interactions can be estimated from measurements of the free energy for the transfer of hydrocarbons from water to a nonpolar solvent. For example, standard free energies of −0.7 kcal/mole and −2 kcal/mole have been estimated for the transfer of a methylene group and aromatic ring, respectively, from water to ethanol.[5] If two hydrophobic molecules are present in water, they tend to associate, not because of the strong interactions between the hydrophobic molecules, but because of the alteration in water structure: Some of the oriented water molecules are released because of the

TABLE II. Representative Time Constants for Solvation-Desolvation of Hydrophobic Structures

Species	$k(s^{-1})$	Reference
$(PhCH_3)_2NCH_3 \cdot H_2O$	2.7×10^{9} [a]	6
$Dioxane(H_2O)_2$	2.8×10^{8} [a]	7
$(Dioxane)_2(H_2O)_2$	1.0×10^{8} [a]	7
Glycine or diglycine or triglycine$\cdot H_2O$	4×10^{8} [b]	8
Polyethylene glycol	$\sim 10^{8}$ [b]	9

[a] Water dissociation rate constant, 25°C.

[b] Sum of rate constants for solvation and desolvation; 25°C for polyethylene glycol, 4°C for glycine, diglycine, triglycine.

decreased surface area formed on association; this has an unfavorable enthalpy change (i.e., positive), but a very favorable entropy change (i.e., positive). Thus in a protein molecule the hydrophobic residues tend to associate inside the structure—as far as possible from water. The dynamics of these hydrophobic interactions can be studied in model systems by measuring the rate of water dissociation from hydrophobic molecules or the ultrasonic water-solute relaxation rate in solutions of a polymer, such as polyethylene glycol, which has many hydrocarbon groups. Some representative data are presented in Table II. The rate constants for the water dissociation from hydrophobic molecules are typically $\geqslant 10^8 \text{ s}^{-1}$.

III. SINGLE POLYPEPTIDE CHAINS

As indicated by the preceding discussion, the noncovalent interactions involved in the stabilization of a catalytically active polypeptide structure are individually quite weak. However, the polypeptide chain has hundreds of these weak interactions, and the summation of these interactions produces a stable and unique structure. Conceptually the hydrophobic interactions can be thought of as the dominant force leading to the establishment of structure, with hydrogen bonding and electrostatics being especially important in generating specificity. An important factor not yet discussed is the role of cooperativity in creating and regulating the correct structure. The effect of cooperativity is most readily observed in studies of protein denaturation where the native structure is converted to a random structure over a very small range of an external parameter such as temperature, pH, and so on. These transitions are associated with large changes in thermodynamic parameters such as enthalpy and entropy. In other words, the hundreds of weak noncovalent interactions are *coordinated;* this cooperative behavior means that a limited number of structures have a high stability, and therefore the multitude of possible structures predicted by the random probabilities of interaction is not found.

Cooperativity also has a profound influence on the dynamics of functionally important conformational changes within a polypeptide chain. Some typical rates of conformational transitions in enzymes are presented in Table III. These transitions are believed to occur within a single polypeptide chain, although not all the enzymes consist of single polypeptide chains. The operational definition of a conformational change is a change in state that occurs within the protein molecule (or ligand-protein molecule) as judged either by measurement of a property of the protein or ligand or by kinetic criteria. Changes in covalent bonds are not

TABLE III. Rates of Representative Conformational Changes in Enzymes

Enzyme	Approximate rate (s^{-1})	Reference
Ribonuclease	10^3–10^4	10–13
Chymotrypsin	10^2	14
Lysozyme	10^2–10^3	15
Creatine kinase	10^4	16
Lactate dehydrogenase	10^3	17
Aspartate aminotransferase	10^2–10^3	18–20
Aspartate transcarbamylase	10^3–10^4	21
Glyceraldehyde 3′-phosphate dehydrogenase (yeast)	10^4	22

involved. The actual changes in structure are almost certainly very small and can be observed directly only in rare cases. Conformational transitions are inferred to be of importance in catalysis if they occur more rapidly than the overall catalytic process; such transitions are usually triggered by the binding of substrates. Although the rates of all the elementary steps are $>10^8\ s^{-1}$, as previously discussed, the rates of the conformational changes are considerably slower. This is undoubtedly because the conformational changes are cooperative in nature, and the coordinated changes in the noncovalent interactions are considerably slower than any of the elementary steps. This phenomenon is well illustrated by studies of a model system, polyglutamic acid. This polymer undergoes a cooperative transition from a helical to random coil configuration that is triggered by small changes in pH. This change involves the making and breaking of hydrogen bonds, both intramolecular and with the solvent, and has a rate of about $10^6\ s^{-1}$,[23] considerably slower than the rate constants for the elementary steps. This transition is highly cooperative only for relatively large polymers.[24] An important point to note is that cooperative changes are possible only for large molecules where a large number of interacting elements are present.

The discussion thus far indicates that a polypeptide chain can fold into specific structures through cooperative processes involving many weak noncovalent interactions and that cooperative conformational transitions can occur rapidly when triggered by events such as ligand binding, pH changes, and temperature changes. How such a macromolecule can be used to control chemical reactivity is now considered. The initial event in enzyme catalysis is the combination of enzyme and substrate. The known X-ray structures of enzymes and many kinetic investigations have established that the combining site created by the enzyme is exquisitely specific. The interactions stabilizing the enzyme-substrate complex are similar in nature to the noncovalent interactions within the enzyme framework, and again it is a multitude of interactions that leads to a stable enzyme-substrate molecule. Typical binding constants are about $10^4\ M^{-1}$, which corresponds to a standard free energy change of about -5 kcal/mole. The combination of enzyme and substrate is characterized by a quite large second order rate constant, typically $10^7\ M^{-1}\ s^{-1}$, which is close to that expected for a diffusion controlled process. Some typical rate constants for enzyme-substrate reactions are presented in Table IV. The binding of substrate has been found to trigger a conformational transition for many enzymes, and in fact, such a conformational change can be regarded as a general phenomenon.[29] In essence the binding energy is used to produce a macromolecular structure that is better optimized for catalysis. Probably one of the most important functions of this conformational change is to place the substrate in a hydrophobic pocket. In such a hydrophobic pocket the strength of hydrogen bonding and electrostatic interactions is enhanced; the exclusion of water also permits efficient acid-base catalysis to occur.

Acid-base catalysis appears to be an important factor in virtually all enzymatic reactions. The rates of proton transfer reactions have been well studied in model systems,[30] but not during the course of enzyme catalysis. The protonation and deprotonation of acids and bases can be represented as

$$B + H^+ \rightleftharpoons BH^+ \tag{3}$$

$$BH^+ + OH^- \rightleftharpoons B + H_2O \tag{4}$$

The rate constants in the forward direction are diffusion controlled for normal acids and bases and have typical rate constants of $10^{10}\ M^{-1}\ s^{-1}$. This implies that the actual event of proton transfer is faster than the diffusion apart of reactants; therefore the specific rate constant for intramolecular proton transfer in water is about $10^{12}\ s^{-1}$. Water is a special solvent for proton transfer reactions because proton conduction can occur rapidly in the hydrogen-bonded network of structured water. In fact, if the water structure is perturbed (e.g., by internal hydrogen bonding), the forward reactions of (3) and (4) are no longer diffusion controlled. If the forward reactions are diffusion controlled, the rate constants for the reverse reactions can be calculated from the known equilibrium constants: The rate constants are approximately $10^{10}\ K_a\ s^{-1}$ and $10^{10}\ K_w/K_a\ s^{-1}$ for the reactions of (3) and (4), respectively, where K_a is the acid ionization constant of B and K_w is the ionization constant of water. For catalysis, both of the reactions represented by (3) and (4) must

TABLE IV. Representative Rate Constants for Enzyme-Substrate Complex Formation

$$E + S \underset{k_r}{\overset{k_f}{\rightleftharpoons}} ES$$

Enzyme (E)	Substrate (S)	$10^{-7}\,k_f$ ($M^{-1}\,s^{-1}$)	$k_r(s^{-1})$	Reference
Ribonuclease	Cytidine 3′-phosphate	4.6	4.2×10^3	10
	Uridine 3′-phosphate	7.8	1.1×10^4	10
	Cytidine 2′,3′-cyclic phosphate	2–4	$1–2\times10^4$	11
	Uridine 2′,3′-cyclic phosphate	1	2×10^4	12
	Cytidylyl 3′,5′-cytidine	1.4	7×10^3	13
Chymotrypsin	Furylacryloyl-L-tryptophan amide	0.62	2.7×10^3	14
Lactate dehydrogenase	NADH	5.46	39	17
Glyceraldehyde 3′-phosphate dehydrogenase (yeast)[a]	NAD^+	1.1	1.1×10^3	22
		0.032	8×10^2	
Creatine kinase	ADP	2.2	1.8×10^4	16
	MgADP	0.53	5.1×10^3	16
	CaADP	0.17	1.2×10^3	16
	MnADP	0.74	4.1×10^3	16
Pyruvate carboxylase-Mn^{2+}	Pyruvate	0.45	2.1×10^4	25
Pyruvate kinase-Mn^{2+}	Fluorophosphate	0.13	3.4×10^4	26
Peroxidase	H_2O_2	0.9	<1.4	27
	Methyl H_2O_2	0.15	<2.2	27
	Ethyl H_2O_2	0.36	–	27
Catalase	H_2O_2	0.5	–	28

[a] Two types of binding sites.

occur to complete a catalytic cycle so that the maximum catalytic rate constant, $10^3\,s^{-1}$, occurs when the pK_a is about 7. However, acid-base catalysis need not be mediated by water. A general formulation of intramolecular proton transfer is

$$DH + A \rightleftharpoons HA + D \tag{5}$$

where D and A denote a proton donor and acceptor, respectively. If the pK of the acceptor is much higher than that of the donor, the specific rate constant for proton transfer is approximately $10^{12}\,s^{-1}$, whereas the specific rate constant for the reverse reaction is approximately $10^{12}\,K_A/K_D\,s^{-1}$, where K_A and K_D represent the acid ionization constants of the acceptor and donor, respectively. Again for catalysis, a cycle of both the forward and reverse reactions must occur. The difference between pK values associated with protein ionizable groups involved in catalysis and common substrates is typically very large. For example, if the pK difference is seven units, the maximum catalytic rate constant is about 10^5 s^{-1}. This large pK difference also implies that the concentration of one of the intermediates would be only 10^{-7} ($10^5/10^{12}$) of the total enzyme concentration, which in turn requires very large rate constants for further reactions if a high catalytic rate is to be maintained. Finally, the possibility of concerted acid-base catalysis should be considered. A simultaneous proton acceptance and donation by the substrate could readily be part of a cooperative conformational change of the macromolecule-substrate framework. Such a reaction would eliminate the necessity for forming a reaction intermediate in very low concentrations. An upper bound for the rate constant associated with such a process is the rate constant for the direct proton transfer between the acid and base groups on the enzyme involved in the catalysis. A typical pK difference is two units, which gives an upper bound of 10^{10} s^{-1}. This is obviously too high an estimate, since substrates are generally not good proton acceptors and donors and might be expected to reduce this maximum rate constant several orders of magnitude.

The maximum turnover numbers of enzymes are about 10^5 to 10^6 s^{-1}, with typical values being 10^2 to 10^3 s^{-1} (cf. Ref. 31). This is surprisingly close to the maximum possible rate constant predicted from consideration of proton transfer reactions. However, the maximum rates are derived from consideration of proton transfer rates between good proton acceptors and donors. These rates are considerably decreased for poor acceptors and donors, such as typical substrates. Therefore, the protein must alter the effective pK values of the substrate. Several plausible mechanisms can be envisaged. The formation of a hydrophobic pocket for the substrate permits proton transfer reactions to occur without the competition of water; a small number of structured water molecules even might assist in proton transfer. Cooperative conformational transitions coordinated with proton transfer could lower the required activation energies. Such an effect is seen in simple reactions where the simultaneous breaking of one bond and formation of another has a lower activation energy than breaking of the same bond alone.[32] In a protein, the multitude of noncovalent interactions acting cooperatively could act in a similar manner to lower the activation energy for a process such as concerted proton transfer. Another way of viewing the role of the protein is as in the "rack" or "strain" hypothesis:[33,34] The protein is proposed to distort the structure of the substrate through conformational changes until it closely resembles the transition state, thus effectively lowering the activation energy for formation of the transition state of the rate-determining step. The driving energy for the "strain" is the binding energy that would be gained by the distortion of the substrate to a more tightly bound species.

In a number of enzymatic reactions, the substrate forms a covalent intermediate either with the enzyme or with a tightly, often covalently, bound coenzyme.[34] For example, a group of proteases and esterases catalyze the hydrolysis and transfer of acyl groups with an intermediate acyl transfer to a sulfhydryl or hydroxyl group of cysteine or serine in the enzyme. Evidence exists for the formation of phosphoryl enzymes with covalent linkages with a serine, imidazole, or carboxyl group. Enzymes utilizing pyridoxal phosphate as a coenzyme frequently form covalent intermediates in which the substrate is bound through an imine (or Schiff base) linkage. The advantage of forming covalent intermediates generally is to break down a kinetically difficult reaction into kinetically easier steps. The enzyme is able to create a unique environment favorable for formation of the covalent intermediate and an equally unique environment favorable for its decomposition. Again conformational changes must play an important role in establishing the appropriate environment and may also be utilized as in the dynamic mechanism discussed previously.

An enzymatic reaction has important advantages over a similar reaction in solution (see Refs. 34–37). Several different functional groups important in catalysis can simultaneously interact with substrates. This is a considerable entropic gain for the system, or in more simplistic terms, raises the effective concentrations of the reactants. The enzyme also restricts the substrates to a specific structure and orientation with respect to the functional groups of the enzyme. This too is basically an entropic effect in that the number of possible conformations (or isomers) of the system is restricted. The catalytic activity of an enzyme then can be depicted as being derived from the creation of a specific polypeptide structure through cooperative interactions that binds the substrate (or substrates) very specifically and quite tightly. This binding event initiates a series of cooperative conformational changes that optimize the polypeptide structure for catalysis of the reaction as discussed earlier. This flexibility of structure, which only can be obtained with a macromolecule, is an essential part of the modulation of enzymic activity. As might be expected, modulation of structure, and hence of activity, also can be achieved in some instances by pH changes, temperature changes, and the binding of ligands other than the substrate.

A. Ribonuclease A

As a specific example the mechanism of action of bovine pancreatic ribonuclease A is now considered. Ribonuclease A is a relatively small enzyme consisting of a single polypeptide chain of molecular weight 13,680. It has been extensively studied by many methods: The amino acid sequence and three-dimensional structure are known; and extensive equilibrium and kinetic studies have been carried out as well as other chemical and physical studies (see Ref. 38 for an extensive review). This enzyme catalyzes the breakdown of ribonucleic acid in two steps, as shown in Fig. 1: The diester linkage is first broken and a pyrimidine 2′,3′-cyclic phosphate is formed; the cyclic phosphate is subsequently hydrolyzed to give the pyrimidine-3′-monophosphate and purine oligonucleotides with a terminal pyrimidine 3′-phosphate. Most of the kinetic studies have utilized dinucleotides and pyrimidine 2′,3′-cyclic phosphates as substrates rather than ribonucleic acid, which gives rise to a heterogeneous mixture of substrates and products. Ribonuclease is a compact, kidney-shaped molecule with the active site located in a groove.[38] Substrates are bound in the groove near two histidine residues (numbers 12 and 119) that are believed to act as proton acceptors and donors during catalysis. A third histidine residue (number 48) is located at the top of the hinge of the groove. A conformational change of the enzyme has been observed that can be associated with an opening and closing of the groove, which alters the pK of the imidazole of histidine 48.[39] The interaction of the enzyme with dinucleosides, pyrimidine 2′,3′-cyclic phosphates and pyrimidine 3′-phosphates involves two distinct steps: a bimolecular combination of enzyme and substrate that is almost diffusion controlled (see Table IV) followed by a conformational change of the enzyme-substrate complex characterized by typical rate constants of 10^3 to 10^4 s^{-1}. The suggestion has been made that this conformational change is associated with a closing of the groove that brings a lysine residue (number 41) close to the catalytic site;[10] the substrates also may be put into a more hydrophobic environment.

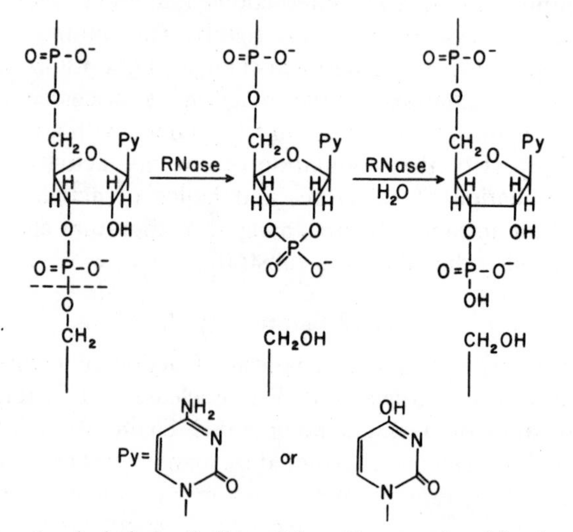

Fig. 1. The two-step hydrolysis of ribonucleic acid as catalyzed by bovine pancreatic ribonuclease A.

For ribonuclease A the occurrence of conformational changes and the occurrence of acid-base catalysis has been well documented. The overall mechanism can be envisaged as follows. The enzyme exists in dynamic equilibrium between two forms differing in the structure of the active site groove. The substrate is bound almost as rapidly as it can diffuse to the active site. Binding of the substrate induces a conformational change that

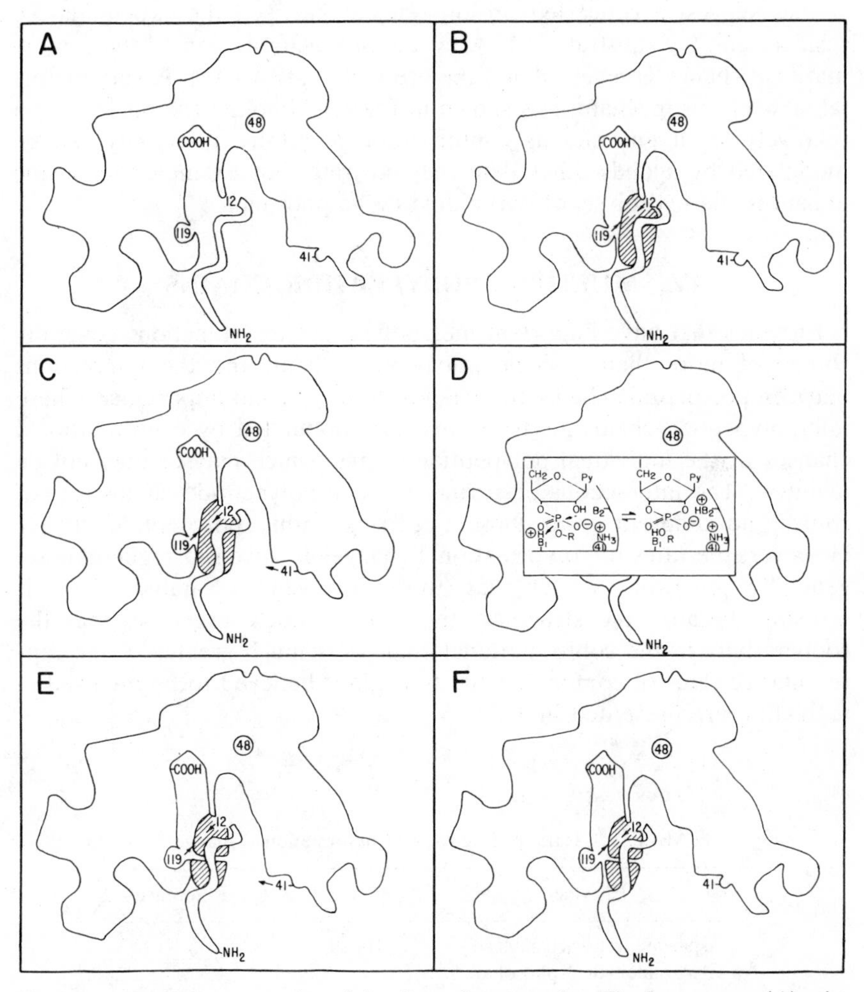

Fig. 2. A pictorial representation of the ribonuclease reaction. The free enzyme (*A*) exists in two conformational states differing by small movements of the hinge region joining the two halves of the molecule. The substrate is bound (*B*) and a conformational change occurs closing the hinge (*C*). Concerted acid-base catalysis then occurs (*D*); products are formed (*E*); the conformational change is reversed (*F*); and product(s) dissociate to give free enzyme.

closes the groove somewhat, thereby swinging lysine 41 closer to the substrate, making the catalytic site more hydrophobic and orienting the imidazole residues on histidine 12 and 119 precisely for acid-base catalysis of the chemical reaction. The details of the proton transfer process cannot be studied directly, probably because reaction intermediates are not present in sufficiently large concentrations. However, consideration of the three-dimensional structure and stereochemical studies suggest a concerted proton transfer between the two imidazole residues and the substrate.[40–43] After the product(s) is formed the conformational change is reversed and the product(s) dissociates. A very pictorial view of this mechanism is shown in Fig. 2. Although ribonuclease is a relatively small enzyme, its conformation and thus its activity can be modulated by ligands other than the substrate. For example, adenosine enhances the hydrolysis of pyrimidine cyclic phosphates.[44]

IV. MULTIPLE POLYPEPTIDE CHAINS

Enzymes that have important metabolic regulatory functions generally consist of more than a single polypeptide chain, and the interactions between polypeptide chains are believed to be of great importance. These interpolypeptide chains modulate and are modulated by conformational changes of the individual polypeptide chains, which may or may not be identical. The interactions occurring between polypeptide chains are, of course, no different from those occurring within polypeptide chains. However, the rates of conformational changes involved in regulation are generally somewhat slower than those involved in catalysis. This is probably because the structures tend to be much larger, so that the cooperativity in the conformational changes is much greater. Some representative rates of conformational transitions believed to be involved in regulation are presented in Table V.

TABLE V. Rates of Regulatory Conformational Changes

Enzyme	Approximate Rate	Reference
Aspartate transcarbamylase	10–$10^4\ s^{-1}$	45–47
Glyceraldehyde 3-phosphate dehydrogenase (yeast)	1–$10\ s^{-1}$	22
Homoserine dehydrogenase (*E. coli*)	min^{-1}	48
Threonine deaminase	min^{-1}	49

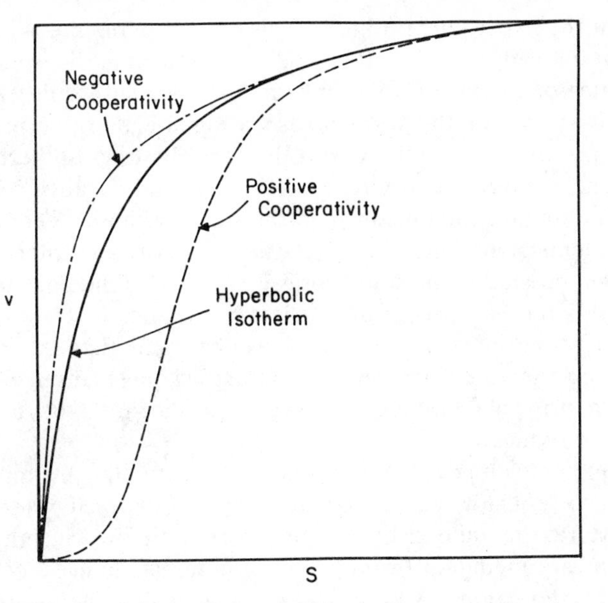

Fig. 3. A schematic plot of the steady-state initial velocity v, versus the substrate concentration S, illustrating a hyperbolic saturation isotherm (——), positive cooperativity (– – –), and negative cooperativity (– · – ·).

Regulation of enzymic activity occurs via two modes (cf. Ref. 50): alteration of the substrate binding process and/or alteration of the catalytic efficiency (turnover number) of the enzyme. The initial rate of a simple enzymatic reaction v is governed by the Michaelis–Menten equation

$$v = \frac{V_m}{1 + K_m/(S)} \tag{6}$$

where V_m is the maximal velocity, K_m is the Michaelis constant, and (S) is the substrate concentration. The hyperbolic rate isotherm generated by this equation is shown in Fig. 3. The simplest regulation occurs by utilizing nonhyperbolic isotherms. Two limiting possibilities are included in Fig. 3. With the sigmoidal initial rate-concentration isotherm, a very small change in substrate concentration produces a larger change in the rate of the reaction than a hyperbolic isotherm over a certain concentration range. The sigmoidicity suggests the binding of substrate is very weak at low substrate concentrations and becomes increasingly strong at high substrate concentrations. In the second limit, the binding is very strong at low substrate concentrations and becomes weaker as the substrate concentration increases. In this case the range of concentration over which the substrate concentration can alter the reaction rate is greater than in the case of the hyperbolic isotherm. This change in binding with substrate concentration can be due either to an interaction between identical binding sites or to several different binding sites with different affinities for the substrate. In the former case a sigmoidal isotherm is said to represent positive cooperativity, and the other limiting isotherm is said to represent negative cooperativity. Both positive and negative cooperativity also may be found simultaneously in the same isotherm. The interactions between identical binding sites are called homotropic interactions. Although the possibility of nonidentical substrate binding sites is quite possible, it is rarely invoked for regulatory enzymes. A third explanation of nonhyperbolic initial rate-substrate isotherms is that special relationships among the rate constants of a complex mechanism exist; this is possible in principle but has not yet been shown to be true for any regulatory enzymes.

Metabolic activators and inhibitors are structurally dissimilar to substrates. These effectors exert regulatory control over catalysis by binding at an allosteric site quite distinct from the catalytic site. Such heterotropic interactions are mediated through conformational changes, often involving subunit interactions. Allosteric effectors can alter the catalytic rate by changing the apparent substrate affinity (*K* system) or by altering the

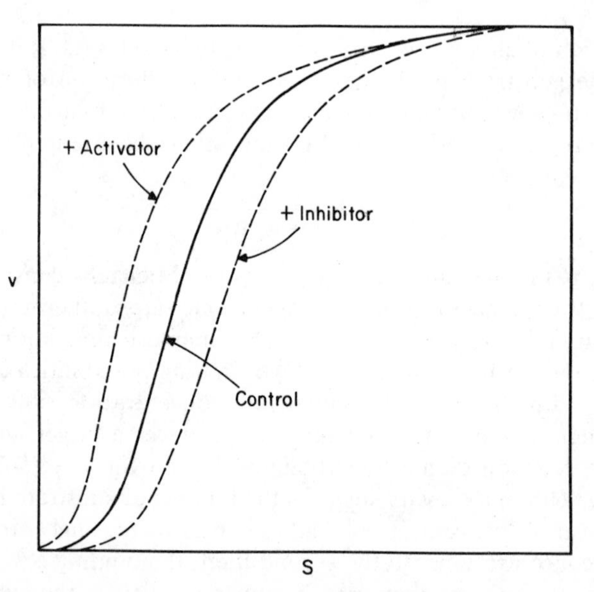

Fig. 4. A schematic plot of the steady-state initial velocity v versus the substrate concentration S, illustrating the effect of an activator and an inhibitor on a K system displaying positive cooperativity.

turnover number (V system) or both. For K systems an inhibitor generally causes the initial velocity-substrate isotherm to become more sigmoidal, whereas an activator makes it less sigmoidal, as shown in Fig. 4. In both cases, regulation is provided only over a restricted range of substrate concentrations.

Two limiting models have been proposed to describe the regulatory phenomena represented in Figs. 3 and 4. One is due to Monod, Wyman, and Changeux (MWC);[51] the other is due to Adair, Koshland, Nemethy, and Filmer (AKNF).[52,53] The MWC model has three underlying assumptions: (*1*) The enzyme consists of two or more identical subunits, each containing a site for the substrate and/or effector; (*2*) at least two different conformational states (usually designated as R and T states) are in equilibrium and differ in their affinities for substrate and/or effector; and (*3*) the conformational changes of all subunits occur in a concerted manner (i.e., all subunits have an identical conformation). This model is illustrated for a four-subunit enzyme in Fig. 5, with circles and squares denoting different subunit conformations. In the absence of substrates

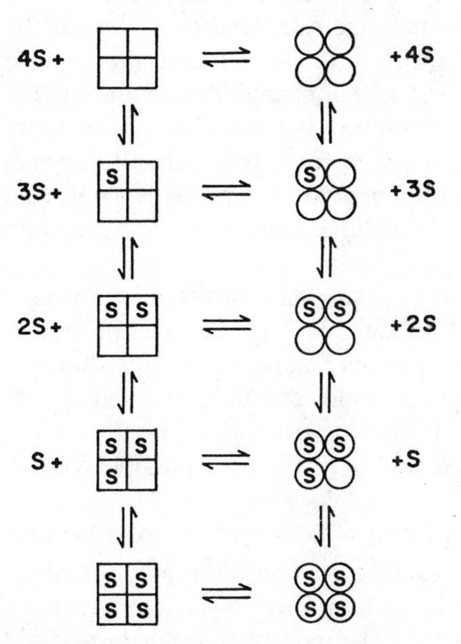

Fig. 5. Schematic representations of the MWC and AKNF models for a four-subunit enzyme. The squares and circles designate different subunit conformations and S is the substrate. The free substrate has been omitted from the AKNF model for the sake of simplicity.

(and activators) the enzyme exists largely in T states, but substrates bind preferentially to the R states. The binding of substrate is weak at low substrate concentrations, since essentially only the T state is present, but the preferential binding of substrate to the R state shifts the enzyme to the R state so that the apparent affinity of the substrate for the enzyme increases markedly as the substrate concentration is raised. This generates a sigmoidal binding isotherm. Activators and inhibitors, by binding preferentially to the R and T states, respectively, can reduce and enhance the sigmoidicity of the binding isotherm, as illustrated in Fig. 4. With this model, however, only positive cooperativity can occur.

The basic premises of the AKNF model are that (*1*) two conformational states are available to each subunit, (*2*) only the subunit to which the ligand is bound changes its conformation, and (*3*) the ligand-induced conformational change in one subunit alters its interactions with neighboring subunits. This model is also illustrated in Fig. 5 for a four-subunit enzyme. Each liganded state has different subunit interactions; therefore each liganded state can have a different effective binding constant for adding the next ligand. Thus binding isotherms displaying positive or negative cooperativity or both can be generated with this model. Also activators and inhibitors can alter the substrate binding constants by changing the subunit conformation, which alters the subunit interactions. The AKNF model utilizes a strictly sequential change of subunit conformations, in contrast to the concerted nature of the MWC model. A more general model, such as that shown in Fig. 6 for a four-subunit enzyme, can be generated by allowing both sequential and partially concerted conformational changes. In practice the various possible models are often extremely difficult to distinguish.

A quite different mechanism for altering subunit interactions is through polymerization-depolymerization of subunits.[54,55] If different polymeric states of the enzyme have different turnover numbers and/or different affinities for substrates and effectors, a model can be generated that is similar to the MWC model except that the cooperativity is also dependent on the enzyme concentration. Both K and V systems are possible with all the models.

Covalent modifications of enzymes also play an important regulatory role.[56] For example, enzyme catalyzed phosphorylation-dephosphorylation and adenylation-deadenylation are used to regulate phosphorylase and glutamine synthetase. The molecular mechanisms used for regulation are presumably similar to those associated with reversible ligand binding except that the ligand binding is now covalent. The metabolic conditions determine when the protein-modifying enzymes are active, and therefore when the regulatory machinery is set into motion.

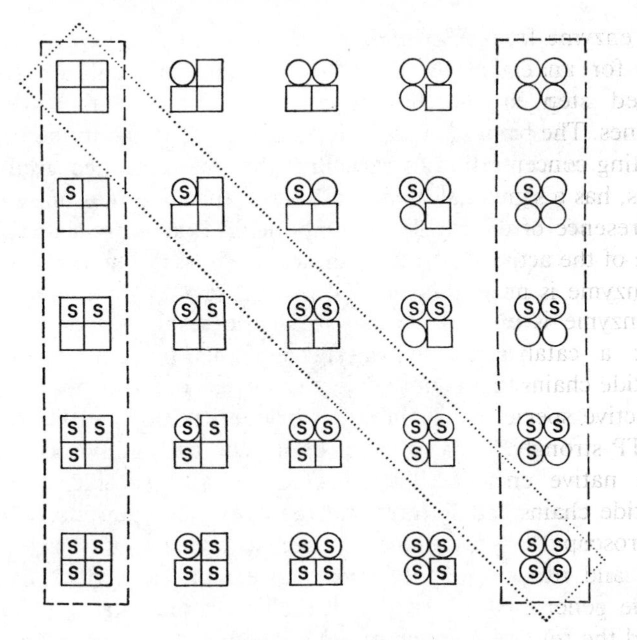

Fig. 6. A general allosteric model for the binding of substrate *S* to a four-subunit enzyme. The squares and circles represent different conformations of the subunits. The MWC model is shown by dashed lines and the AKNF model by dotted lines. The free substrate and arrows connecting the states are omitted for the sake of clarity.

It should be noted that regulatory conformational changes need not be rapid. In fact, in some cases these transitions are extremely slow (see Table V), and an observable time lag in the enzyme response may occur. The molecular basis for regulation is not fundamentally different for slow and rapid conformational transitions, but slowly responding systems are often termed *hysteretic*.[57]

A. Aspartate Transcarbamylase

Aspartate transcarbamylase catalyzes the reaction

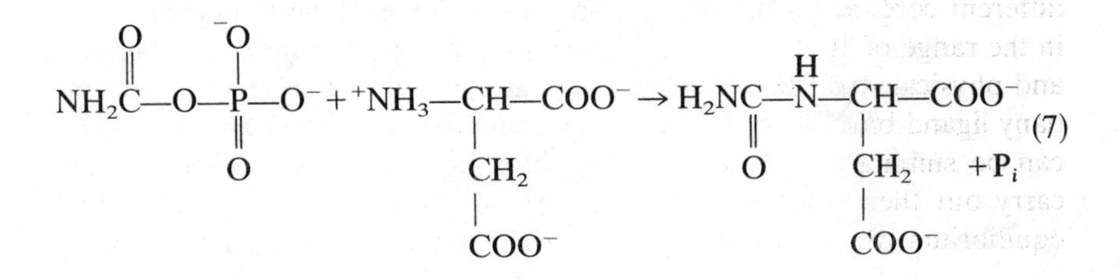

$$NH_2C(=O)—O—P(=O)(O^-)—O^- + {}^+NH_3—CH(CH_2COO^-)—COO^- \rightarrow H_2NC(=O)—NH—CH(CH_2COO^-)—COO^- + P_i \quad (7)$$

and the enzyme from *Escherichia coli* has been extensively studied (see Ref. 58 for an extensive review). The reaction catalyzed is the first committed step in the biosynthetic pathway for the synthesis of pyrimidines. The binding of aspartic acid to the enzyme in the presence of a saturating concentration of carbamyl phosphate, as measured by initial velocities, has a sigmoidal binding isotherm that becomes more sigmoidal in the presence of the feedback inhibitor CTP and less sigmoidal in the presence of the activator ATP, as shown in Fig. 4.[59] The turnover number of the enzyme is unaltered by CTP and ATP.

The enzyme is unusual in that it can be resolved into two types of proteins: a catalytically active species consisting of three identical polypeptide chains that is not subject to nucleotide control and a catalytically inactive species consisting of two identical polypeptide chains that binds CTP strongly.[60] These two proteins can be readily reconstituted to give the native enzyme. The enzyme contains six identical catalytic polypeptide chains and six identical regulatory polypeptide chains. Electron microscopy,[61] X-ray studies,[62] fluorescence energy transfer measurements,[63] and nuclear magnetic resonance measurements[64] have established the general nature of the three-dimensional structure of the enzyme and the relative location of the six regulatory and six catalytic sites: Two catalytic trimers are connected by three regulatory dimers, and the regulatory and catalytic sites are quite far apart (>40 Å), thus clearly establishing the allosteric nature of the regulation. A metal ion, Zn, is required to bind each regulatory polypeptide chain to a catalytic polypeptide chain. A very schematic model of the structure is shown in Fig. 7.

The binding of nucleotides to the regulatory subunits displays negative cooperativity:[65–67] the binding of an effector molecule to one dimer site considerably reduces the binding affinity of the second regulatory site for an effector molecule. The binding of succinate, an aspartate analogue, to the native enzyme in the presence of saturating carbamyl phosphate displays a sigmoidal binding isotherm.[68] Therefore aspartate transcarbamylase utilizes both positive and negative cooperativity in its regulatory mechanism.

Extensive kinetic studies have been made of the binding of ligands to aspartate transcarbamylase using fast reaction techniques, and several different conformational changes have been detected, with time constants in the range of 10^{-3} to 10^{-1} s.[69,70] In addition, a large number of chemical and physical studies have suggested that conformational changes accompany ligand binding.[58] The overall mechanism for catalysis and regulation can be summarized as follows. The effector molecules, ATP and CTP, carry out their function by altering a two-state (on-off) conformational equilibrium that occurs essentially independently in each regulatory

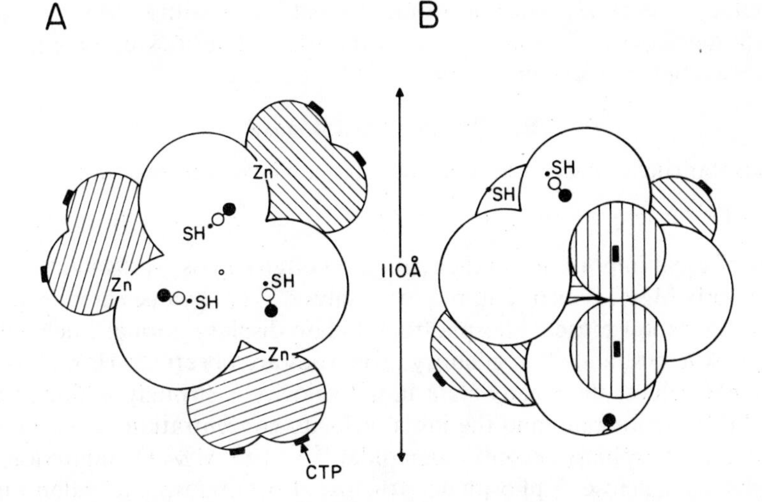

Fig. 7. A pictorial representation of two views of the structure of aspartate transcarbamylase. The hatched portions are regulatory subunits and the white portions are catalytic subunits. The binding sites for CTP (■), carbamyl phosphate (○), and aspartate (●) are indicated along with an SH group that is near the active site and a Zn atom that is required for the binding between regulatory and catalytic subunits.

chain. These local conformational changes in the regulatory polypeptide chains alter the interactions between the regulatory and catalytic polypeptide chains, changing the affinity of the enzyme for aspartate. The binding of carbamyl phosphate and succinate to the catalytic subunit also induces local conformational changes that are probably of importance in catalysis. The suggestion has been made that a conformational change enhances catalysis by compressing the reactants together.[71] (This is essentially the "strain" theory of catalysis previously discussed.) In addition to the conformational changes associated with catalysis, two conformational changes are observed (one associated with carbamyl phosphate binding, the other with succinate binding) that appear to be quite distinct from each other and from that induced by CTP and ATP binding. These transitions appear to be involved in regulation and are concerted in nature. Although the molecular nature of the concerted conformational transitions is not known, it may involve rotation of the catalytic subunits with respect to each other around the threefold symmetry axis of the molecules (Fig. 7). The overall control mechanism appears to be a combination of several different conformational transitions, each of which can lead to inhibition or enhancement of enzymic activity. A multiplicity and coupling of conformational changes may be a general phenomenon in regulatory processes, since it would provide a sensitive and versatile control mechanism. It is analogous to the use of interlocking switches to control complex electronic circuits.

B. Phosphofructokinase

Phosphofructokinase is an enzyme that catalyzes the reaction

$$\text{ATP} + \text{fructose 6-phosphate} \rightleftharpoons \text{ADP} + \text{fructose 1,6-bisphosphate} \quad (8)$$

and is of great importance in the regulation of glycolysis. A divalent metal ion, usually Mg^{2+}, is also required for catalysis. At pH values greater than about 7.5, rabbit muscle phosphofructokinase displays normal Michaelis–Menten kinetics, and no regulatory properties are observed. However, at lower pH values, the steady-state initial velocity is strongly inhibited by MgATP (the substrate) and the initial velocity-concentration isotherm for fructose 6-phosphate becomes sigmoidal.[72,73] The MgATP inhibition is relieved by fructose 6-phosphate, fructose 1,6-bisphosphate, adenosine nucleotides, and phosphate. The reaction is also strongly inhibited by citrate, which could be a physiologically important control mechanism. The inhibition by MgATP and activation by other ligands appear to be primarily due to changes in the affinity of the enzyme for fructose 6-phosphate, whereas citrate appears primarily to lower the turnover number;[74] therefore the citrate and MgATP inhibition are quite distinct and different. The regulation of phosphofructokinase activity at pH 7 (and below) thus occurs through alterations of both the binding and the turnover number.

The enzyme contains identical polypeptide chains of molecular weight 80,000 that aggregate to form dimers, tetramers, octamers, and higher aggregates.[75–77] The structure of the enzyme is not known, but electron microscope studies suggest the monomer can be approximated as a prolate ellipsoid with a long axis of 67 Å and a short axis of 25 Å.[78] The dimer appears to be formed by aggregation along the long axes, whereas the tetramer is formed by end-to-end aggregation of the dimers, as illustrated in Fig. 8. Higher aggregates exist as sheetlike structures and fibers. The aggregation behavior of the enzyme is of interest because the monomer and dimer have essentially no enzymatic activity, whereas the tetramer and higher aggregates are fully active. Furthermore, polymerization of the dimer to the tetramer is induced by the binding of substrates and activators, whereas citrate depolymerizes the enzyme in a concentration range that may be physiologically significant.[79] The rates of polymerization and depolymerization are quite rapid in the presence of substrates, activators, and citrate (i.e., complete in a few minutes at 5°C), whereas depolymerization caused by dilution is rather slow. In principle,

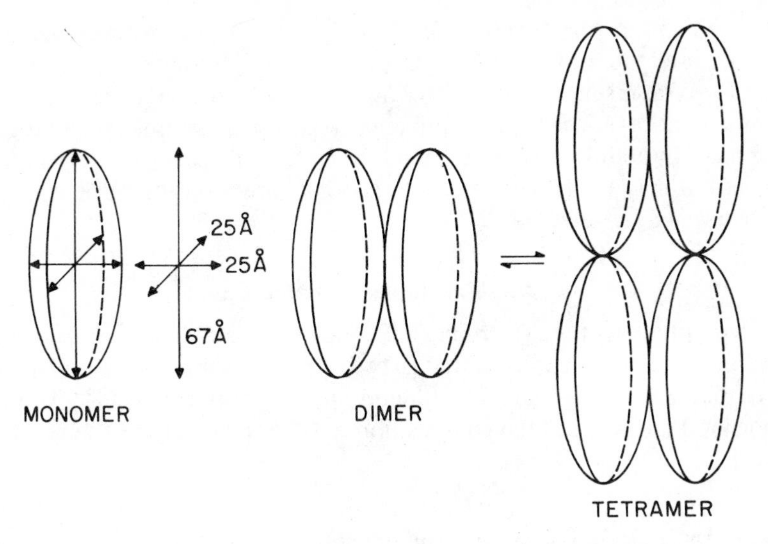

Fig. 8. A pictorial representation of the monomer, dimer, and tetrameric structures of rabbit muscle phosphofructokinase as deduced from electron microscopy.

the sigmoidal initial velocity-fructose 6-phosphate isotherm might be due to the dimer-tetramer equilibrium and preferential binding of fructose 6-phosphate to the tetramer. However, a detailed study of the binding of fructose phosphates to the enzyme indicates this is not the case:[80] The binding displays either negative cooperativity or no cooperativity, depending on the aggregation state. Only in the presence of MgATP (or more precisely, its analogue Mg-adenylyl imidodiphosphate) is a sigmoidal binding isotherm observed. Therefore homotropic and heterotropic cooperative interactions are found in the tetrameric structure; aggregation phenomena also may play a regulatory role, especially in the effects of citrate. Studies of the pH dependence of the reaction velocity and of protein aggregation also suggest regulation occurs within the tetrameric structure.[81] Detailed kinetic studies of ligand binding and elucidation of the allosteric conformational changes remain as future goals, but already it is clear that interactions of a single type of polypeptide chain can be used for regulation with several different types of binding sites and probably more than a single regulatory mechanism.

V. MULTIENZYME COMPLEXES

Another mode of regulating enzymic activity is through the formation of multienzyme complexes.[82] These complexes are aggregates of enzymes that catalyze two or more steps in a metabolic sequence. The chemical intermediates in the reaction sequence go directly from one enzyme to another and do not appear free in solution. Multienzyme complexes, therefore, represent a mode of synchronizing a sequence of reactions that is much more efficient than utilizing a group of soluble unaggregated enzymes. In addition to the efficient utilization of reaction intermediates, regulation may be achieved by interactions between different enzymes in the complex.

A. Tryptophan Synthetase

Tryptophan synthetase from *Escherichia coli* is a simple example of a multienzyme complex: It contains two types of subunits, α and β, that have molecular weights of 29,500 and 54,000, respectively.[83,84] The fully associated enzyme has the composition $\alpha_2\beta_2$[85] and catalyzes the reaction

$$\text{InGP} + \text{Ser} \rightleftharpoons \text{Trp} + \text{GP} \tag{9}$$

where InGP is indole-3-glycerol phosphate, Ser is L-serine, Trp is L-tryptophan and GP is glyceraldehyde 3-phosphate. The isolated α subunit catalyzes the reaction

$$\text{InGP} \rightleftharpoons \text{In} + \text{GP} \tag{10}$$

and the β_2 dimer catalyzes the reaction

$$\text{In} + \text{Ser} \rightleftharpoons \text{Trp} \tag{11}$$

The coenzyme pyridoxal 5′-phosphate is required for the reactions in both (9) and (11). The sum of (10) and (11) gives (9), so that tryptophan synthetase is indeed a multienzyme complex catalyzing a sequence of reactions. The intermediate indole cannot be detected when the overall reaction is carried out, although the native enzyme will catalyze the partial reactions [(10) and (11)] 50 to 100 times more efficiently than the isolated subunits.[86–88]

Numerous equilibrium and kinetic studies have been made with tryptophan synthetase and its subunits, and considerable information has been obtained about the reaction pathway and reaction intermediates (cf. Refs. 89–92). For the purposes of this review, the principal conclusion reached is that the interaction of the α and β subunits appears to restrict the conformations of the α and β subunits to those that bind the substrates tightly and catalyze the reaction efficiently. The basic mechanism is not altered by the subunit interactions; instead stabilization of particular conformations and binding sites is the important advantage gained in formation of the multienzyme complex.

B. Pyruvate Dehydrogenase Complex

The pyruvate dehydrogenase complex from *Escherichia coli* is considerably more complex than tryptophan synthetase. It has a molecular weight of approximately 4.6 millon and contains three enzymes: pyruvate dehydrogenase (E_1), dihydrolipoyl transacetylase (E_2), and dihydrolipoyl dehydrogenase (E_3).[82] The overall reaction catalyzed by the complex is

$$\begin{aligned} &CH_3COCOOH + CoA-SH + NAD^+ \\ &\qquad \rightarrow CH_3CO-S-CoA + CO_2 + NADH + H^+ \end{aligned} \quad (12)$$

The overall reaction is postulated to proceed by the following sequence of reactions:

$$\begin{aligned} &CH_3COCOOH + E_1[\text{-TPP}] \xrightarrow{Mg^{2+}} CO_2 + E_1[\text{hydroxyethyl-TPP}] \\ &E_1[\text{hydroxyethyl-TPP}] + E_2[\text{Lip-S}_2] \rightleftharpoons E_1[\text{-TPP}] + E_2[\text{HS-Lip-S-acetyl}] \\ &E_2[\text{HS-Lip-S-acetyl}] + \text{CoA} \rightleftharpoons E_2[\text{Lip-(SH)}_2] + \text{acetyl-CoA} \\ &E_2[\text{Lip-(SH)}_2] + E_3[\text{FAD}] \rightleftharpoons E_2[\text{Lip-S}_2] + E_3[\text{FAD(red)}] \\ &E_3[\text{FAD(red)}] + NAD^+ \rightleftharpoons E_3[\text{FAD}] + NADH + H^+ \end{aligned} \quad (13)$$

where TPP, FAD, and Lip-S_2 represent the cofactors thiamine pyrophosphate, flavin adenire dinucleotide, and lipoic acid, respectively. Each of the enzymes is a single polypeptide chain, but the stoichiometry of the polypeptide chains in the complex still is a matter of controversy: Values for $E_1{:}E_2{:}E_3$ of 24:24:12,[93] 16:16:16,[94] and 24:24:24[95] have been proposed. The symmetry requirements imposed by electron microscopy studies appear to require 24 transacetylase polypeptide chains as the core of the complex with the E_1 and E_3 enzymes combining with the E_2 core.[82] Stable E_1-E_2 and E_2-E_3 complexes are formed, but an E_1-E_3 complex is not readily formed.[96] The isolated enzymes have essentially the same catalytic activity as in the complex when assays for the individual enzymes are utilized.[96] The binding of substrates and other ligands to E_1 and E_2 is not markedly different in the intact complex and the E_1 and E_2-E_3 subcomplexes, and the binding of ligands to one enzyme is not altered appreciably by ligand binding to other enzymes.[97,98] Therefore the formation of the multienzyme complex does not appear to cause large changes in conformation of the polypeptide chains, and the interactions between different types of polypeptide chains do not seem to play a major role in regulation.

Free intermediates have not been observed in the overall reaction so that the primary function of the multienzyme complex in this case appears to be a "channeling" of intermediates between enzymes. A mechanism has been proposed in which the lipoic acid, which is covalently bound to E_2, carries the substrate from site to site by forming covalent intermediates, as outlined in (13).[82] This mechanism requires all the catalytic sites to be within a circle of about 30 Å, since lipoic acid has a maximum extension of about 15 Å from the lysine residue to which it is linked. A schematic illustration of this mechanism is presented in Fig. 9. Thus far no experimental proof of this mechanism is available; in fact, recent fluorescence energy transfer measurements suggest the catalytic sites of E_1 and E_3 and of E_2 and E_3 are quite far apart, >45 Å, so that the mechanism clearly is more complex than proposed.[97,98] Regardless of the detailed mechanism, this multienzyme complex enhances catalysis by directing the intermediates between different enzymes with little influence of enzyme-enzyme interactions.

The pyruvate dehydrogenase enzyme (E_1) of the complex is also subject to allosteric regulation:[82] It is inhibited by acetyl-CoA, a product of pyruvate inhibition, and GTP. The acetyl-CoA inhibition is reversed by nucleoside monophosphates, and the GTP inhibition is reversed by GDP. Apparently acetyl-CoA and GTP act at different sites. The complex is also inhibited by NADH through binding to E_3. The mammalian pyruvate dehydrogenase complexes contain two additional regulatory enzymes, a kinase and a phosphatase, and further regulation occurs by phosphorylation-dephosphorylation.[82] Thus both allosterism and complex formation play an important role in regulating the overall reaction (12).

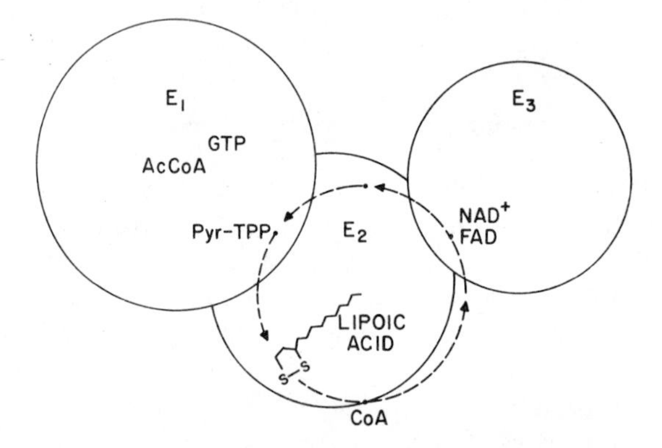

Fig. 9. A schematic drawing of a possible mechanism for the reaction catalyzed by the pyruvate dehydrogenase complex. The three enzymes E_1, E_2, and E_3 are located so that lipoic acid covalently linked to E_2 can rotate between the active sites containing thiamine pyrophosphate (TPP) and pyruvate (Pyr) on E_1, CoA on E_2, and FAD on E_3. Acetyl-CoA and GTP are allosteric effectors of E_1, and NAD^+ is an inhibitor of the overall reaction.

VI. MEMBRANE-BOUND ENZYMES

The final mode of regulating enzymic activity to be discussed is the coupling of an enzyme with a membrane. Several different types of regulation are possible: (1) Specific interactions between the protein and phospholipid may be required; (2) a general requirement for a hydrophobic type of environment may exist; (3) the enzyme can be immobilized by a membrane and can be localized in a particular place where it is needed; (4) the function of the enzyme can be coupled with another membrane function, such as transport (this coupling may require a closed membranous structure); and (5) the enzymic activity can be modulated by interaction of the enzyme with other membrane proteins (e.g., by coupling to other enzymes or to receptors). A few examples illustrating these possibilities are now considered.

A. β Hydroxybutyrate Dehydrogenase

The enzyme $D(-)\beta$ hydroxybutyrate dehydrogenase catalyzes the reaction

$$\text{Acetoacetate} + \text{NADH} + \text{H}^+ \rightleftharpoons D\text{-}\beta\text{-hydroxybutyrate} + \text{NAD}^+ \quad (14)$$

The enzyme is tightly bound to the mitochondrial membrane but can be solubilized by use of phospholipase A or detergents.[99–102] The enzyme, which is a single polypeptide chain of molecular weight 32,000,[102] is totally inactive in the absence of lecithin or lecithin analogues.[103,104] The specificity of the enzyme-lecithin interaction has been probed by utilizing lecithins with different fatty acid side chains and related compounds.[105,106] Lecithins, lysolecithins, and stearylphosphorylcholine form active complexes with the enzyme. A hydrophobic chain followed sequentially by a negative and positive charge, as in stearylphosphorylcholine, is the minimal structural requirement of an activator. The dependence of the rate on the chain length of saturated fatty acids in lecithin is shown in Fig. 10. The nature of the fatty acid side chain clearly alters the activity, but this is a quantitative rather than a qualitative effect; both short-chain soluble lecithins and long-chain vesicle-forming lecithins are good activators. The degree of saturation of the side chains also influences the activation. Although short-chain lecithins are good activators, an enzyme-lecithin complex that is stable over long time periods only forms with lecithins that are capable of forming lamellar vesicular structures. The temperature dependence of the rate follows normal Arrhenius behavior over the temperature range of approximately 5 to 30°C, and the activation energy is essentially the same for a series of lecithins, so that the nature of the catalytic process appears to be essentially the same for all activators.

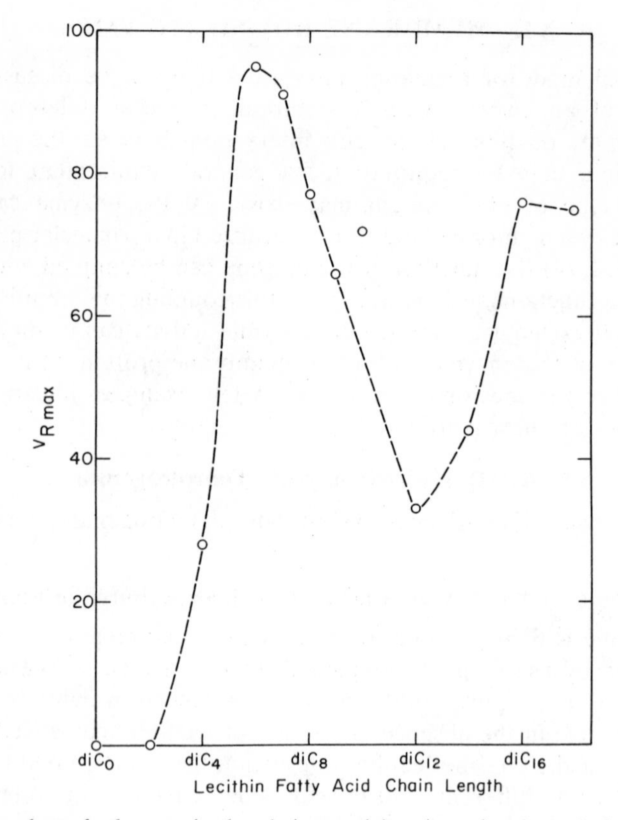

Fig. 10. A plot of the maximal relative activity ($v_{R\,max}$) of β hydroxybutyrate dehydrogenase-lecithin mixtures versus the number of carbon atoms in the saturated fatty acid side chains of the lecithins.

Phospholipid vesicles (and bilayers) composed of phospholipids with well-defined fatty acid side chains undergo a sharp transition from a crystallinelike state to an amorphous state as the temperature is raised.[107] The transition temperature depends on the nature of the fatty acid side chains. For example, for C_{12} saturated fatty acid chains on lecithin the transition temperature is 0° and for C_{18} saturated fatty acid chains it is 58°C; for unsaturated lecithins the transition temperature is below zero.[107] For real membranes sharp phase transitions are not observed, because of the heterogeneous composition of the membrane. In the case of β hydroxybutyrate dehydrogenase, the enzymic activity apparently is not influenced by this phase transition as judged by the temperature dependence of the reaction rate. However, for some membrane-bound proteins, a plot of the reaction rate versus the reciprocal temperature displays sharp breaks;[108] this is frequently attributed to local phase transitions (or phase separations) within the membrane. Such an interpretation is largely speculative at the present time. Nevertheless, phase transitions and separations of phospholipids in membranes might play an important role in regulating enzyme catalysis.

B. Coupling Factor 1

The coupling factor 1 of mitochondria is an ATPase when isolated as a soluble enzyme. In mitochondria it is bound to the inside of the mitochondrial membrane and synthesizes ATP from ADP and P_i when coupled to an oxidation chain of enzymes. A similar coupling factor is found to be operative in photophosphorylation and oxidative phosphorylation in chloroplasts and bacteria (see Refs. 109 to 112 for recent reviews). Coupling factor 1, which is readily solubilized, has a molecular weight of about 360,000 and contains five different types of polypeptide chains in an unknown stoichiometry. For the chloroplast enzyme, a minimal subunit stoichiometry of $\alpha_2\beta_2\gamma\delta\varepsilon_2$ has been determined from chemical cross-linking studies.[113] The subunits are generally identified by Greek letters in order of descending molecular weight. The function of the individual subunits is not yet resolved. However, the ATPase active site is associated with the β subunit,[114] and additional nucleotide binding sites, which may serve an allosteric regulatory function, are found on the α-β subunits.[115] Which of these sites is directly involved in phosphorylation is still a matter of controversy. The γ subunit is required for phosphorylation and is believed to play a regulatory role.[114,116] The δ subunit is required for binding to the membrane,[117] and the ε subunit is an inhibitor of ATPase activity in some coupling factors and may play an

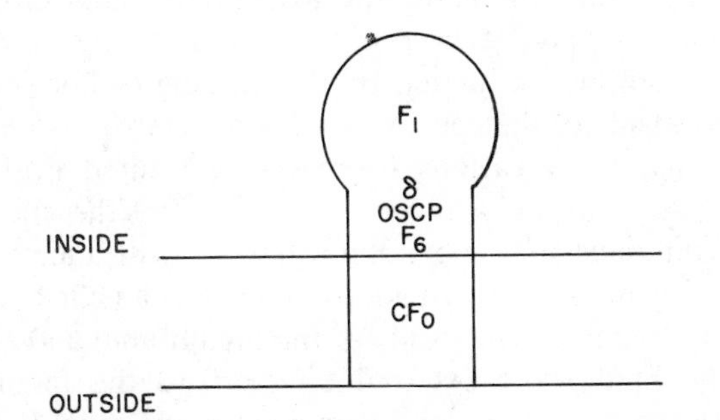

Fig. 11. A pictorial representation of the mitochondrial ATP synthesizing coupling factor interacting with the mitochondrial membrane. F_1 contains five polypeptide chains α, β, γ, δ, and ϵ and is readily solubilized. The stalk is probably made up of three polypeptide chains, δ, OSCP, and F_6, which interact with a small group of hydrophobic polypeptides, CF_0, embedded in the membrane.

important role in regulation.[118] In order for coupling factor 1 to be functionally coupled to the membrane, two additional soluble polypeptide chains and hydrophobic polypeptide chains (CF_0), which are embedded in the membrane, are required.[109,119] A schematic illustration of how this coupling system may be situated in the membrane is shown in Fig. 11. Electron microscopy studies show ball-like structures attached to the membrane; these structures are inferred to be the portion of the coupling factor that is easily solubilized.[120]

The complete coupling system synthesizes ATP when coupled to other enzymes, but for many years a controversy has centered around the source of energy for this synthesis. At the present time the most likely explanation is that the phosphorylation is coupled with the production of a proton gradient across the membrane that provides the energy needed for ATP synthesis (the Mitchell hypothesis[121]). The molecular mechanism, of course, undoubtedly involves conformational changes within the coupling system. However, the important point to note is that the enzymic activity is modulated by coupling it with a proton gradient (or proton pump), a process that obviously requires an intact membranous structure.

C. Adenylate Cyclase

As a final example of a membrane-bound enzyme, adenylate cyclase is considered (see Refs. 122 and 123 for reviews of this system). This enzyme has not yet been prepared in any reasonable state of purity, so that its chemical and physical characteristics are not known. However, it is of crucial importance in metabolic regulation: It synthesizes cyclic 3′,5′-adenosine phosphate, which regulates the activity of protein kinase. Protein kinase in turn regulates the activity of many enzymes through phosphorylation. Adenylate cyclase is located on the inside of the cell, but its activity can be modulated by the binding of hormones and other agents to the outside of the cell. A list of some activators and inhibitors of adenylate cyclase from various sources is presented in Table VI. The external ligands bind to a receptor protein on the outer membrane surface that interacts in some unknown way with adenylate cyclase to regulate the enzymic activity. In addition, GTP is an allosteric activator of the enzyme (probably on the inside of the membrane) and apparently also modulates the binding of external effectors to the membrane.[124] For many systems structurally different hormones are capable of modulating adenylate cyclase activity, and it has been shown that hormones bind to different protein receptors. Apparently, a variety of receptor molecules are capable of interacting with a single adenylate cyclase molecule. The effects of hormones and GTP are not additive: The

TABLE VI. Some Hormone Effectors of Adenylate Cyclase Systems[a]

Hormone	Tissue source
Catecholamines	Muscle, liver, spleen, kidney, brain pineal gland, adipose, pancreatic islets, erythrocytes, leukocytes, thymocytes, fat cell, lung fibroblasts, parotid gland, endothelium, peritoneal macrophages
ACTH	Fat cell
Glucagon	Liver, fat cell, pancreatic islets
Secretin	Fat cell
Thyrotropin	Thyroid
Prostaglandin	Thyroid, kidney, platelets
Oxytocin	Bladder
Corticotropin	Adrenal, fat cell
Calcitonin	Kidney
Parathyroid	Kidney

[a] Refs. 124 and 125.

adenylate cyclase has some maximal activity, and when this maximal activity is achieved, the addition of activators results in more binding to receptors but not more enzymic activity. This suggests that the system may be viewed as a simple two-state allosteric model where one state is active, the other is inactive, and ligand binding alters the equilibrium between the active and inactive states.[126] Once the system is in the active state, further binding of an activator does not enhance the activity. Each state includes the enzyme, receptors, and perhaps the membrane phospholipids and other proteins.

Thus far solubilization of the adenylate cyclase systems has been difficult. Although the enzyme has been solubilized in several instances, the specific activity is generally low relative to that in the membrane and an unambiguous hormone response similar to that found in membranous preparations is lacking.[123,127] A schematic model showing how the adenylate cyclase system (enzyme and receptors) may be situated in the membrane is presented in Fig. 12. Many questions remain to be answered. What is the mode of interaction of receptor and enzyme? Do the proteins interact directly or is the interaction modulated by lipids and other proteins? Perhaps the enzyme itself extends through the membrane. A receptor might not interact with the enzyme until it contains a ligand. This would allow an excess of receptor molecules to exist relative to the number of enzyme molecules; an enzyme molecule might then only

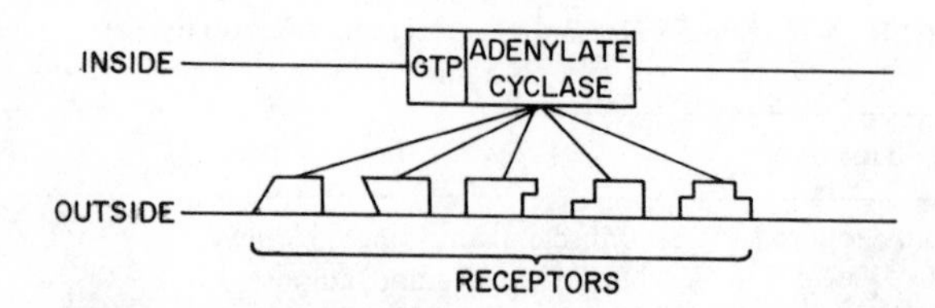

Fig. 12. A pictorial representation of adenylate cyclase and hormone receptors interacting with a membrane. The GTP control component is also shown. The different hormone receptors may not interact directly with the adenylate cyclase and may be diffusing freely in the membrane until a hormone is bound.

interact with a single liganded receptor. If this model is correct, the receptor and enzyme must diffuse freely in the membrane phase, since activation is quite rapid. Cooperative ligand binding may also play an important role in the regulatory process. In spite of these many uncertainties, it is apparent that the adenylate cyclase system involves the interaction of many different components, probably both proteins and lipids. The membrane structure itself is of obvious importance, and the molecular details of the mode of communication between the outside and inside of a cell is an intriguing problem.

VII. CONCLUSION

This review has tried to present an overview of the control of enzymic activity in complex polyatomic frameworks. The examples discussed are intended to be representative; obviously many other examples could be cited. The elementary interactions involved in modulating enzymic activity are well understood in terms of thermodynamics, kinetics, and structure. A considerable amount of information is also available for the simplest type of macromolecular framework, enzymes consisting of a single polypeptide chain, although a considerable amount of work remains to be done.

Regulatory enzymes containing multiple polypeptide chains are just beginning to be understood in molecular terms. Considerably more thermodynamic, kinetic, and structural information is required. Several multienzyme complexes are available in a reasonably pure state, but the molecular characterization of their mechanisms is still in a rather primitive state. The situation is even more difficult with membrane-bound enzymes. A few of these enzymes can be obtained as well-defined entities, but in many cases purification of the enzyme system and all its components is quite far off in the future. The small quantity of material usually available is also a great problem with these systems. As might be anticipated, a wide range of molecular interactions are used to regulate enzymic activity in biological systems, but the central theme appears to be the modulation of macromolecular conformation; for multienzyme complexes and membrane systems, the localization of specific structures also is of primary importance.

Acknowledgment

I am indebted to the National Institutes of Health and the National Science Foundation for their support of my research.

References

1. G. G. Hammes and A. C. Park, *J. Am. Chem. Soc.*, **91,** 956 (1969).
2. G. G. Hammes and P. L. Lillford, *J. Am. Chem. Soc.*, **92,** 7578 (1970).
3. G. G. Hammes and H. O. Spivey, *J. Am. Chem. Soc.*, **88,** 1621 (1966).
4. M. Eigen, *Pure Appl. Chem.*, **6,** 97 (1963).
5. C. Tanford, *J. Am. Chem. Soc.*, **84,** 4240 (1962).
6. E. Grunwald and E. K. Ralph, III, *J. Am. Chem. Soc.*, **89,** 4405 (1967).
7. G. G. Hammes and W. Knoche, *J. Chem. Phys.*, **45,** 4041 (1966).
8. G. G. Hammes and N. C. Pace, *J. Phys. Chem.*, **72,** 2227 (1968).
9. G. G. Hammes and T. B. Lewis, *J. Phys. Chem.*, **70,** 1610 (1966).
10. G. G. Hammes and F. G. Walz, Jr., *J. Am. Chem. Soc.*, **91,** 7197 (1969).
11. J. E. Erman and G. G. Hammes, *J. Am. Chem. Soc.*, **88,** 5067 (1966).
12. E. J. del Rosario and G. G. Hammes, *J. Am. Chem. Soc.*, **92,** 1750 (1970).
13. J. E. Erman and G. G. Hammes, *J. Am. Chem. Soc.*, **88,** 5614 (1966).
14. G. P. Hess, J. McConn, E. Ku, and G. McConkey, *Phil. Trans. Roy. Soc.*, **B257,** 89 (1970).
15. E. Holler, J. Rupley, and G. P. Hess, *Biochemistry*, **14,** 2377 (1975).
16. G. G. Hammes and J. K. Hurst, *Biochemistry*, **8,** 1083 (1969).
17. G. Czerlinski and G. Schreck, *Biochemistry*, **3,** 89 (1963).
18. G. G. Hammes and J. L. Haslam, *Biochemistry*, **7,** 1519 (1968).
19. G. G. Hammes and J. L. Haslam, *Biochemistry*, **8,** 1591 (1969).
20. P. Fasella and G. G. Hammes, *Biochemistry*, **6,** 1798 (1967).
21. G. G. Hammes, R. W. Porter, and G. R. Stark, *Biochemistry*, **10,** 1046 (1971).
22. K. Kirschner, E. Gallego, I. Schuster, and D. Goodall, *J. Mol. Biol.*, **58,** 29 (1971).
23. A. F. Barksdale and J. E. Stuehr, *J. Am. Chem. Soc.*, **94,** 3334 (1972).
24. J. Applequist and P. Doty, Abstracts, 135th Meeting of American Chemical Society, 1959.
25. A. S. Mildvan and M. C. Scrutton, *Biochemistry*, **6,** 2978 (1967).
26. A. S. Mildvan, J. S. Leigh, and M. Cohn, *Biochemistry*, **6,** 1805 (1967).
27. B. Chance, *Arch. Biochem.*, **22,** 224 (1949).
28. B. Chance, in *Currents in Biochemical Research*, D. E. Green, Ed., Wiley-Interscience, New York, 1956, p. 308.
29. G. G. Hammes and P. R. Schimmel, *The Enzymes* (3rd ed.) **2,** 67 (1970).
30. M. Eigen, *Angew. Chem.*, **75,** 489 (1963).
31. M. Eigen and G. G. Hammes, *Adv. Enzymol.*, **25,** 1 (1963).
32. G. G. Hammes, *Nature*, **204,** 342 (1964).
33. R. Lumry, *The Enzymes* (2nd ed.), **1,** 157 (1959).
34. W. P. Jencks, *Catalysis in Chemistry and Enzymology*, McGraw-Hill, New York, 1969.

35. T. C. Bruice, *Ann. Rev. Biochem.*, **45,** 331 (1976).
36. D. E. Koshland, Jr., *J. Theoret. Biol.*, **2,** 75 (1962).
37. D. E. Koshland, Jr., and K. E. Neet, *Ann. Rev. Biochem.*, **37,** 359 (1968).
38. F. M. Richards and H. W. Wyckoff, *The Enzymes*, **4,** 647 (1971).
39. T. C. French and G. G. Hammes, *J. Am. Chem. Soc.*, **87,** 4669 (1965).
40. D. A. Usher, D. I. Richardson, Jr., and F. Eckstein, *Nature*, **228,** 663 (1970).
41. G. C. K. Roberts, E. A. Dennis, D. H. Meadows, J. S. Cohen, and O. Jardetzky, *Proc. Nat. Acad. Sci.* (*U.S.*), **62,** 1151 (1969).
42. D. Findlay, D. G. Herries, A. P. Mathias, B. R. Rabin, and C. A. Ross, *Biochem. J.*, **85,** 152 (1962).
43. D. A. Usher, E. S. Ehrenrich, and F. Eckstein, *Proc. Nat. Acad. Sci.* (*U.S.*), **69,** 115 (1972).
44. H. J. Wieker and H. Witzel, *Eur. J. Biochem.*, **1,** 251 (1967).
45. L. W. Harrison and G. G. Hammes, *Biochemistry*, **12,** 1395 (1973).
46. G. G. Hammes and C.-W. Wu, *Biochemistry*, **10,** 1051 (1971).
47. G. G. Hammes and C.-W. Wu, *Biochemistry*, **10,** 2151 (1971).
48. E. D. Barber and H. J. Bright, *Proc. Nat. Acad. Sci.* (*U.S.*), **60,** 1370 (1968).
49. G. W. Hatfield and H. E. Umberger, *J. Biol. Chem.*, **245,** 1742 (1970).
50. G. G. Hammes and C.-W. Wu, *Adv. Biophys. Bioeng.*, **3,** 1 (1974).
51. J. Monod, J. Wyman, and J. P. Changeux, *J. Mol. Biol.*, **12,** 88 (1965).
52. G. S. Adair, *J. Biol. Chem.*, **63,** 529 (1925).
53. D. E. Koshland, Jr., G. Nemethy, and D. Filmer, *Biochemistry*, **5,** 365 (1966).
54. C. Frieden and R. Colman, *J. Biol. Chem.*, **242,** 1705 (1967).
55. L. W. Nichol, W. J. H. Jackson, and D. J. Winzor, *Biochemistry*, **6,** 2449 (1967).
56. E. R. Stadtman, *The Enzymes*, **1,** 398 (1970).
57. C. Frieden, *J. Biol. Chem.*, **245,** 5788 (1970).
58. G. R. Jacobson and G. R. Stark, *The Enzymes*, **9,** 225 (1973).
59. J. C. Gerhart and A. B. Pardee, *J. Biol. Chem.*, **237,** 891 (1964).
60. J. C. Gerhart and H. K. Schachman, *Biochemistry*, **4,** 1054 (1965).
61. K. E. Richards and R. C. Williams, *Biochemistry*, **11,** 3393 (1972).
62. S. H. Warren, B. P. Edwards, D. R. Evans, D. C. Wiley, and W. N. Lipscomb, *Proc. Nat. Acad. Sci.* (*U.S.*), **70,** 1119 (1973).
63. S. Matsumoto and G. G. Hammes, *Biochemistry*, **14,** 214 (1975).
64. S. Fan, L. W. Harrison, and G. G. Hammes, *Biochemistry*, **14,** 2219 (1975).
65. C. C. Winlund and M. J. Chamberlin, *Biochem. Biophys. Res. Commun.*, **40,** 43 (1970).
66. S. Matsumoto and G. G. Hammes, *Biochemistry*, **12,** 1388 1973).
67. C. W. Gray, M. J. Chamberlin, and D. M. Gray, *J. Biol. Chem.*, **248,** 6071 (1973).
68. J. P. Changeux, J. C. Gerhart, and H. K. Schachman, *Biochemistry*, **7,** 538 (1968).
69. G. G. Hammes and C.-W. Wu, *Science*, **172,** 1205 (1971).
70. C. Tondre and G. G. Hammes, *Biochemistry*, **13,** 3131 (1974).
71. K. D. Collins and G. R. Stark, *J. Biol. Chem.*, **244,** 1869 (1969).
72. H. Hofer and D. Pette, *Z. Physiol. Chem.*, **349,** 1378 (1968).
73. J. V. Pasonneau and O. H. Lowry, *Biochem. Biophys. Res. Commun.*, **7,** 10 (1962); **13,** 372 (1963).
74. P. M. Lad and G. G. Hammes, *Biochemistry*, **13,** 4530 (1974).
75. K. R. Leonard and I. O. Walker, *Eur. J. Biochem.*, **26,** 442 (1972).
76. M. J. Pavelich and G. G. Hammes, *Biochemistry*, **12,** 1408 (1973).
77. C. J. Coffee, R. B. Aaronson, and C. Frieden, *J. Biol. Chem.*, **248,** 1381 (1973).

78. J. N. Telford, P. M. Lad, and G. G. Hammes, *Proc. Nat. Acad. Sci.* (*U.S.*), **72,** 3054 (1975).
79. P. M. Lad, D. E. Hill, and G. G. Hammes, *Biochemistry,* **12,** 4303 (1973).
80. D. E. Hill and G. G. Hammes, *Biochemistry,* **14,** 203 (1975).
81. C. Frieden, H. R. Gilbert, and P. E. Bock, *J. Biol. Chem.*, **251,** 5644 (1976).
82. L. J. Reed and D. J. Cox, *The Enzymes,* **1,** 213 (1970).
83. C. Yanofsky and I. P. Crawford, *The Enzymes,* **7,** 1 (1972).
84. I. P. Crawford and C. Yanofsky, *Proc. Nat. Acad. Sci.* (*U.S.*), **44,** 1161 (1958).
85. M. E. Goldberg, T. E. Creighton, R. L. Baldwin, and C. Yanofsky, *J. Mol. Biol.*, **21,** 71 (1966).
86. D. A. Wilson and I. P. Crawford, *J. Biol. Chem.*, **240,** 4801 (1965).
87. M. Hatanaka, E. A. White, K. Horibata, and I. P. Crawford, *Arch. Biochem. Biophys.*, **97,** 596 (1962).
88. G. M. Hathaway, S. Kido, and I. P. Crawford, *Biochemistry,* **8,** 989 (1969).
89. M. E. Goldberg, S. York, and L. Stryer, *Biochemistry,* **7,** 3662 (1968).
90. E. J. Faeder and G. G. Hammes, *Biochemistry,* **9,** 4043 (1970); **10,** 1041 (1971).
91. W. O. Weischet and K. Kirschner, *Eur. J. Biochem.*, **65,** 365, 375 (1976).
92. T. E. Creighton, *Eur. J. Biochem.*, **13,** 1 (1970).
93. L. J. Reed, F. H. Petit, M. H. Eley, L. Hamilton, J. H. Collins, and R. M. Oliver, *Proc. Nat. Acad. Sci.* (*U.S.*), **72,** 3068 (1975).
94. O. Vogel, B. Hoehn, and U. Henning, *Proc. Nat. Acad. Sci.* (*U.S.*), **69,** 1615 (1972).
95. D. L. Bates, R. A. Harrison, and R. N. Perham, *FEBS Lett.* **60,** 427 (1975).
96. M. Koike, L. J. Reed, and W. R. Carroll, *J. Biol. Chem.*, **238,** 30 (1963).
97. O. A. Moe, Jr., D. A. Lerner, and G. G. Hammes, *Biochemistry,* **13,** 2552 (1974).
98. G. B. Shepherd, N. Papadakis, and G. G. Hammes, *Biochemistry,* **15,** 2888 (1976).
99. I. Sekuzu, P. Jurtshuk, Jr., and D. E. Green, *J. Biol. Chem.*, **238,** 975 (1963).
100. G. S. Gotterer, *Biochemistry,* **6,** 2139 (1967).
101. H. G. Bock and S. Fleischer, *J. Biol. Chem.*, **250,** 5774 (1975).
102. H. M. Menzel and G. G. Hammes, *J. Biol. Chem.*, **248,** 4885 (1973).
103. P. Jurtshuk, I. Sekuzu, and D. E. Green, *Biochem. Biophys. Res. Commun.*, **6,** 76 (1961).
104. G. S. Gotterer, *Biochemistry,* **6,** 2147 (1967).
105. A. K. Grover, A. J. Slotboom, G. H. de Haas, and G. G. Hammes, *J. Biol. Chem.*, **250,** 31 (1975).
106. P. Gazzotti, H. G. Bock, and S. Fleischer, *J. Biol. Chem.*, **250,** 5782 (1975).
107. B. D. Ladbrooke and D. Chapman, *Chem. Phys. Lipids,* **3,** 304 (1969).
108. J. K. Raison, *Bioenergetics,* **4,** 285 (1973).
109. E. Racker, in *Molecular Oxygen in Biology: Topics in Molecular Oxygen Research,* (O. Hayaishi, Ed., North-Holland, New York, 1974, pp. 339–361.
110. H. S. Penefsky, *The Enzymes,* **10,** 375 (1974).
111. A. E. Senior, *Biochim. Biophys. Acta,* **301,** 249 (1973).
112. P. L. Pederson, *Bioenergetics,* **6,** 243 (1975).
113. B. A. Baird and G. G. Hammes, *J. Biol. Chem.*, **251,** 6953 (1976).
114. D. W. Deters, E. Racker, N. Nelson, and H. Nelson, *J. Biol. Chem.*, **250,** 1041 (1975).
115. L. C. Cantley, Jr., and G. G. Hammes, *Biochemistry,* **14,** 2976 (1975).
116. N. Nelson, D. W. Deters, H. Nelson, and E. Racker, *J. Biol. Chem.*, **248,** 2049 (1973).
117. M. Futai, P. C. Sternweis, and L. A. Heppel, *Proc. Nat. Acad. Sci.* (*U.S.*), **71,** 2725 (1974).

118. N. Nelson, H. Nelson, and E. Racker, *J. Biol. Chem.*, **247,** 7657 (1972).
119. R. Serrano, B. I. Kanner, and E. Racker, *J. Biol. Chem.*, **251,** 2453 (1976).
120. Y. Kagawa and E. Racker, *J. Biol. Chem.*, **241,** 2461 (1966).
121. P. Mitchell, *Biol. Rev. Camb. Phil. Soc.*, **41,** 445 (1966).
122. G. A. Robinson, R. W. Butcher, and E. W. Sutherland, *Ann. Rev. Biochem.*, **37,** 149 (1969).
123. J. P. Perkins, *Adv. Cyclic Nucleotide Res.*, **3,** 1 (1973).
124. M. Rodbell, M. C. Lin, Y. Salomon, C. Londos, J. P. Harwood, B. R. Martin, M. Rendell, and M. Berman, *Adv. Cyclic Nucleotide Res.*, **5,** 3 (1975).
125. R. J. Lefkowitz, L. E. Limbird, C. Mukherjee, and M. G. Caron, *Biochim. Biophys. Acta*, **457,** 1 (1976).
126. G. G. Hammes and M. Rodbell, *Proc. Nat. Acad. Sci.* (*U.S.*), **73,** 1189 (1976).
127. E. J. Neer, *J. Biol. Chem.*, **249,** 6527 (1974).

LIST OF INTERVENTIONS

1. Welch
   - 1.1 Hammes
2. Hess
   - 2.1 Hammes
   - 2.2 Ubbelohde
   - 2.3 Hammes
3. Ubbelohde
   - 3.1 Hammes
   - 3.2 Ubbelohde
   - 3.3 Klotz
   - 3.4 Ubbelohde
   - 3.5 Hammes
4. Carafoli
   - 4.1 Hammes
   - 4.2 Damjanovich
   - 4.3 Ubbelohde
   - 4.4 Danijanovich
   - 4.5 Hammes
   - 4.6 Thomas
   - 4.7 Hammes
   - 4.8 Simon
   - 4.9 Korenstein
5. Welch
   - 5.1 Hammes
6. Caplan
   - 6.1 McConnell
   - 6.2 Eisenman
   - 6.3 Hammes
   - 6.4 McConnell
   - 6.5 Bangham
7. Williams
   - 7.1 Ross
   - 7.2 Somorjai
   - 7.3 Ubbelohde

1. Intervention of Welch

I am one of those of the "extreme" opinion that most enzymes of intermediary metabolism are organized in some manner, among themselves and/or with membranous matrices, in situ. In your presentation you indicated the various improvements in efficiency afforded by structural organization. With this in mind, may we not approach the question of organization in vivo according to the adage "Absence of evidence is not necessarily evidence of absence"? Considering the advantages (e.g., better overall efficiency) provided by enzyme organization, would you accept the preceding adage as regards the possibility of heretofore undetected structures in vivo?

1.1 Intervention of Hammes

I don't quite know how to answer this question, as it raises a philosophical point rather than scientific fact. Certainly the fact that many enzymes are isolated as soluble species is not proof that these same enzymes do not exist as part of multienzyme complexes in the cell. Nevertheless, I think it is unlikely that this is the case. The formation of multienzyme complexes presumably was due to evolutionary selection. Unless some appreciable advantage is gained by the formation of a multienzyme complex, no reason exists for its formation. Based on this line of reasoning, I believe many enzymes exist as soluble species in the cell rather than as part of specifically organized structures.

2. Intervention of Hess

What is the principal difference between the ATPase activity of F_1 in solution and the membrane-bound ATPase activity apart from the protontransfer function?

2.1 Intervention of Hammes

A number of cases are known in which the properties of an enzyme are markedly altered by interaction with a membrane. Of course, in some cases the normal function of an enzyme is destroyed when it is removed from the membrane. For example, the mitochondrial coupling factor cannot synthesize ATP when removed from the membrane, since coupling to a proton gradient is required. The portion of the coupling factor that is easily solubilized (F_1) is an ATPase. The steady-state kinetic properties of this solubilized ATPase are appreciably changed when it is reconstituted with mitochondrial membranes: The turnover numbers and pH dependencies are different; the solubilized enzyme is strongly inhibited by ADP, whereas the reconstituted enzyme is not; and the reconstituted enzyme is inhibited by oligomycin, whereas the solubilized enzyme is not.

2.2 Intervention of Ubbelohde

Can one give any general interpretation of this difference?

2.3 Intervention of Hammes

The oligomycin inhibition requires interaction of F_1 with other polypeptide chains of the coupling factor that are associated with the membrane used for reconstitution. In a sense this is quite analogous to allosteric enzymes, where regulation is achieved by modulation of intersubunit interactions.

3. Intervention of Ubbelohde

Should one regard the enzymes as integrally bound up with the membrane or as merely associated with it?

3.1 Intervention of Hammes

The question of whether an enzyme is membrane bound or membrane associated is to some extent a matter of semantics. However, it is certainly true that some proteins are readily dissociated from membranes whereas others require quite drastic conditions before they can be dissociated from the membrane. As limiting cases, the former can be designated as membrane associated and the latter as membrane bound. Enzymes that are generally considered membrane bound are firmly embedded in the membrane structure. For example, the mitochondrial coupling factor is strongly coupled to the bilayer structure by hydrophobic polypeptides. The Na^+-K^+ ATPases that have been purified have a small patch of associated phospholipids; when the enzyme is delipidated, enzymatic activity is lost. In fact, membrane-bound enzymes appear to be an integral part of the membrane structure; perhaps this should be used as a basis for the definition of a membrane-bound enzyme.

3.2 Intervention of Ubbelohde

Is there evidence for the conditioning of enzyme behavior by the membrane?

3.3 Intervention of Klotz

No work has been done with our polymers that corresponds closely to placing an enzyme in a membrane. I did mention earlier that we have considered immobilizing synzymes by coupling to a solid support, which would constrain the polymer. Also most of our polymers have some apolar side chains, which give these macromolecules some lipidlike character.

3.4 Intervention of Ubbelohde

If the enzyme is located in deep clefts, might its behavior be conditioned or restricted by this?

3.5 Intervention of Hammes

The conformation of membrane-bound enzymes is undoubtedly restricted by the membrane. However, the mechanism of action of these enzymes appears to be similar to that of soluble enzymes, so that the presence of clefts and conformational flexibility is to be expected. The mitochondrial coupling factor apparently contains both the ATP synthesizing enzyme and a proton channel; conformational changes undoubtedly play a role in the function of this system. A large movement of polypeptide chains has been proposed in the functioning of this system (and for other membrane-bound enzymes), but no convincing experimental evidence is available to support such a hypothesis.

4. Intervention of Carafoli

One striking characteristic of the coupling ATPase of energy-transducing membranes, apart from the extraordinarily large number of different polypeptide subunits, is the existence of two different polypeptides involved in the response of the enzyme to the inhibitor oligomycin. One binds the inhibitor, the other, separated in space from the former by possibly as much as 10 to 15 Å, confers oligomycin sensitivity to the entire enzyme complex. How could the transfer of information between these two polypeptide subunits and their concerted interaction with the ATPase proper be visualized?

4.1 Intervention of Hammes

This interesting observation is another example of how intersubunit interactions can control the reactivity of specific ligand-binding sites.

As with other multisubunit enzymes (e.g., allosteric enzymes), the structural integrity of a membrane-bound enzyme primarily is maintained by noncovalent interactions such as hydrogen bonding, electrostatics, and hydrophobic interactions. Hydrophobic polypeptides (or hydrophobic portions of polypeptides) apparently are used to anchor the enzymes to the membrane through interactions with phospholipids. Therefore, I would characterize the interaction between the enzyme and membrane as "chemical" in nature rather than as "geometric."

4.2 Intervention of Damjanovich

Enzyme regulation and action are highly influenced by the near aqueous environment. The protein fluctuation is generated by the translational thermal coupling of the environment. Since translational $(T) \rightarrow$ vibrational (V) transitions during collisions between enzymes and the environment are regulated by quantum-physical conditions, among them the velocity threshold of the small molecules, an enzyme model can be elaborated using this idea. We assume [*J. Theor. Biol.*, **51,** 393–401 (1975)] a direct interference of the mass distribution of the enzyme environment and the enzyme action and regulation. There is a preferential, near simultaneous collision pattern between the protein and the environment influencing the catalysis. The mathematical elaboration of this model gives a deeper physical meaning of the phenomenological kinetic constants, such as k_{+1}, k_{-1}, k_2, and so on. These assumptions predict a new kind of enzyme regulation depending on the environmental mass distribution. To support the existence of this regulation, a combination of spectroscopic methods (microwave, nmr, etc., spectroscopies) carried out during the enzyme action, using different environmental mass distribution, was recommended.

4.3 Intervention of Ubbelohde

Is the regulation to be associated with the bulk environment or with the proximity of the membrane?

4.4 Intervention of Damjanovich

The energetic coupling between environment and enzymes is more general. It must occur also in membranes and free aqueous solutions as well. In view of the overall occurrence of the translational thermal coupling and the experimentally proved fluctuations of proteins, our model must be valid in case of any enzymes.

4.5 Intervention of Hammes

The idea that statistical fluctuations play a role in the mechanism of enzyme catalysis is intriguing. However, as far as I know, no experimental evidence exists to support such a hypothesis.

4.6 Intervention of Thomas

There are two kinds of effects of the membrane on the enzyme behavior: a specific interaction between the enzyme and the lipid membrane and a nonspecific interaction of the membrane structure by itself on the enzyme kinetics. In the case of ATPase, the enzyme in solution is working in homogeneous and isotropical conditions. At the opposite extreme, in the membrane the enzyme is working under asymmetrical boundary conditions. In the last case there is a coupling between a scalar process and the vectorial transport effect. In conclusion, the effect of the membrane on the enzyme behavior is not only a chemical effect, but also a geometrical one.

4.7 Intervention of Hammes

I, of course, agree that proper orientation of a membrane-bound enzyme is essential. However, the orientation is achieved mainly through noncovalent interactions between polypeptide chains and between polypeptide chains and phospholipids. With this in mind, I classified the enzyme-membrane interaction as "chemical" rather than as "geometric." This is simply a matter of semantics.

4.8 Intervention of Simon

It is a heavy oversimplification to compare the activity of an enzyme in a membrane with its capability to bind a substrate in a bulk system. A more relevant parameter is the free energy of transfer for a substrate molecule to pass from an aqueous phase into the membrane phase and the binding sites of an enzyme. As an example, the ion selectivity of some ionophores changes drastically when comparing the selectivity of a carrier-modified membrane with the selectivity of the same carrier in bulk systems such as water.

4.9 Intervention of Korenstein

The insertion of a protein within a membrane not only stabilizes the optimal conformation for the enzyme activity, but also limits the possible conformations it may undergo. Thus, assuming a mechanism for the enzymatic activity, where the underlying assumption is that the conformational fluctuation controls the chemical reaction, a decrease in the number of degrees of freedom for the enzyme increases its probability for obtaining the active fluctuation needed for catalysis. Moreover, lipid-protein or protein-protein interactions within the membrane may decrease the rates of protein conformational changes. In those cases, where the conformational-change rates are decreased to the same order of magnitude as those of the chemical rates, a modulation of the chemical reaction by the conformational change is produced. Thus when an electric potential difference is established across the membrane, a conformational change is induced in the protein, so that the chemical reaction is now modulated by the electric field.

5. Intervention of Welch

I have a question that might be posed to both Professor Hammes and Professor Hess. This concerns the nature of the hydrophobic microenvironment surrounding many membrane-associated enzyme systems. For example, what do we know of the dielectric properties of these phases?

I was thinking particularly of electrostatic interactions between enzyme residues and substrate molecules. Let us compare the hydrophilic cytoplasmic phase (say, with dielectric constant $\epsilon = 80$) and the hydrophobic regions within membranes (say, with $\epsilon = 2$). Is it possible that protein-substrate interactions may be enhanced in certain membrane-associated enzyme schemes? That is, might specific intermolecular forces play a more significant role in influencing the site-to-site migration of intermediate substrates, as compared to the same system in the hydrophilic phase? [R. Coleman, *Biochim. Biophys. Acta*, **300,** 1 (1973); P. A. Srere and K. Mosbach, *Ann. Rev. Microbiol.*, **28,** 61 (1974); and H. Fröhlich, *Proc. Nat. Acad. Sci.* (*U.S.*), **72,** 4211 (1975).]

5.1 Intervention of Hammes

The question as to what the dielectic constant is of a membrane is difficult to answer because a membrane is of molecular dimensions whereas a dielectric constant is well defined only for a macroscopic system. Perhaps the essential point to note is that electrostatic interactions clearly are of great importance in membrane-protein interactions.

In general, the strength of substrate binding to membrane-bound enzymes spans a range similar to that encountered with soluble enzymes. The presence of a membrane can either enhance or inhibit substrate binding. The solubilized ATPase and the reconstituted enzyme have only slightly different Michaelis constants.

6. Intervention of Caplan

Perhaps it is worthwhile reemphasizing at this point, although this has been touched on before, the role of the membrane as a coupling device between enzymes in certain multienzyme systems. For example, the proton pump associated with the respiratory enzymes is coupled to the proton-driven ATPase in mitochondria by means of the membrane to which they are mutually attached, although at different points. The proton is pumped out at a site in one region of the membrane and reenters at a site in another region of the membrane. Thus it actually "circulates" between the two sites, and this provides the coupling. The same is true of photophosphorylation in *H. halobium*, where protons circulate between purple membrane patches containing bacteriorhodopsin and apomembrane regions containing the ATPase.

6.1 Intervention of McConnell

Van Deenen and colleagues have shown (*Biochim. Biophys. Acta* **406,** 169, 1975) that the activity of a phospholipase against phosphatidylcholine liposomes has a sharp temperature maximum at the phase transition temperature. In addition, Linden et al. (*Proc. Natl. Acad. Sci. US* **70,** 2271, 1973) showed a marked temperature dependence of sugar transport into *E. Coli* fatty acid auxotrophs at the temperature corresponding to the onset of phase separations of the lipids.

6.2 Intervention of Eisenman

One important effect of a membrane on an embedded enzyme has not been mentioned, namely, the effect of perturbing the concentration presented to the enzyme at the membrane-solution interface from the concentration present in the bulk solution. This is particularly clear for (1) Gouy-Chapman surface potential effects due to charged polar head groups that affect the concentrations of such charged species as H^+ and ATP; (2) dipolar potential jumps across the interface, which can be substantially different between neutral lipids (e.g., 160 mV between phosphatidyl ethanolamine and glycerol monooleate); and (3) repulsive "image potential" effects due to the energy of transfer of charged species into the less polar environment inside the membrane, as just mentioned by Simon.

6.3 Intervention of Hammes

I agree with Professor McConnell that phospholipid phase transitions may play a role in controlling the activity of a membrane-bound enzyme. However, the case cited is somewhat ambiguous, since porcine phospholipase A_2 is a soluble enzyme acting on a phospholipid surface. The major effect of the phase transition in this case is to alter the nature of the substrate rather than the intrinsic catalytic activity of the enzyme.

6.4 Intervention of McConnell

Metcalf and collaborators (*Biochemistry* **15,** 4145, 1976) have shown that the activity of the Ca^{2+} ATPase from rabbit sarcoplasmic reticulum is dependent on lipid composition.

6.5 Intervention of Bangham

Dawson and I (*Biochim. Biophys. Acta* **59,** 103, 1961) studied the interaction of two phospholipases with their common substrate, lecithin, and concluded that the enzyme must extend an active site ("masculine" enzyme) into the planar lecithin monolayer or bilayer, respectively, in order to reach the ester bond. The phospholipase A_2 from penicillin notaton required a net negative charge at the substrate surface, the α toxin from *Cl. welchii,* a positive one, to facilitate penetration.

7. Intervention of Williams

I am disturbed that following Professor Hammes' presentation the language in the discussion has changed. While discussing enzymes we looked at molecule properties closely related to the discussion of small molecules (e.g., atom position and motions). Now in the discussion of complex enzymes, especially in membranes, we have started to use bulk properties (e.g., we talk of phases, dielectric constants, Chapman-Gouy theory, etc.). Is it the view of the discussants that events in membrane–coupled-enzyme systems cannot be described by molecular events because of the complexity of the system resulting from extensive cooperativity within the membrane (e.g., between lipids and proteins)?

7.1 Intervention of Ross

In principle one should not, of course, use a macroscopic concept like dielectric constant in a complex of molecular dimensions. But in the theory of electrolyte solutions this has been done for some time and will continue to be done until we have something better and more convenient.

7.2 Intervention of Somorjai

I think Dr. Williams' avowed pessimism is inevitable if one is trying to explain all biological phenomena in terms of molecular, microscopic concepts. But molecular description can be a Procrustean bed when dealing with complex, intrinsically macroscopic phenomena, because simple interpretability may not be feasible. In fact, the selection of a few essential macroscopic variables from a vast number of microscopic variables (or their combinations) is crucial not only for understanding via *simplification,* but also because collective variables (order parameters) tend to obey qualitatively different rules or laws that are not obvious in, or even deducible from, a microscopic description. The power of alternate description, signaling or leading to qualitatively different behavior, has been convincingly demonstrated by the work of Professor Prigogine's group. Macroscopic descriptions may be phenomenological and must be approximate, yet they lead to understanding of experimental observations that would be inconceivable in terms of more exact microscopic theories. Theoretical physics, where such theories are very much in vogue, nevertheless realized the importance and inevitability of macroscopic descriptions, as evidenced by the success of the theory of superconductivity, to mention but one example. Biology, an inherently more complex discipline, needs the power of macroscopic description even more (whenever it is appropriate), despite the success of molecular theories.

7.3 Intervention of Ubbelohde

Membranes certainly introduce cooperative processes, so that a merely molecular approach will not be enough, particularly with reference to boundary conditions. Whether a "cell" is large enough, on the other hand, to justify statistical averaging as implied by such terms as *phase* and *dielectric* field may involve quite a profound distinction. As a speculation, a cell diameter might be conditioned by the natural mode interval in diffusive systems and *phase* is not a justifiable term. A related question is whether the thickness of a membrane measured in molecular dimensions can play an important role structurally or whether a membrane behaves merely as an indefinitely thin boundary.

8. Intervention of Williams

Is it possible for Professor McConnell to give a description of his observed λ-point transition on a molecular basis, or does he believe that the complexity is so great that he can only describe the system in general functional thermodynamic terminology? It is sometimes possible to describe λ-point transitions in pure solids (e.g., NH_4Cl) in a molecular fashion, but can we hope to do this for a biological membrane?

8.1 Intervention of Ross

It seems we have the same problem with any phase transition, and certainly in some a molecular description has been useful.

8.2 Intervention of McConnell

Drs. J. Owicki and M. S. Springate and I are currently attempting to apply liquid crystal continuum models to account for some of the effects of temperature on the activity of membrane-associated biochemical processes. This approach is also being used to estimate how close the observed temperatures are to critical temperatures. Few if any of the observed transitions have been shown to be second-order or λ-transitions. Information about molecular changes during these phase transitions is already available. For leading references, see H. M. McConnell, "Molecular Motion in Biological Membranes," in *Spin Labeling: Theory and Applications*, L. Berliner, Ed., Academic Press, New York, pp. 525–560, 1976.

9. Intervention of Eisenman

I wish to respond to Professor Ubbelohde's question regarding what thinness, per se, of biological membranes could be important. For ion movements across membranes as mediated catalytically by carriers or channels, the thinness permits local deviations from the electroneutrality constraint that, for example, enables neutral molecules to carry cations across the membrane as charged species, leaving their counterions behind in the aqueous solutions. This is not possible when the thickness of the system becomes large.

10. Intervention of Somorjai

I would like to emphasize the importance of statistical fluctuations in enzymes. The point is that an individual enzyme molecule, large as it may be when viewed as a *single molecule*, is finite and indeed very small when considered as a *statistical ensemble* of atoms. Thus transient fluctuations in solution are frequent, large, and inevitable, even at thermodynamic equilibrium. In fact, root mean square fluctuations of internal energy or volume can be estimated (A. Cooper, Proc. Nat. Acad. Sci. US **73,** 2740, 1976) in individual enzyme molecules and turn out to be surprisingly large, comparable to changes observed on thermal denaturation. Furthermore, the *most probable* value for the internal energy of a single enzyme is estimated to be higher than its *mean* energy! Now if these fluctuations are or can be confined to a small region of the enzyme (active site?), then they could produce appreciable conformational and/or energy changes there that could be crucial for enzyme action. The important question to be settled is the specific molecular form these giant fluctuations take, in particular their correlation pattern in both space and time. To put it differently, we have to find a plausible mechanism by which an enzyme converts the random influx of translational energy from its surroundings into coherent vibrations that might channel this energy into the active site region (Damjanovich's energy funnels). We are in the process of formulating such a mechanism, based on concepts that derive from nonlinear wave propagation theory (Solitons).

10.1 Intervention of Welch

With regard to the comments just made by Professor Somorjai, it may be appropriate to note that some experimental and theoretical evidence does suggest a role for low-frequency, thermal fluctuations in protein structure in enzyme catalysis.

References

1. K. G. Brown, S. C. Erfurth, E. W. Small, and W. L. Peticolas, *Proc. Nat. Acad. Sci. (U.S.)*, **69,** 1467 (1972).
2. G. Kemeny, *Proc. Nat. Acad. Sci. (U.S.)*, **71,** 2655 (1974); *J. Theor. Biol.*, **48,** 231 (1974).
3. J. L. Shohet and S. A. Reible, *Ann. N.Y. Acad. Sci.*, **227,** 641 (1974).
4. H. Fröhlich, *Proc. Nat. Acad. Sci. (U.S.)*, **72,** 4211 (1975).
5. R. Lumry and R. Biltonen, in *Structure and Stability of Biological Macromolecules*, S. N. Timasheff and G. D. Fasman, Eds., Marcel Dekker, New York, 1969.

10.2 Intervention of Ross

Are these fluctuations in an equilibrium situation or under nonequilibrium conditions?

10.3 Intervention of Somorjai

I like to look at enzymes as open systems and consider the active site as analogous to a dissipative structure. This structure is triggered (excited parametrically) by the approach of the appropriate substrate.

10.4 Intervention of Welch

A *molecular enzyme-kinetic model*, formulated recently by Professor Damjanovich and his colleague Dr. Somogyi [*Acta Biochim. Biophys. Acad. Sci. Hung.*, **6,** 353 (1971); *J. Theor. Biol.*, **41,** 567 (1973): *J. Theor. Biol.*, **51,** 393 (1975)], takes into account specific properties of protein structure responsible for the uptake of thermal energy (e.g., by translational-vibrational transitions) and its "channeling" to the active center. Their "energy funnel" mechanism supposes a coupled oscillator system, depending on the structure of the protein, which conveys in an anisotropic fashion some part of the collisional energy from exterior sites to the active center. Green and Ji (*Proc. Nat. Acad. Sci. (U.S.)*, **70,** 904 (1973); *Ann. N.Y. Acad. Sci.*, **227,** 6 (1974)] developed a similar idea along different lines and applied it specifically to mitochondrial electron transport. According to the latter model, the catalytic process requires that an internal protein coordinate (e.g., the separation of two charged moieties) at the active center reach a critical extension. This crucial coordinate is assumed coupled specifically to low-frequency, collective vibrational modes of the protein structure [K. G. Brown, S. C. Erfurth, E. W. Small, and W. L. Peticolas, *Proc. Nat. Acad. Sci.* (*U.S.*), **69,** 1467 (1972)]. Certain structural properties of proteins (e.g., "fluidity" in the region of the active center) are consistent with such models; compare the following references:

References

1. R. Lumry and R. Biltonen, in *Structure and Stability of Biological Macromolecules*, S. N. Timasheff and G. D. Fasman, Eds., Marcel Dekker, New York, 1969.
2. F. M. Richards, *J. Mol. Biol.*, **82,** 1 (1974).
3. C. Chothia, *Nature*, **254,** 304 (1975).
4. J. L. Shohet and S. A. Reible, *Ann. N.Y. Acad. Sci.*, **227,** 641 (1974).

11. Intervention of Hess

In biological membranes, according to the Mitchell hypothesis, the generation of a proton gradient might well be a property for the production of ATP by ATPase. Therefore membrane-bound systems, which are able to generate a gradient, are of great interest. Recently, a relatively simple bacteriorhodopsin was discovered by Oesterhelt and Stoeckenius [D. Oesterhelt and W. Stoeckenius, *Nat. New Biol.*, **233,** 149 (1971)]. This pigment is located in the cellular wall of salt bacteria Halobacterium halobium. The pigment was isolated in the form of membrane fragments and could be shown to release and bind protons in the isolated form as well as in the intact bacteria [D. Oesterhelt, and B. Hess, *Eur. J.*

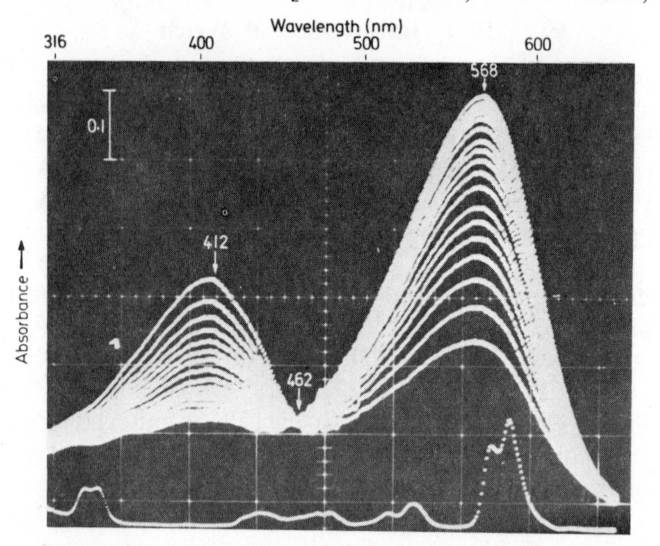

Fig. 1 Time series of spectra taken immediately after bleaching the purple membrane with white light with a rapid scanning photometer. The lower spectrum represents a filter spectrum for calibration. (Courtesy of *Eur. J. Biochem.*)

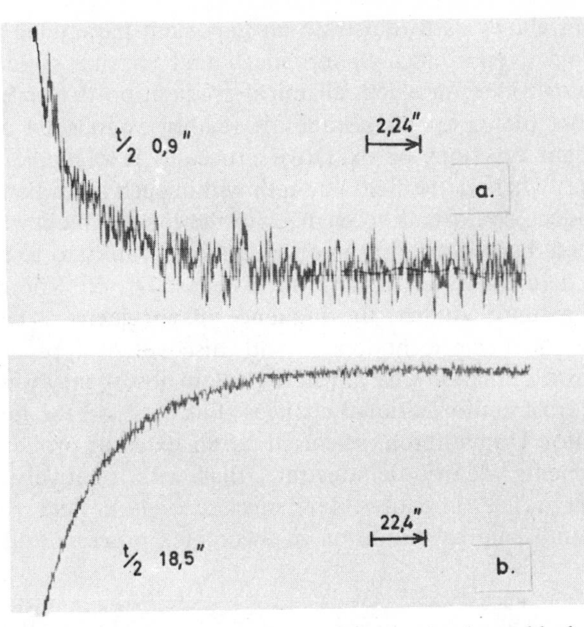

Fig. 2. Deprotonation and protonation of bacteriorhodopsin triggered by laser-excitation at 580 nm in a pulse of 100 nsec length. Methylumbilliferon was used as pH indicator. (*a*) deprotonation recorded at −53°C, (*b*) protonation recorded at −34°C.

Biochem., **37,** 316–326 (1973), D. Oesterhelt, and W. Stoeckenius, *Proc. Nat. Acad. Sci.* (*U.S.*), **70,** 2835–2857 (1973)].

The proton cycle is coupled to the photochemical cycle in an approximately one-quantum process with two major components shown in Fig. 1 (from D. Oesterhelt, and B. Hess. Courtesy of *Eur. J. Biochem.*). The purple pigment absorbing at 568 nm at maximum is the protonated component occurring in the dark. On illumination the sample is bleached with the appearence of a composed pigment absorbing at 412 nm in a process passing through several intermediates and releasing a proton. In the dark the purple pigment is regenerated, coupling to a protonation process. The protonation and deprotonation reaction can fluorometrically be recorded and time resolved as a two-exponential process, as shown in Fig. 2. It is interesting to note that the intramembrane protein structure is a prerequisite of the deprotonation-protonation cycle, and the functional and structural stability of the rhodopsin membrane system is independent of the physical and chemical state of the bulk medium. Indeed, the intramolecular buffering of the system allows the photochemical reaction to occur between pH 9 and 2 without losing overall properties (B. Hess and D. Oesterhelt, The Photochemical Cycle of Bacteriorhodopsin in *Halobacterium halobium*, 10th FEBS Meet. Paris, Abstr. 1207, 1975).

This system allows us to illustrate an important parameter that might well affect and even control membrane-bound enzyme systems. It has long been known that an electrochemical gradient on the order of 50 to 200 mV across biological membranes is readily established as soon as energy via light reactions or oxidative processes is available. Here the problem arises whether the field strength within such a gradient is strong enough to induce conformation changes in the located membrane-bound species. Indeed, bacteriorhodopsin offers an opportunity to test the effect of electrical field pulses on membrane patches. Dr. R. Korenstein and myself have recently studied the influence of an electrical field on the conformation of bacteriorhodopsin and discovered appreciable field effects as shown in Fig. 3*a* and *b*, where protein absorbancy was recorded as an indicator of conformational changes (Fig. 3*a*) and the fluorescence of the indicator Umbilliferon was used as an external probe (Fig. 3*b*). Both experiments clearly demonstrate that with relatively low field strength after a lag time of a few microseconds a field response is established almost up to saturation in a complex process. Following the

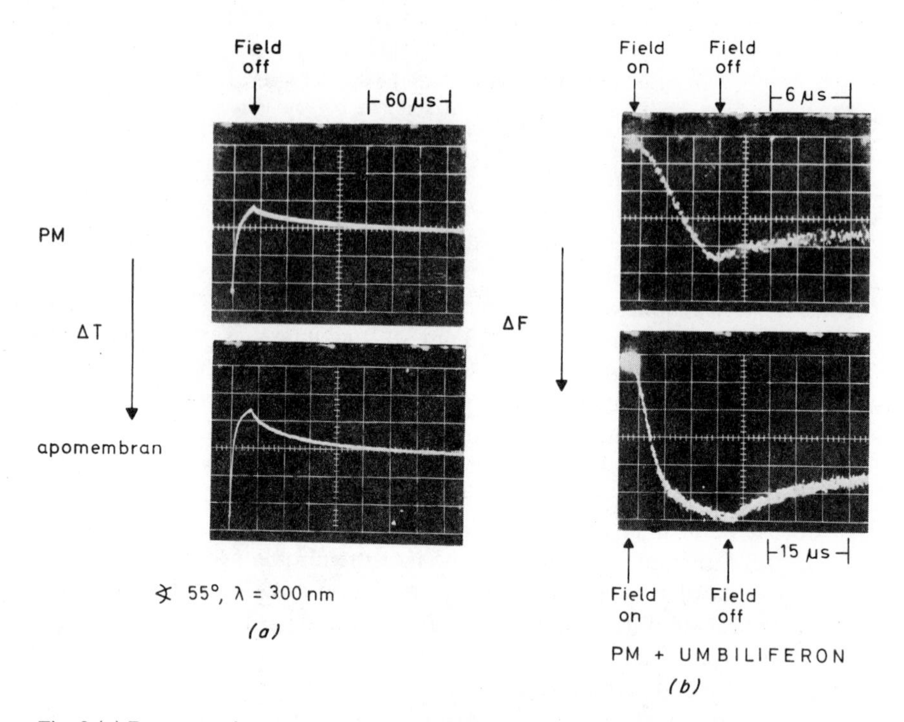

Fig. 3.(*a*) Response of the purple membrane as well as apomembrane following an electrical field pulse of 20 kV/cm, measuring light 300 nm with polarization angle 55°. (*b*) See above, fluorescence difference was recorded as indicator of umbilliferon.

field pulse the system relaxes with a series of exponentials over a rather long period of time to its original state. It is obvious that such a condition would strongly affect a mechanism of proton conductivity as well as proton binding via the generation of induced dipoles within the molecule. As shown in Fig. 3*a*, the field effect not only can be demonstrated with the intact purple membrane system, but also is observed in the absence of the retinal moiety from the protein, which was prepared according to Oesterhelt in form of a so-called apo-membrane (Hess and Korenstein, Katzir-Conference, Paris 1976).

11.1 Intervention of Ubbelohde

What is the potential gradient across the membrane?

11.2 Intervention of Caplan

We have actually measured the electrical potential differences in intact purple membrane bacteria, using the distribution of a lipid-soluble cation as an indicator, and found values of up to 100 mV [E. Bakker, H. Rottenberg, and S. R. Caplan, *Biochim. Biophys. Acta,* **440,** 557 (1976)]. These values obviously correspond to enormous potential gradients across the membrane.

11.3 Intervention of McConnell

About 10 years ago Wayne Hubbel and I examined the "fluidity" of a large variety of membranes using spin labels. Excitable membranes were very fluid by our criteria; the most rigid membrane we ever studied was the "purple" membrane supplied to us by Walter Stoeckenius, which was like *concrete* on our scale of fluidity. At that time I felt that this membrane could only translocate electrons or protons, because of this rigidity (see Ref. 31 in my report and earlier references contained therein).

DIFFUSION-REACTION IN STRUCTURED MEDIA AND MEMBRANES BEARING ENZYMES

DANIEL THOMAS

Laboratoire de Technologie Enzymatique Compiègne, France

Within the living cell the great majority of the enzymes are attached to membrane structures or contained in cell organelles. When enzymes are isolated, they are removed from their natural state and quite often are highly unstable. The artificial binding of enzymes into membranes makes possible study of the interaction between diffusion and enzyme reaction within a well-defined context.[1]

In this way Goldman et al.[2–5] incorporated papain and phosphatase in collodion membranes, and other authors[6–7] have done likewise with several enzymes in cellophane membranes. It is important to note the work of Suzuki[8] on collagen membranes as well as the work on enzyme reactors[9,10]

These methods produce active artificial membranes, but the active site distribution is not well defined and the theoretical treatment is difficult.

In order to give a homogeneous distribution of enzyme molecules inside the membrane, it was necessary to synthesize the membrane and to incorporate the enzymes at the same time. The co-cross-linking of enzyme molecules with an inert protein appears to be a proper solution. Purely active proteic films were created by using this procedure.[11–12] These artificial enzyme membranes can be used in the study of heterogeneous enzyme kinetics and for modeling biological membranes. The phenomena in the enzyme membranes can be classified in two parts.

I. THE EFFECT OF THE COMPOSITION AND STRUCTURE OF THE MEMBRANE ITSELF ON THE ENZYME MEMBRANE BEHAVIOR

In reference to this, it is possible to describe examples of facilitated transport and active transport.

Facilitated transport of CO_2 with a membrane-bearing carbonic anhydrase was described by Broun et al.[13] In this system a hydrophobic membrane separates two compartments containing buffer solutions. This membrane is permeable to gases like CO_2 but impermeable to water and electrolytes. CO_2 diffusion velocity was measured for a membrane without enzyme, with grafted enzyme on one side, and with grafted enzyme on both sides. With enzyme the apparent permeability increased two times and four times, respectively. The concentration profiles of CO_2 with and without enzyme can explain this phenomenon.

An active glucose transport effect occurs with a bienzyme membrane composed of two active protein layers and two selective films (14–15). The active enzyme films carry, respectively, hexokinase and phosphatase co-cross-linked with an inert protein. Both are impregnated with ATP and covered on their external sides by two selective films permeable to glucose but impermeable to glucose-6-phosphate. In this asymmetrical membrane glucose is temporarily phosphorylated and the system behaves chemically as a simple ATPase. In the first layer glucose is a substrate and G-6 P diffuses along its own concentration gradient into the second layer, where glucose is a product. The glucose and glucose-6-phosphate concentration profiles explain how this diffusion reaction is converted into a pumping phenomenon.

When the concentration in the donor compartment remains constant during the experiment, the concentration in the receptor compartment increases regularly at first, then attains a plateau. For an ideal system with the thickness of a biological membrane a maximum increase in concentration of 130 times can be predicted by computer simulation. This system gives an experimental physicochemical example of a transformation of scalar chemical energy into a "vectorial catalysis effect."

II. THE EFFECT OF THE REACTANT CONCENTRATION DISTRIBUTIONS ON THE ENZYME MEMBRANE BEHAVIORS

First of all, the behavior of the enzymes in the membrane differs markedly from the behavior of the unbound enzymes in solution. It is pertinent to note that the medium in which the enzyme bound to a membrane acts might be determined not only by the composition and structure of the membrane itself, but also by the local concentration distribution of substrate and products. The microenvironment in the membranes is the result of a balance between the flow of matter and enzyme reactions. The substrate and product concentrations in the membrane differ from point to point across the membrane and also from those at the outer solution. By electron microscopy this was experimentally demonstrated beyond doubt with the DAB-peroxidase system by Barbotin and Thomas.[16] The effects of these profiles were studied with monoenzyme membranes by Goldman et al.[2–5] and Thomas et al.[11] A study of activity as a function of inhibitor concentration for enzyme membrane with different Thiele moduli was done. An enzyme in a membrane is less sensitive to the effect of an inhibitor. In contrast, the effect of the reaction reversibility on an enzyme reaction is increased by the diffusion limitation. These problems were studied experimentally and by simulation on computer.[17] The profiles are still more important when nonlinearity of the enzyme reactions can produce memory, oscillation, and spontaneous structuration.

A. Memory in Enzyme Membranes[18]

A coating bearing one enzyme (papain) is produced on the surface of a glass pH electrode by the method previously introduced (co-cross-linking). The papain reaction decreases the pH, and the pH-activity variation gives an autocatalytic effect for pH values greater than the optimum; under zero-order kinetics for the substrate (benzoyl arginine ethyl ester) the pH inside the membrane is studied as a function of the pH in the bulk solution in which the electrode is immersed. A hysteresis effect is observed and *the enzyme reaction rate depends not only on the metabolite concentrations, but also on the history of the system.*

B. Oscillatory Behavior of Enzyme Membranes[19]

When the same kind of electrode is introduced in a solution with a high pH (i.e., pH = 10) and a lower substrate concentration (first order kinetics), an oscillation in time of the measured pH inside the membrane spontaneously occurs. This enzyme, which has been extensively studied, does not give oscillation for any conditions of pH and substrate concentration. The period of oscillation is around one-half minute, and the oscillation is abolished by introducing an enzyme inhibitor. The phenomenon can be explained by the autocatalytic effect and by a feedback action of OH^- diffusion in from the outside solution. The diffusion of this ion is quicker than the diffusion of the substrate. There is a qualitative agreement between the computer simulation and the experimental results.

C. Spontaneous Structuration in Enzyme Membrane[20]

An artificial membrane bearing two different enzymes (glucose oxidase and urease) in a spatially homogeneous fashion is produced by using the method previously described. The glucose oxidase reaction decreases the pH, and the urease increases the pH. The pH activity profiles show an autocatalytic effect for the glucose oxidase in the range of pH values greater than the optimum, and for the urease, smaller than the optimum pH. When the two enzymes are mixed together the global pH variation is zero for one well-defined pH value.

The active membrane separates two compartments and it is possible to get this pH value throughout the system, in presence of the two substrates, by the transient use of a buffer. The pH values outside are controlled and H^+ fluxes measured by pH-stat systems. After small asymmetrical perturbations of the pH values at the boundaries (0.05), an inhomogeneous pH distribution arises spontaneously inside the membrane. The initial perturbations are amplified and the pH values in the compartments tend to evolve in opposite directions. The H^+ fluxes entering and leaving the membrane can be determined by pH-stat measurements. If the boundary pH values are not maintained constant by a pH stat, the system evolves to a new stationary state characterized by a pH gradient of two pH units across the membrane.

D. Interaction between Enzyme Activity and the Potential Difference at the Boundaries of the Membrane

For the first time, the use of artificial enzyme membranes allows the study of the interaction between enzyme activity and membrane potential in a well-defined context. Before the recent progress in manufacturing artificial membrane bearing immobilized enzyme, Blumenthal et al.[21] described a system in which a papain solution was sandwiched between two cation and anion exchange membranes. Under short-circuit conditions the system was able to generate a current. A nonequilibrium thermodynamic analysis was developed by the authors.

David et al.[22] have described recently some studies dealing with a urease membrane. The enzyme transforms a neutral molecule, urea, into two ions, ammonium carbonate. The reaction gives both a modification of the ion concentration and an increase of the pH. The membrane was studied between two compartments containing, respectively, 10^{-3} *M* and 10^{-2} *M* phosphate solution at pH 5.1. The potential is recorded at the boundaries of the membrane after introducing the substrate; in both compartments a jump of the potential difference is observed. After introduction of an enzyme inhibitor the potential difference returns to the initial value. The phenomenon of potential modification is truly linked to the enzyme reaction. The steady-state membrane potential was studied both at constant substrate concentration as a function of the ion concentration gradient and at constant ion gradient as a function of the substrate concentration. It is especially important to discuss the last point. The relationship between the potential and the substrate concentration is a sigmoid curve recalling the behavior of some biological membranes.[23]

The coupling between a simple enzyme reaction and a structure is able to explain the "cooperative" effect observed with excitable membranes.

References

1. D. Thomas and G. Broun, *Methods in enzymology*, **44,** 901 (1976).
2. R. Goldman, H. I. Silman, S. R. Caplan, O. Kedem, and E. Katchalski, *Science*, **150,** 758 (1965).
3. R. Goldman, O. Kedem, H. I. Silman, S. R. Caplan, and E. Katchalski, *Biochemistry*, **7,** 486 (1968).
4. R. Goldman, O. Kedem, and E. Katchalski, *Biochemistry*, **7,** 4518 (1968).
5. R. Goldman, O. Kedem, and E. Katchalski, *Biochemistry*, **10,** 165 (1971).
6. G. Broun, S. Avrameas, E. Selegny, and D. Thomas, *Biochem. Biophys. Acta*, **185**, 260 (1969).
7. E. Selegny, S. Avrameas, G. Broun, and D. Thomas, *C.R. Acad. Sci. C.* **266,** 1931 (1968).
8. I. Karube and S. Suzuki, *Biochem. Biophys. Res. Commun.*, **47,** 51 (1972).
9. A. Bareli and E. Katchalski, *J. Biol. Chem.*, **238,** 1690 (1963).
10. M. D. Lilly, W. E. Hornby, and E. M. Crook, *Biochem. J.*, **100,** 718 (1966).
11. D. Thomas, G. Broun, and E. Selegny, *Biochimie*, **54,** 229 (1972).
12. G. Broun, D. Thomas, G. Gellf, D. Domurado, A. M. Berjonneau, and C. Guillon, *Biotechnol. Bioeng.*, **15,** 359 (1973).
13. G. Broun, E. Selegny, C. Tran Minh, and D. Thomas, *FEBS Lett.*, **7,** 223 (1970).
14. D. Thomas, C. Tran Minh, G. Gellf, D. Domurado, B. Paillot, R. Jacobsen, and G. Broun, *Biotech. Bioeng. Symp.*, **3,** 299 (1972).
15. G. Broun, D. Thomas, and E. Selegny, *J. Membrane Biol.*, **8,** 373 (1972).
16. J. N. Barbotin and D. Thomas, *J. Histochem. Cytochem.*, **22,** 1048 (1974).
17. D. Thomas, C. Bourdillon, G. Broun, and J. P. Kernevez, *Biochemistry*, **13,** 2995 (1974).
18. A. Naparstek, J. L. Romette, J. P. Kernevez, and D. Thomas, *Nature*, **249,** 490 (1974).
19. A. Naparstek, D. Thomas, and S. R. Caplan, *Biochem. Biophys. Acta*, **323,** 643 (1973).
20. D. Thomas, A. Goldbeter, and R. Lefever, unpublished results.
21. R. Blumenthal, S. R. Caplan, and O. Kedem, *Biophys. J.*, **7,** 735 (1967).
22. A. David, M. Metayer, D. Thomas, and G. Broun, *J. Membrane Biol.*, **18,** 113 (1974).
23. J. P. Changeux and J. Thiery, in *Regulatory Function of Biological Membranes*, Vol. 11, J. Järnefelt, Ed., Elsevier, Amsterdam, 1968.

LIST OF INTERVENTIONS

1.1 Carafoli
1.2 Thomas
1.3 Clementi
1.4 Thomas
1.5 Hess
1.6 Thomas
2. Bangham
3. Sanfeld-Steinchen
3.1 Williams
3.2 Lefever

4. Simon
4.1 Thomas
5. McConnell
5.1 Thomas
5.2 Carafoli
6. Clementi
7. Caplan
8. Ross
8.1 Thomas
9. Goldanski
10. Caplan
11. Cerf
11.1 Mandel
11.2 Williams
11.3 Hammes
12. Ubbelohde
12.1 Thomas
12.2 Goldanski
12.3 McConnell
12.4 Goldanski
12.5 McConnell
12.6 Klotz
12.7 Goldanski
12.8 Ross
12.9 Goldanski

1.1 Intervention of Carafoli

Concerning your experiment on the measurement of intramembrane pH, pH indicators are charged molecules, which could conceivably migrate in or out of the membrane is response to electrical fields. Have you checked whether your indicators really stay in the membrane, and in the same place in the membrane, during the experiment?

1.2 Intervention of Thomas

The absence of oscillation in the bulk solution was checked by using both a second pH electrode and a dye. The oscillation inside the membrane is not an interaction between polyelectrolyte and glass surface; the effect is destroyed in the presence of an inhibitor of the enzyme activity.

1.3 Intervention of Clementi

In the very interesting report by Dr. Thomas, the diffusion aspect of chemicals through artificial membranes has been beautifully analyzed. However, I have some hesitation if one attempts to extend a concept proved for macroscopic (bulk) phenomena to microscopic (surface) ones. Let us start with the simple consideration that a biological membrane to a first approximation is built by a double layer of oriented phospholipids containing enzymes and proteins of nearly the same size as the membrane thickness. The membranes described by Dr. Thomas are of the order of microns in thickness and are made of cellulose or other nonphospholipid materials. Biological membranes are about 150±50 Å thick, and the *main* likely constituents are oriented phospholipids. Thus the diffusion in a biological membrane (*1*) is through the phospholipid bilayer (passive), (*2*) is through the phospholipid bilayer in proximity (or within) to the enzyme, (*3*) has a path (=membrane thickness) at most an order of magnitude larger than the chemical going through, (*4*) involves a chemical that "senses" the entire bilayer during diffusion.

The question is "How much can one infer from Dr. Thomas' diffusion experiments on the diffusion mechanism in large (bulk) artificial bilayer membranes about mechanisms on bilayer membranes of phospholipids with proteins inclusion?" Some part of the formulation may break down because we pass from bulk to surface only, from macro- to microdescription.

1.4 Intervention of Thomas

The scope of my contribution referred to diffusion in enzyme-membrane. The problem of protein-lipids interaction would demand a further lecture. With lipid-protein membranes, there are many phenomena, but diffusion still plays an important role.

1.5 Intervention of Hess

The elegant experiments of Dr. Thomas illustrate the very fact that the whole system is more than the sum of its parts and that chemical reactions coupled to diffusion under nonequilibrium conditions might indeed yield interesting properties, quite different from those observed under near equilibrium conditions. I would like to discuss the relevance of the Thiele modulus σ in terms of the structure and function of biological membrane-bound systems. First of all, with the well-known kinetic constants for soluble enzymes and their substrates, a σ-value of approximately 10^{-4} is computed, well below the order of magnitude reported by Dr. Thomas for his model systems. However, it is obvious that in membrane-bound biological systems the parameters might vary appreciably and should be given a better definition with respect to their physical meaning. With a thickness on the order of 100 Å for a biological membrane the diffusion term might well be modified if the dimensionality of the diffusion coefficient is reduced by a mechanism implying two- or even one-dimensional diffusion (lateral or across the membrane) (see B. Hess, above, also P. H. Richter and M. Eigen, *Biophys. Chem.*, **2,** 255 (1974). In addition, the kinetic term of the equation, although in soluble enzyme systems on the order of 100/sec, might well be different in membrane-bound systems because of the second-order reaction velocity constant as part of K_m being a function of the diffusion coefficient and its dimension. Indeed, second-order reaction velocity constants of some membrane-bound processes were found to be on the order of $10^9\ M^{-1} \times s^{-1}$ and to point even more to the occurrence of membrane-controlled lateral diffusion processes.

1.6 Intervention of Thomas

The value of Thiele modulus in biological membrane was evaluated by Professor Horvath of Yale. The value is not at all negligible. Because of the distances involved the Thiele modulus is still more important in lateral diffusion.

2. Intervention of Bangham

Dr. Thomas drew our attention to the fact that, although biological membranes are thin in comparison to his preparations, their diffusion constants may be long. I have measured permeability coefficients (transmembrane "diffusion") for a number of solutes for artificial phospholipid bilayers (liposomes). The values follow and are to be compared for calculated permeabilities for a solute diffusion across a comparable thickness of water.

	Permeability coefficient cm sec^{-1}	
Solute	*Membrane* (50 Å)	*Water* (50 Å)
H_2O	10^{-3}	~10
Malonamide	10^{-4}	~10
Glucose	10^{-8}	~10
Ions	10^{-13}	~10

These values represent several mechanisms, including simple diffusion through an oil phase (e.g., water and small nonelectrolytes). The values for large hydrophilic molecules may represent the frequency of membrane breakage and/or dislocation.

3. Intervention of Sanfeld-Steinchen

When dealing with two-dimensional formation of patterns in lipidic-proteic membranes (fluid membranes), not only does the coupling between the chemical reaction in the membrane and the surface diffusion have to be considered (i.e., the Thiele modulus), but one must also consider the coupling with the onset of convection (Navier-Stokes equation), through surface boundary conditions, taking into account the variation of the membrane surface tension with the concentrations of the various species reacting in the membrane. The formation of hexagonal patterns in the membrane can be explained by a kind of "Marangoni" effect, after a critical threshold of the concentrations of the reacting species, in far from equilibrium conditions, and for nonlinear mechanisms of reactions. We have performed the calculations for such a coupling between surface chemical reactions, surface diffusion, and surface convection in the frame of a very simplified model of membrane (a Gibbs division surface model), and we have shown the possibility of obtaining such kinds of two-dimensional stationary patterns and time-oscillating behavior as well.

References

1. A. Sanfeld and A. Steinchen-Sanfeld, *Biophys. Chem.*, **3,** 99 (1975).
2. F. Deyhimi and A. Sanfeld, *C.R. Acad. Sci. Paris,* **279,** 437 (1974).
3. M. Hennenberg, I. S. Spreusen, A. Steinchen-Sanfeld, A. Sanfeld, *J. Chim. Phys.*, **72,** 1202 (1975).
4. J. S. Sorensen, M. Hennenberg, A. Steinchen-Sanfeld, and A. Sanfeld, *J. Coll. Int. Sci.*, **56,** 191 (1976).

3.1 Intervention of Williams

Can the internal rotary motions of fluids observed in many cells be related to the convection discussed by Dr. Sanfeld?

3.2 Intervention of Lefever

Dr. Sanfeld refers to phenomena that concern membranes and surface behaviors rather than bulk solutions; whether such behaviors play a role in the motion of fluids in cells remains an open question. On the other hand, the coupling of chemical and hydrodynamic effects is not a necessary condition to the existence of rotations and macroscopic transport of matter. We have seen in the lecture of Professor Prigogine that such effects can result more simply from the coupling of chemical reactions with transport phenomena like diffusion.

4. Intervention of Simon

I can easily understand a thickness of the order of 50 to 100 μm for the unstirred diffusion layers of the flat and thin macromembranes you discussed. In microsystems such as mitochondria with diameters around a few μm these unstirred layers must, however, be considerably smaller. Would you please comment on this substantial difference between the model membranes you studied and actual biological membrane systems?

4.1 Intervention of Thomas

The size of the unstirred layer does not depend on the thickness of the membrane. The water "does not know" what is the thickness of the membrane.

The idea of convection was introduced, which is not in contradiction with Thiele principle; it is possible to take into account not only the diffusion itself, but also the mass transfer phenomena (electric field, convection, etc.).

5. Intervention of McConnell

For bilayer membrane problems it may be a mistake to treat permeation through the membrane as a single diffusion process. In the first place it is extremely unlikely that the distribution of permeant molecules across the membrane is described by the diffusion equation. Second, the permeation may be related to lateral density fluctuations in the membrane, giving a quite nonuniform lateral distribution of the permeant molecules near the membrane surface at any instant.

5.1 Intervention of Thomas

Our work deals with the necessity of creating kinetics laws for heterogeneous enzymology. There was a big gap between the classical enzyme kinetics in solution and highly structured biological systems. All the concepts of diffusion reaction are clear for our thick membrane but are also useful for lipid-protein membranes, even if the process of transport is not only classical diffusion.

5.2 Intervention of Carafoli

On the matter of the unstirred layer, it must be remembered that model membranes are static structures, whereas natural membranes are highly dynamic; that is, they continuously move in their normal environment. This is certainly going to disturb the formation and the maintenance of unstirred layers more than a few molecules thick.

6. Intervention of Clementi

It seems that there is a need to reexamine, some of the basic quantities used in transport processes, like Thiele numbers, attempting to connect them to more chemical quantities. For example, the macroscopic quantity, ϵ the dielectric constant, can be interpreted in terms of dipole moment distribution, and the dipole moment has immediate structural implications. Now to talk of a dielectric constant in the interaction of *two* atoms would be a rather useless exercise, since the dilectric constant is a continuous matter concept, not a discrete matter concept. In the same way there is a "limiting size" criterion that must be used when we use continuous matter terminology. Above the critical size the concept and definitions do make sense; below, they must be translated to a discrete (atomic or molecular) language.

7. Intervention of Caplan

I want to make three brief remarks. First, if we are discussing convective effects I think mention should be made of certain related thermal effects. In membranes in which ion flows are monitored by electrical current measurements, such as black lipid films and excitable tissues, the occurrence of a relatively high level of noise ("flicker noise") is well known. Recent studies by Lifson and Gavish (as yet unpublished) indicate that this noise may be due to the formation of an extremely thin depletion layer on one side of the membrane, with a corresponding enrichment layer on the other. The depletion layer may become completely desalinated, and the resulting rise in resistance leads to substantial local heating. This in turn results in hydrodynamic instability and turbulent local convection, which gives rise to the noise. Second, I refer to the point raised by Dr. Lefever with regard to the possible existence of a critical value for the Thiele modulus beyond which oscillations can take place. I do not know the answer to this in general, but in the case of the papain oscillator it should be realized that the kinetics are not described by a simple Michaelis-Menten relation, and the effective Thiele modulus is pH dependent. We have studied the stability "map" of the system in the substrate concentration versus pH plane and find that oscillations can in fact occur only in an extremely limited region of this plane (Hardt, Segel, and Caplan, unpublished results). Third, with regard to the suggestion by Dr. Hess that the highly regular "pattern" or arrangement of bacteriorhodopsin in the purple membrane represents a dissipative structure, I think one must be cautious about the use of such a description. The purple membrane is extremely stable; one can keep preparations for periods of time of the order of a year, in the absence of any energy source, without any noticeable loss of function. One can prepare functional subbacterial particles as well as artificial liposomes containing purple membrane, and the function remains intact, although no ATP is present. It seems to me that the pattern of the purple membrane is a stable structure that results from a minimization of free energy.

In reply to Dr. Hess I may add that, although energy is admittedly required to construct the bacteriorhodopsin molecules in the first place, the proteins are not assembled inside the membrane. They enter the membrane and form an equilibrium structure. On another point, ordering inside the membrane is analogous to crystallization.

8. Intervention of Ross

With regard to turbulence, in some cases an additional complication may have to be considered in that the temperature may vary locally, and then thermal conduction is important. It is possible to vary the constraints on the systems you have studied (e.g., concentration) so that by variation of one or more of these constraints both hysteresis and no hysteresis (at different constraints) can be observed in the same system. This is important to make sure that the observed hysteresis is due to the nonlinear kinetics and not due to other reasons.

8.1 Intervention of Thomas

With well-defined values for the parameters values of the enzyme membrane itself it is only possible to get hysteresis phenomena. By using inhibitors or activators it is possible to modulate the enzyme activity in the membrane and in this case the hysteresis loop can be generated or be absent.

The behavior of papain, for example, is very well known in solution and no hysteresis or oscillation is possible. The new behavior is only due to the adding of diffusion limitations.

In your catalysis it is a problem of surface; in our case it is a problem of active sites, but in a similar way by poisoning the enzyme sites the system can cross a critical value of the enzyme activity.

9. Intervention of Goldanskii

I wonder whether there were some attempts to obtain the general expression for the ratio of the rates of two possible ways of bimolecular reaction between reactants A and B for the membrane surrounded by the solvent:

$$A \xrightarrow{D_A} B$$

1. Reaction $A+B$ is preceded by lateral (two-dimensional diffusion of A (with a coefficient D_A) to the reactant B within the membrane (i.e., without any participation of environment. ($\alpha_B = n_B \cdot \sigma_B / S$ is a fraction of membrane cross-section S that is occupied by B; σ_B is the cross-section of B; n_B cm^{-2} is its concentration.)

2. Reaction $A \rightarrow a$; $a+B$

$$A \xrightarrow{D_a} a \xrightarrow{D_A} B$$

which proceeds in two steps, with the participation of the environment of membrane, when A leaves the membrane transforming into a (with a volume diffusion coefficient D_a, which can be much larger than D_A) and the particle a hits later the membrane surface reacting with B. Obviously, this ratio should depend on both D_A and D_a (i.e., on the properties of both membrane and its environment) and on the α_B [i.e., on the surface (two-dimensional) concentration of B].

10. Intervention of Caplan

There is another facet of membrane function to which little reference has been made, and that is the ability of the membrane to communicate signals from point to point in its plane. One sees this in such phenomena as cap formation at the cellular level, and of course the classical example is the propagation of an action potential along a nerve axon. It is becoming clear that the enzyme acetylcholine-esterase plays a key role in postsynaptic transmission D. Nachmansohn, *Proc. Nat. Acad. Sci.* (*U.S.*), **73,** 82 (1976)]. In this regard it is worth mentioning that the papain membrane can in principle, at least on the basis of computer "experiments," transmit simple signals S. Hardt, A. Naparstek, L. A. Segel, and S. R. Caplan, in D. Thomas and J. P. Kernevez, Eds., *Analysis and Control of Immobilized Enzyme Systems*, North-Holland, New York, 1975, p. 9). A strip of membrane is hypothetically immersed in a solution that provides the requisite boundary conditions (i.e., pH, substrate concentration), and a local perturbation in pH is brought about, say, by adding acid from a pipette applied directly to one end of the strip. The pipette could be regarded in this context as the analogue of a synapse. The resulting behavior is obtained by solving the equations for the system in two dimensions, rather than in one dimension, as discussed by Dr. Thomas. If the substrate concentration is such that the membrane operates in the regime of first-order kinetics, and initially in what I called earlier a reaction-controlled stationary state, and acid pulse from the "synapse" throws it locally into a diffusion-controlled state. This pulse propagates, but unfortunately attenuates rapidly, since the leading edge travels more slowly that the trailing edge. On the other hand, if the membrane is operated in the regime of zero-order kinetics, a nonattenuated signal can be propagated. If the system is initially in a reaction-controlled steady state in the hysteresis range described by Dr. Thomas, the perturbation shifts it locally to the corresponding diffusion-controlled state. This propagates along the membrane like a row of falling dominoes at a rate of the order of 10 μ/sec, which of course is very slow compared to signal propagation in the nerve. (There is no electrical component to the signal in this model.) An appropriate opposite perturbation returns the system to its orginal state, as though the dominoes rise again.

11. Intervention of Cerf

This remark may have relevance to the cooperative dynamics of protein assemblies.

In collaboration with the laboratory of virology of the Institute of Molecular and Cellular Biology of Strasbourg, our laboratory studied the ultrasonic absorption of the BMV (Brom mosaic virus) capside and of the dissociated system. The capside is made of 180 proteins; it can easily be dissociated into 90 dimers.

Measurements carried out between 2 and 20 MHz show that the capsides absorb significantly more than the dimers. Since one would expect that the absorption is mainly due to protein-solvent interaction, one may wonder whether the excess absorption is due to movements that characterize the protein assembly.

We are studying other self-assembling systems and extending the measurements to lower frequencies. We shall also use the pressure-jump technique for such studies.

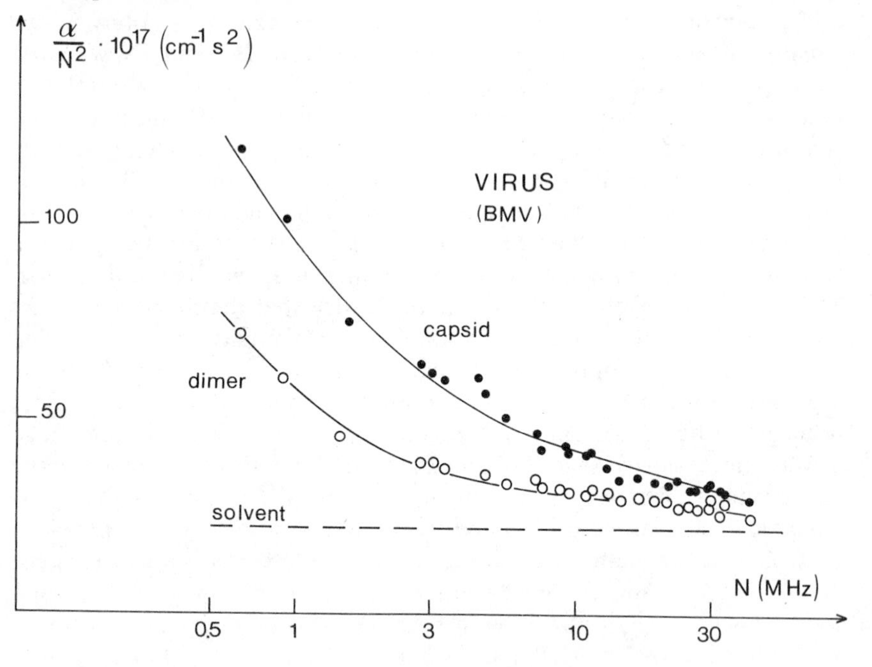

Fig. 1. Results of Michels, Schulz, and Cerf (Laboratory of Acoustics) and Pfeiffer and Hirth (Laboratory of Virology).

The wavelength is of the order of one-tenth to a few tenths of a millimeter. Therefore it is much larger than the diameter of the capside.

One of our aims in using ultrasonics was to study the kinetics of the association-dissociation process. Kinetic effects would produce a maximum of absorption when the pH is changed at constant frequency. No information about the kinetics has been obtained so far. However, the difference of absorption in capsides and in the dissociated system turned out to be of interest, because of the unexpected increase of absorption by capsides.

In capsides the self-assembling process is entropy driven. Of course, it would be of interest to carry out experiments in dissipative structures.

11.1 Intervention of Mandel

I do not know whether my remark is of some relevance to your problem, particularly as I do not know if your capside carries a net charge.

Some time ago we measured the dielectric dispersion of a plant virus particle and found a critical frequency in the kHz region, therefore in a region that may well correspond to the eventual maximum of your ultrasonic absorption. We were able to describe the mechanism leading to this dispersion in terms of the rotation of the complete virus particle and of the motion of associated (or bound) counterions on the elongated surface of the particle.

I wonder if one of these mechanisms can be related to the one responsible for the strong ultrasonic absorption you observe with the capside.

11.2 Intervention of Williams

I wonder if Professor Cerf has considered the examination of the protein hemocyanin. The work of Van Bruggen (Groningen) shows that a variety of polymeric states can be obtained under different conditions. The proteins combine in a helical cylindrical way with a molecular weight of many million. The system is comparable with the packing of the coat protein of virus. However, the cooperativity of interaction in the hemocyanins reveals itself in the cooperativity of oxygen binding. The self-assembly depends on the concentration in a variety of salts. I believe that some extracellular hemoglobins are similar. A review is given by R. J. P. Williams 1975, *Biochem. Biophys. Acta*, **416,** 237–286.

11.3 Intervention of Hammes

Some years ago we studied the absorption of ultrasonic radiation in water solutions of phospholipid vesicles. Excess absorption was seen in a frequency region similar to that observed for the virus capsid. We attributed this to the interaction of an organized water structure with the vesicles. Perhaps such a phenomenon is also of importance in explaining the interesting results of Dr. Cerf.

12. Intervention of Ubbelohde

I think we should now return to the key remarks made by Goldanskii. It appears that he is contrasting thermal effects of diffusion and reaction with quantum effects for which temperature is unimportant.

12.1 Intervention of Thomas

What is the parallel between your work and the work of Professor Douzou on enzymology at low temperature?

12.2 Intervention of Goldanskii

I am not familiar with the results you mentioned. Looking for the biochemical studies that have directly confirmed our observation of molecular tunneling in chemical reactions, one should mention the investigations of the dynamics of ligand rebinding to heme proteins performed by H. Frauenfelder, I. Gunsalus, and their colleagues at the University of Illinois (Urbana, Ill.) [see, e.g., *Biochemistry*, **14,** 5355 (1975) and *Science*, **192,** 1002 (1976)].

Having obtained the data on the kinetics of such rebinding (after the break of Fe—CO bond in hemoglobin by short laser flash) in the broad interval of temperatures (300 to 2°K), these authors found that the rebinding rate was temperature independent below about 10°K. They have explained such independence as an evidence of tunneling of CO and Fe in each other's directions, with the resulting decrease of distance between these two species.

One interesting analogy should be noted here. It is well known that the exponential factor that determines the rate of tunneling contains the product $d\sqrt{mE}$, where d is the barrier width, E is its height, and m is the mass of tunneling particle. In chemical cases in ours and American works, $d \sim 10^{-8,5}$ cm, $m \sim 30$, $E \sim 0, 1$ eV. In the spontaneous fission of nuclei, $d \sim 10^{-12}$ cm, $m \sim 100$, $E \sim 10^6$ eV. Thus the spontaneous fission of nuclei and molecular tunneling in chemical reactions can be treated to some extent as quite similar phenomena.

One should remember in this connection that many cases of spontaneous fission of excited (isomer) nuclear states with lifetimes about 10^{-2} sec (quite close to the time of addition of a new link to the growing polyformaldehyde chain near absolute zero of temperature) were observed and successfully studied after the pioneering works of Dubna scientists.

12.3 Intervention of McConnell

If the formaldehyde polymerization is exothermic, you may have a problem in interpretation. That is, dynamite probably works quite well when immersed in liquid helium in the presence of an appropriate spark.

12.4 Intervention of Goldanskii

We have directly measured the heat capacity of formaldehyde at temperatures close to that of liquid helium (about $2 \cdot 10^{-3}$ cal g^{-1} grad^{-1} at 6 to 7°K) and we can say that the average heating of our sample by radiation itself (at the dose rate of about 50 rad sec^{-1}) could not exceed 0.1°K, whereas the average increase of temperature caused by the heat of polymerization ($Q \sim 0.4$ eV, length of polymerization chain $\nu \sim 10^3$) was not larger than about 0.5°K.

The role of low heat conductivity was found to be out of significance by the comparison of data obtained at various thicknesses of monomer layers in our calorimeters, at various dose rates, by the comparison of irradiation of monomer (both radiation and polymerization can heat the system) and polymer (only external radiation is the source of heat).

Another possible explanation of our results is that the energy released in one single act of addition of a new link to the growing chain remains localized at this end link and is later used as an activation energy for the addition of the next link. However, it seems highly improbable to keep the excitation energy of about 0.1 eV being localized for about 10^{-2} s, not dissipated and used just for the addition of a new link to a polymer chain with a probability $(\nu-1)/\nu \approx 1$.

All these questions are described in more detail in my surveys in *Russian Chemical Reviews* (*Uspekhi Khimii*), in English, **44,** 1019 (1975) and *Ann. Rev. Phys. Chem.*, **27** (1976).

12.5 Intervention of McConnell

The critical points are the thermal conductivity and heat capacity of the formaldehyde crystal at the low temperature. If these parameters are low enough relative to the rate of reaction and heat release, the reaction may not be occurring at low temperatures.

12.6 Intervention of Klotz

Since Dr. Goldanskii says he is not familiar with Douzou's work, perhaps I can make some response to the question of Dr. Thomas. Douzou's observations definitely do not illustrate a "Goldanskii effect." In my understanding, the essential point from Douzou experiments, focusing on enzyme kinetics at temperatures from above 0°C to substantially below, is that the kinetic parameters change in a continuous way.

Thus the enzyme mechanism operating in solution above 0°C seems to be retained at much lower temperatures.

12.7 Intervention of Goldanskii

What I emphasize is that you can have no care at all about the decrease of entropy in chemical reactions as long as they proceed near absolute zero.

The equilibria of all reactions under such conditions are displaced toward exothermic processes, even those that lead to the formation of highly ordered systems. Furthermore, one should bear in mind the possibility of a kind of autoregulation of the predominant direction of such spontaneous reactions; processes with a relatively small heat release (closer to "resonance processes") could proceed with higher probability and, as the complexity of the molecules formed increases, the probability of the dissipation of the evolved energy among the intramolecular degrees of freedom becomes more pronounced. Therefore it seems possible that at very low temperatures under the conditions of initiation by cosmic rays, even most complex molecules can be formed with a small, but still measurable, rate, and that slow exothermic low-temperature reactions can play some part in the processes of chemical and biological evolution.

12.8 Intervention of Ross

We should not restrict ourselves to equilibrium conditions.

12.9 Intervention of Goldanskii

Polymers of formaldehyde were found recently in interstellar space by N. Wickramasinghe [*Nature*, **252,** 462 (1974)]. It is well known that polyformaldehyde is thermodynamically unstable already at not very high temperatures (close to room temperature), but it should be stable versus depolymerization near absolute zero. Therefore the formation of polyoxymethylene near absolute zero is not a thermodynamic but a kinetic problem.

Cosmic ray particles can work as a trigger of the chain of polymerization of formaldehyde adsorbed (e.g., at the surface of interstellar silicate dust). However, the triggering of the polymerization chain is a necessary but still not sufficient condition for obtaining the interstellar polymers. If the addition of any new link of the chain would require a hit of adsorbed monomer layer by another cosmic ray particle (i.e., if there is no spontaneous growth of chains after they are started by some external factor), the formation of polymers in interstellar space would be highly improbable. Therefore the spontaneous growth of polymer chains near absolute zero caused by the quantum tunneling mechanism seems to play a decisive role in the interstellar polymerization.

In answer to Professor Hess, I would add that the experiments of B. Chance and R. de Vault [see, for example, *Biophys. J.*, **6,** 825 (1966)] were probably the first to demonstrate the high importance of electron tunneling in low-temperature chemical (even biochemical) processes.

A plateau of oxidation rate found in these experiments was observed for a broad temperature interval from about 4°K up to 120°K.

Although a tunneling mechanism does not necessarily lead to a plateau in the temperature dependence of the chemical reaction rate, it can be believed that the existence of such a plateau means the validity of a tunneling mechanism.

The experiments on the radiation-induced solid-state formaldehyde polymerization at 140 to 4°K were the first to demonstrate the molecular tunneling (i.e., the tunneling of whole molecules and/or molecular groups).

The polymerization of formaldehyde involves the complete reorganization of molecules, including the changes in both lengths and angles of valence bonds. Contrary to the simple transfer of an electron, it can be called a chemical reaction in the full sense of these words.

The next example of molecular tunneling near absolute zero was the rebinding of ligands to heme proteins that I have already mentioned. The typical distances of electron tunneling in various (e.g., radiation-chemical) oxidation reduction processes in solid state at 100 to 140°K are of the order of tens of angstroms. Distances of molecular tunneling at about 4°K are equal to 0.3 to 0.5 Å.

DYNAMIC PROPERTIES OF MEMBRANES; MEMBRANE IMMUNOCHEMISTRY

HARDEN M. McCONNELL

Stauffer Laboratory for Physical Chemistry Stanford University, Stanford, California

Abstract

A very brief description of biological membrane models, and model membranes, is given. Studies of lateral diffusion in model membranes (phospholipid bilayers) and biological membranes are described, emphasizing magnetic resonance methods. The relationship of the rates of lateral diffusion to lipid phase equilibria is discussed. Experiments are reported in which a membrane-dependent immunochemical reaction, complement fixation, is shown to depend on the rates of diffusion of membrane-bound molecules. It is pointed out that the lateral mobilities and distributions of membrane-bound molecules may be important for cell surface recognition.

I. INTRODUCTION

Membranes play essential roles in the functions of both prokaryotic and eukaryotic cells. There is no unicellular or multicellular form of life that does not depend on one or more functional membranes. A number of viruses, the "enveloped" viruses, also have membranes. Cellular membranes are either known or suspected to be involved in numerous cellular functions, including the maintenance of permeability barriers, transmembrane potentials, active as well as specific passive transport across the membranes, hormone-receptor and transmitter-receptor responses, mitogenesis, and cell-cell recognition. The amount of descriptive material that might be included under the title of "biological membranes" is encyclopedic. The amount of material that relates or seeks to relate structure and function is less, but still large. For introductory references see Refs. 53, 38, 12, 47, 34, 13. Any survey of this field in the space and time available here is clearly out of the question. For the purposes of the present paper we have selected a rather narrow, specific topic, namely, the lateral diffusion of molecules in the plane of biological membranes.[38,12,43,34] We consider this topic from the points of view of physical chemistry and immunochemistry.

II. MEMBRANE MODELS AND MODEL MEMBRANES

Over the years the accumulation of chemical, structural, and functional information on biological membranes has led to a number of generalizations that have been incorporated in one or more theoretical "membrane models." Some of these generalizations are enumerated as follows: (*1*) Membranes are composed of proteins and lipids (including glycoproteins and glycolipids), and the relative mass proportions of these two components are equal to one another, to within a factor of 10. (*2*) Many, if not all, membranes involve lipid bilayers as one of the structural features. The bilayer is essential for the maintenance of the cellular permeability barrier and provides a structure to which membrane proteins can be attached (surface proteins) or inserted (penetrating proteins, including *trans*membrane proteins). (*3*) Membranes are typically unsymmetrical in composition with respect to the inner and outer surfaces (i.e., the surfaces that face the inside and outside of the cell). (*4*) In some regions of some biological membranes, the lipids, or lipids and proteins, can undergo lateral diffusion in the plane of the membrane.

Generalizations such as those previously described have been incorporated in schematic drawings or membrane models of biological membranes by a number of investigators. For examples see Refs. 53 and 38. These membrane models have greatly stimulated research and influenced thinking in membrane and cellular biology. Such models do have two drawbacks: They are sometimes so general that they have little predictive value; they also can be somewhat hypnotic, leading a naive onlooker to take them literally. A number of serious workers in the membrane field have recognized this latter problem by employing membrane model drawings that have deliberately included a touch of humor, so as to forewarn the interested but uninformed reader that such drawings are not to be taken too literally (see, for example, Ref. 49).

Figure 1 gives illustrative drawings of membrane models; the reader is referred to the legend for a description of the nature and supposed functions of the various components.

Fig. 1. (*Upper*) An early "schematic and speculative" drawing of a biological membrane. The purpose of the diagram and associated discussion was to point out that membrane fluidity, particularly as inferred using spin labels, should make possible molecular motions of membrane proteins such as rotation (given an appropriate hydrophobic surface) and translation. The membrane is based on a fluid lipid bilayer (lipids denoted *A*), surface-bound proteins (*B*), glycolipids (*C*), a presumed permease protein (*D*) with a hydrophobic surface, and a binding site (*E*) (see Ref. 31). (Reproduced with permission of Plenum Publishing Corporation.) (*Lower*) Adapted from the Singer-Nicolson "fluid mosaic" membrane, showing fluid lipids, peripheral proteins, and integral proteins (see Ref. 48). Proteins known or suspected *at present* to be involved in transmembrane transport appear to be of the channel-forming type rather than the rotator type as in the upper sketch.

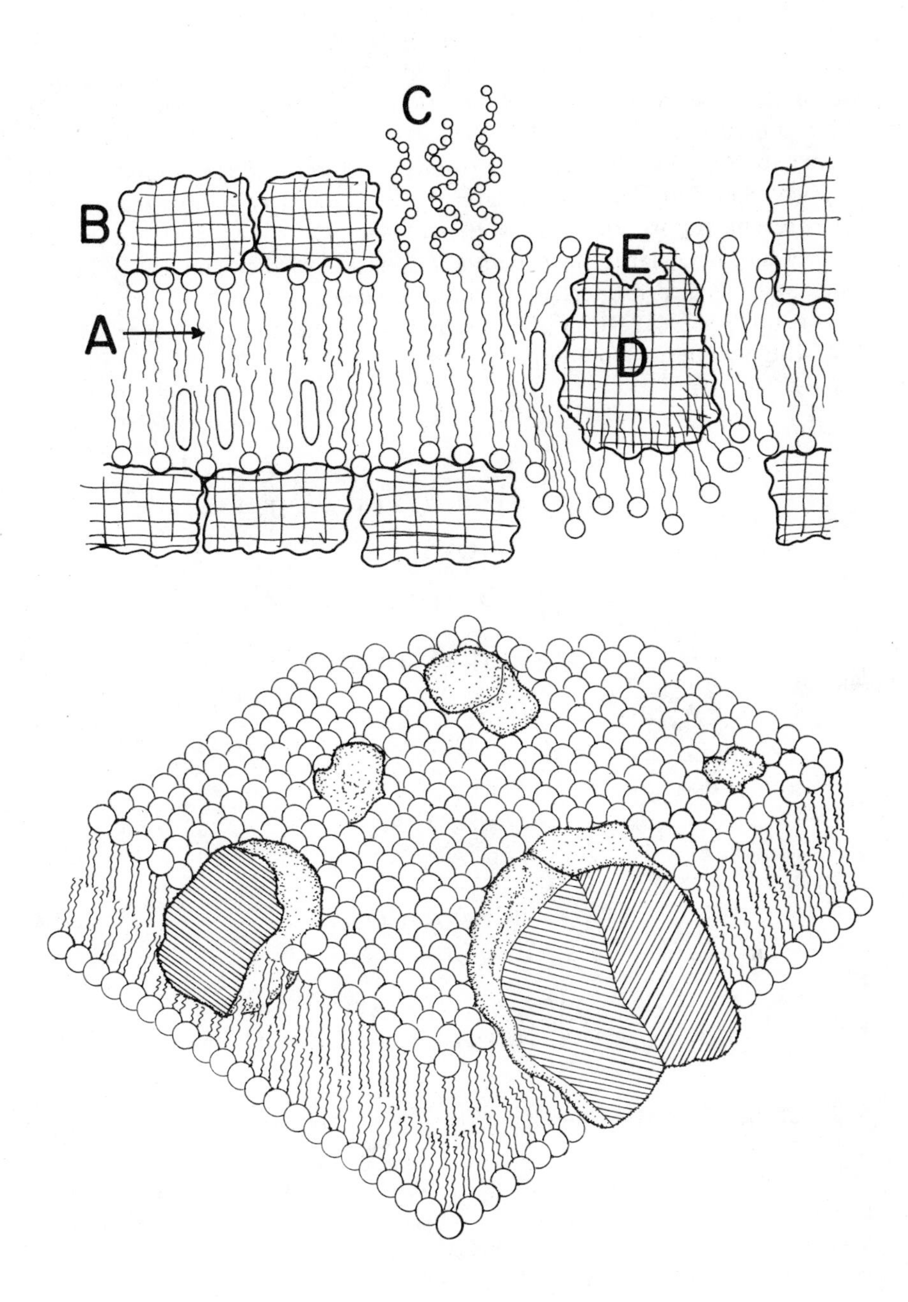
C
B
E
A
D

An important approach to the study of biological membranes has been the preparation and study of *model membranes*. According to current usage, model membranes include lipid bilayers and lipid bilayers into which have been incorporated additional components such as one or more membrane proteins. It is through the study of such model membranes that one has the best opportunity to isolate and study fundamental physical chemical and biophysical processes, and it is for this reason that the present report emphasizes these systems. A discussion of model membranes necessarily starts with a description of the chemical compositions and physical properties of lipid molecules.

For brevity we restrict the discussion in the present paper to the following simple phospholipids.

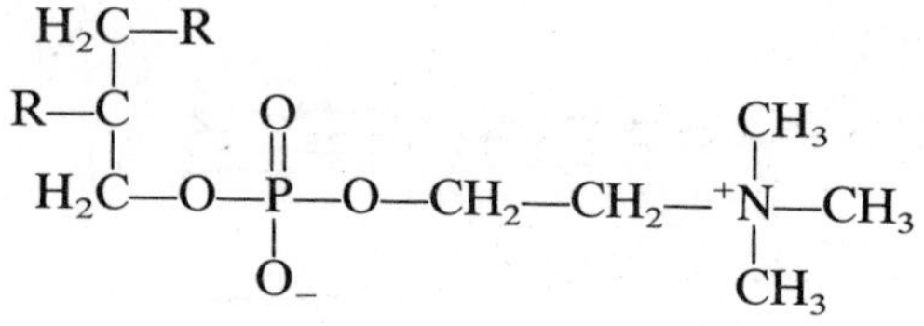

$$
\begin{array}{l}
H_2C{-}R \\
\;\;| \\
R{-}C \qquad\quad O \qquad\qquad\qquad\qquad\quad CH_3 \\
\;\;| \qquad\qquad\;\; \| \qquad\qquad\qquad\qquad\quad\;\; | \\
H_2C{-}O{-}P{-}O{-}CH_2{-}CH_2{-}{}^{+}N{-}CH_3 \\
\qquad\qquad\;\; | \qquad\qquad\qquad\qquad\quad\;\; | \\
\qquad\qquad\;\; O^{-} \qquad\qquad\qquad\qquad\; CH_3
\end{array}
$$

$$R = O{-}\overset{\displaystyle O}{\overset{\|}{C}}{-}(CH_2)_{12}{-}CH_3 \qquad \text{(I)}$$

$$R = O{-}\overset{\displaystyle O}{\overset{\|}{C}}{-}(CH_2)_{14}{-}CH_3 \qquad \text{(II)}$$

$$R = O{-}\overset{\displaystyle O}{\overset{\|}{C}}{-}(CH_2)_{16}{-}CH_3 \qquad \text{(III)}$$

$$R = \textit{trans}\; O{-}\overset{\displaystyle O}{\overset{\|}{C}}{-}(CH_2)_9{-}\overset{H}{C}{=}\overset{H}{C}{-}(CH_2)_5{-}CH_3 \qquad \text{(IV)}$$

Some of the physical properties of these phospholipids are summarized in Table I.

Many phospholipids, including (I) through (IV) above, form lipid bilayers when dispersed in water. These bilayers may take on various shapes, depending on their treatment. When strongly sonicated in aqueous solutions, phosphatidylcholines such as (II) form quite stable single-compartment vesicles, with outer diameters of the order of 200 to 350 Å.

TABLE I

Phospholipid	Chain melting transition		Pretransition	
	Temperature	ΔH (kcal/mole)	Temperature	ΔH (kcal/mole)
I	23.9[a] 23.2[b] 23.5[c]	5.4[a]	14.2[a] 10.1[b] 13.4(H), 9.6(C)[3]	1.0[a]
II	41.4[a] 40.5[b] 40.7[c]	8.7[a]	35.3[a] 29.2[b] 34.2(H), 30.2(C)[c]	1.8[a]
III	54.9[a] 54.0[b]	10.6[a]	51.5[a] 46.1[b]	1.8[a]
IV	10.5–13.5[d]		0(?)[d]	very small

[a] See Ref. 30.
[b] See Ref. 44.
[c] See Ref. 29.
[d] See Ref. 54.

Often when phospholipids such as (I) through (IV) are dispersed in water, they form so-called liposomes, which have a large number (e.g., 10 or more) of concentric bilayer shells. When phospholipids are dissolved in detergents and the detergents are removed slowly by dialysis, the phospholipids form relatively large vesicles containing one or only a few bilayer shells. Lipid bilayers have also been studied as supported by black lipid membranes [Fettiplace et al. (1975)].

When hydrated phospholipids (containing about 50% water) are squeezed between two flat quartz plates, one can frequently obtain highly ordered, uniaxial, parallel arrays of about 100 to 10,000 bilayers (see, for example, Ref. 33). These lipid multilayers are ideal for a number of spectroscopic studies of phospholipids. Most of the membrane work described in the present work has been carried out using liposomes and oriented multilayers.

The lateral motions of phospholipid molecules in phospholipid bilayer membranes can give rise to lateral phase separations. A number of phase diagrams describing these lateral phase separations have been derived using spin labels[44–46,54] as well as differential scanning calorimetry.[36] A number of the phase diagrams derived from spin-label paramagnetic resonance data have been verified by other techniques, such as scanning calorimetry (J. H. Sturtevant, private communication) and freeze-fracture electron microscopy. For electron microscopic studies that have identified

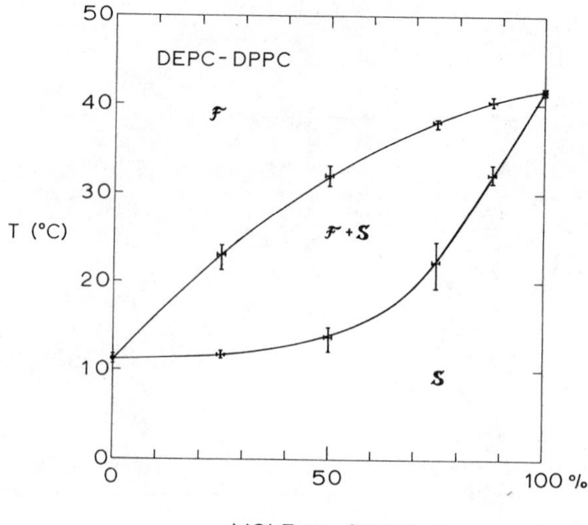

Fig. 2. Phase diagram describing lateral phase separations in the plane of bilayer membranes for binary mixtures of dielaidoylphosphatidylcholine (DEPC) and dipalmitoylphosphatidylcholine (DPPC). The two-phase region ($F+S$) represents an equilibrium between a homogeneous fluid solution F (L_α phase) and a solid solution phase S presumably having monoclinic symmetry ($P_{\beta'}$ phase) in multilayers. This phase diagram is discussed in Refs. 19, 18, 4. The phase diagram was derived from studies of spin-label binding to the membranes.

coexisting phases expected on the basis of spin-label-derived phase diagrams, see Refs. 46, 19, 23, 24, 18.

Figure 2 shows a spin-label-derived phase diagram for binary mixtures of (II) and (IV), dipalmitoylphosphatidylcholine and dielaidoylphosphatidylcholine. It will be seen that the diagram describes miscibility of these two lipids in both the solid and solution phases. (Other binary mixtures of lipids show immiscibility in the solid as well as the fluid phases.[45,54])

Figure 3 shows a freeze-fracture microphotograph of a 50:50 binary mixture of (II) and (IV) quenched from a temperature (~25°C) where roughly equal proportions of solid and fluid phases are expected to coexist. The banded region corresponds to the crystalline, monoclinic $P_{\beta'}$ solid solution phase and the nonbanded region to the fluid phase. For a recent crystallographic study of the $P_{\beta'}$ phase of phosphatidylcholines, see Ref. 22. For a freeze-fracture study of this phase, see Ref. 28.

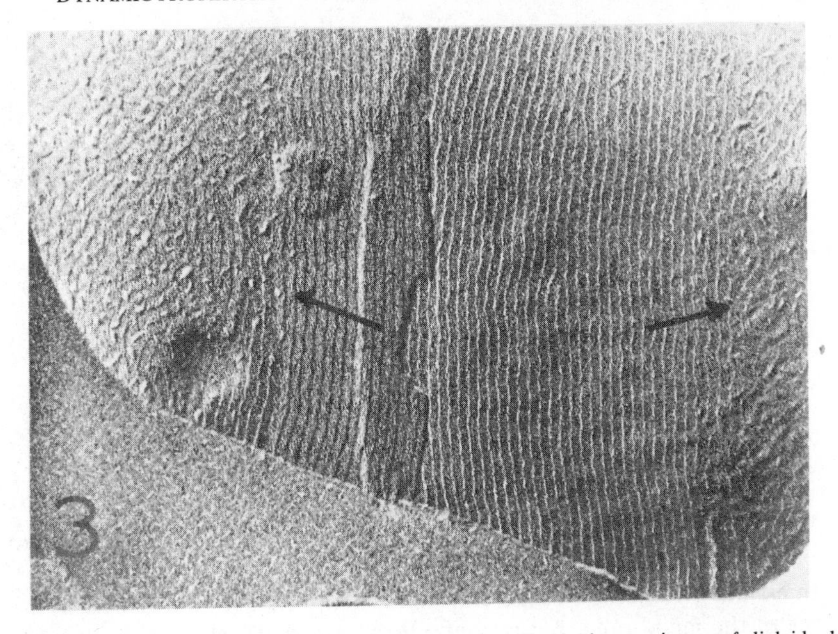

Fig. 3. Freeze-fracture electron microphotograph of a 50:50 binary mixture of dielaidoylphosphatidylcholine and dipalmitoylphosphatidylcholine quenched from a temperature (~23°C) where roughly equal proportions of the fluid and solid phases coexist according to the phase diagram in Fig. 2. The region showing a regular banded pattern is the "solid" phase ($P_{\beta'}$ phase) and the regions showing no bands or irregular bands are due to the fluid domains (L_α phase). The regular bands are about 250 Å apart. The "bumps" in the fluid phase lipids are due to a membrane protein (glycophorin) that was included at a low concentration in this mixture of lipids. Note that the glycophorin is preferentially soluble in fluid phase lipids. Arrow denotes domain boundaries between fluid and solid phase lipids. (From Ref. 18.)

III. LATERAL DIFFUSION

The rates of lateral diffusion of phospholipids in lipid bilayer membranes, and in biological membranes, were first measured using spin-labeled lipids.[26,50,10,11,9] In general, these rates have been determined by incorporating spin-labeled lipids such as (V) and (VI) in phospholipid bilayers, or multilayers. The paramagnetic resonance spectra of labels such as (V), as well as the nuclear resonance spectra of other lipids in membranes containing (V), depend on the concentration c of the label in the membrane and the rate of lateral motion of the lipids. Two methods have been used to deduce the diffusion constants, transient methods and steady-state methods.

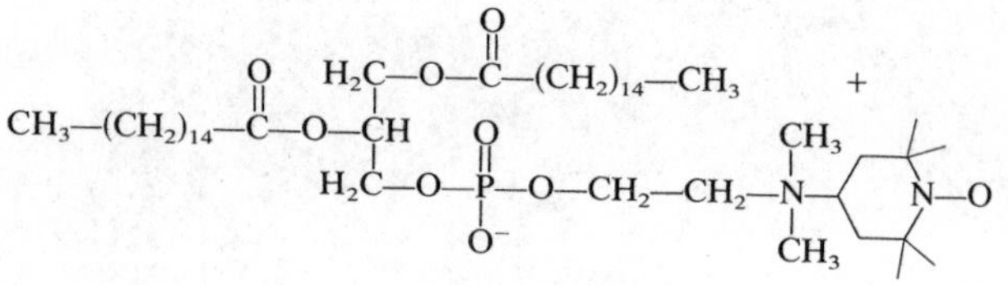

$$\begin{array}{l} \quad\quad\quad\quad\quad\quad\quad\quad O \quad H_2C{-}O{-}\overset{O}{\overset{\|}{C}}{-}(CH_2)_{14}{-}CH_3 \\ CH_3{-}(CH_2)_{14}{-}\overset{\|}{C}{-}O{-}CH \quad\quad O \quad\quad\quad\quad\quad CH_3 \\ \quad\quad\quad\quad\quad\quad\quad\quad\quad H_2C{-}O{-}\overset{\|}{P}{-}O{-}CH_2{-}CH_2{-}\overset{+}{N}{-}\text{(tetramethylpiperidine N–O)} \\ \quad\quad\quad\quad\quad\quad\quad\quad\quad\quad\quad\quad\quad O^- \quad\quad\quad\quad\quad\quad\quad CH_3 \end{array} \qquad \text{(V)}$$

$$\begin{array}{l} \quad\quad\quad\quad\quad\quad\quad\quad\quad\quad\quad\quad\quad\quad\quad\quad\quad\quad O \\ \quad\quad\quad\quad\quad O\;\; N{\rightarrow}O \quad O \quad\quad CH_2{-}O{-}\overset{\|}{C}{-}(CH_2)_{m+n+1}{-}CH_3 \\ CH_3{-}(CH_2)_m{-}C{-}(CH_2)_n{-}\overset{\|}{C}{-}O{-}CH \quad\quad O \\ \quad\quad\quad\quad\quad\quad\quad\quad\quad\quad\quad\quad CH_2{-}O{-}\overset{\|}{P}{-}O{-}CH_2{-}CH_2{-}N^+{-}(CH_3)_3 \\ \quad\quad\quad\quad\quad\quad\quad\quad\quad\quad\quad\quad\quad\quad\quad\quad O^- \end{array} \qquad \text{(VI)}$$

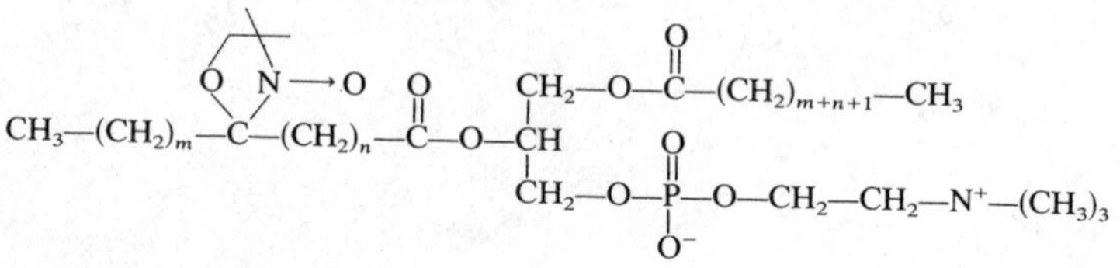

In a transient method one creates an initial distribution of labels in the plane of the membrane that is nonuniform, $c(x, y, 0)$, where c is the mole % of the spin-label lipid (e.g., at point x, y in the membrane, at time $t=0$). For times $t \geqslant 0$ the lateral distribution of label $c(x, y, t)$ is then determined from a solution of the diffusion equation

$$\frac{\partial c}{\partial t} = D\left(\frac{\partial^2 c}{\partial x^2} + \frac{\partial^2 c}{\partial y^2}\right) \qquad (1)$$

Devaux and McConnell[9] took advantage of the fact that in fluid membranes such as egg phosphatidylcholine, the resonance spectra of spin labels such as V and VI depend strongly and monotonically on the label concentration c when $c \geqslant 5$ mole %. The normalized paramagnetic resonance spectra $S_0(H, c)$ of a series of samples, all of uniform concentration c, were determined experimentally.[9] The observed time-dependent spectra are then obtained from the equation

$$S(H, t) = \frac{1}{c_0 XY} \int_0^{X,Y} c(x, y, t) S_0(H, c(x, y, t))\, dx\, dy \qquad (2)$$

for a rectangular sample of dimensions X, Y. By comparisons of the observed time-dependent spectra with the normalized reference spectra, and the solutions of the diffusion equation, it was possible to measure the diffusion constant for lateral motion (e.g., $D = 2 \times 10^{-8}$ cm^2/s for label V in egg phosphatidylcholine multibilayers at room temperature). In this work by Devaux and McConnell[9] the initial label distribution was created by preparing (under a microscope) small patches of V and then surrounding these patches with unlabeled phospholipid (see Fig. 4).

In an effort to improve on the accuracy and versalility of this method, Sheats and McConnell (unpublished) have recently developed a new

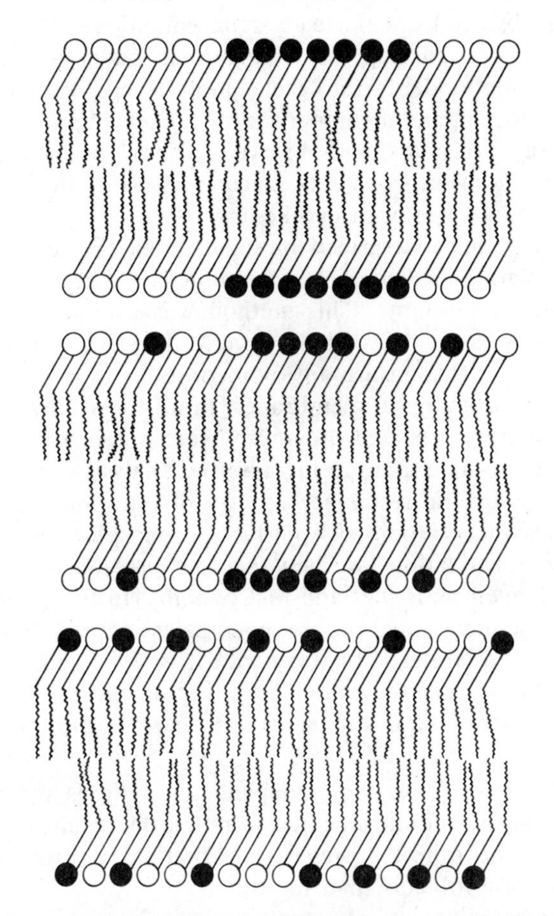

Fig. 4. Schematic representation of transient method employed by Devaux and McConnell[9] to measure the rates of lateral diffusion of phospholipids in model membranes. The upper diagram represents a concentrated patch of labels at the beginning of the experiment, time $t=0$. At later times $t>0$, the molecules diffuse laterally, as shown in the lower two drawings. The paramagnetic resonance spectra depend on the spin-label concentration in the plane of the membrane, and an analysis of the time dependence of these spectra yielded the diffusion constant. [Reprinted with permission from P. Devaux and H. M. McConnell, *J. Am. Chem. Soc.*, **94,** 4475 (1972). Copyright by American Chemical Society.]

method, which offers the possibility of greater precision for these determinations of diffusion constants. In this new method the initial nonuniform distribution of labels is produced by a laser photochemical destruction of labeled lipids in localized regions of multibilayers. The method offers the possibility of great precision in determining the initial distribution function $c(x, y, 0)$.

The steady-state methods involve theoretical analysis of magnetic resonance spectra observed under steady-state conditions. This typically involves assumptions regarding the adequacy of magnetic resonance line shape theory, some model for molecular motions and distances of closest approach on collision, and a comparison of calculated spectra for various assumed diffusion constants, and observed spectra. In general, the agreement between diffusion constants calculated using the transient and steady-state methods has been excellent.

Devaux et al.[11] have described in some detail the use of the Bloch equations to relate electron spin-spin interactions between spin-labeled lipids to diffusion constants. This method was originally employed by Träuble and Sackmann,[50] Scandella et al.,[42] Devaux and McConnell,[10] and Devaux et al.[11] to measure diffusion constants in multibilayer model membranes and in biological membranes. In all cases diffusion constants of the order of those reported previously were obtained.

A second steady-state method involves the analysis of the broadening of the nuclear magnetic resonance spectra of phospholipids in bilayers containing low concentrations of spin-labeled phospholipids. A theoretical analysis of the relation between this line broadening and diffusion rates has been given by Brûlet and McConnell.[3] [In this paper (6) is not correct; the subsequent equations are nonetheless correct. For an alternative derivation, see Brûlet.[2]] In this paper it is shown that a number of measurements of nuclear relaxation rates T_1^{-1} of nuclei in phospholipids are consistent with lateral diffusion constants in the range 10^{-7} to 10^{-8} cm^2/s.

The availability of these two independent methods for determining diffusion rates has been quite significant, in that the results of one method have confirmed the other; however, one method is often more easily adapted to special problems than the other.

Rates of lateral diffusion of membrane components have also been determined using optical methods. The early experiments of Frye and Ediden[16] demonstrated lateral motion of fluorescent-labeled surface antigens in heterokaryons of mouse and human cells. They observed intermixing of fluorescent-labeled antibodies against mouse cell and human cell antigens. Optical methods may also be characterized as either transient or steady state. The use of fluorescence correlation spectroscopy as described by Elson and Webb[14] is an example of a steady-state method. Photobleaching methods involve the bleaching of a chromophoric group in a localized region of a biological membrane and the observation of the time course of recovery of optical absorption due to diffusion of molecules into the target region (transient methods). For this and other optical transient methods, see Refs. 37, 35, 8, 43, and 17.

At the present time, the rates of lateral diffusion of phospholipids and membrane proteins in the solid phase of pure phospholipids is not known. It is hoped that such diffusion constants can be obtained by one of the transient methods mentioned earlier. It is likely that these diffusion rates will be found to be quite low.

A more interesting problem from both the experimental and theoretical point of view is the lateral diffusion of phospholipids in mixtures of lipids, when both solid and fluid phases coexist. At least three questions arise in connection with this problem. (*1*) What is the rate of lateral diffusion of phospholipids in solid solution domains? (*2*) To what extent do solid solution domains act as obstacles to the lateral diffusion of lipid molecules in fluid domains? (*3*) To what extent are there composition and density fluctuations present in "fluid" lipid bilayers, and to what extent do these fluctuations affect lateral diffusion? Let us consider these questions one at a time, bearing in mind that these questions may to some extent be interrelated.

1. At the present time, nothing is known about the rate of lateral diffusion of phospholipids in solid solution phases ($P_{\beta'}$ or $L_{\beta'}$). However, the freeze-fracture electron microphotographs of binary mixtures of phospholipids are of particular interest in connection with this problem. Consider the phase diagram in Fig. 2. When a 50:50 binary mixture of dielaidoylphosphatidylcholine and dipalmitoylphosphatidylcholine is maintained at a temperature (e.g., 25°C) where the system is roughly 50% fluid and 50% solid, the coexisting fluid and solid domains can be visualized by rapid temperature quenching followed by freeze-fracture electron microscopy. In such electron microphotographs the solid phase lipids (at 25°C) can be identified by a regular band pattern, such as the one seen in Fig. 3. This strongly indicates that the composition of the solid phase lipids is *homogeneous* over the entire solid domain (i.e., over distances of the order of 1 to 5 μ). This is in contrast to the "coring" effect often observed in alloys, where different regions of the solidified alloy have different compositions, because of low diffusion rates in these solids.[51] The nonuniform band patterns sometimes seen in freeze-fracture microphotographs may be due to coring (see Fig. 3*b* in Ref. 19). However, many if not most of the micrographs of binary mixtures we have observed show uniform band patterns for the solid solution phases. A

simple interpretation of our results in these cases is that the rates of lateral diffusion in the solid solution phases are appreciable. An alternative mechanism for homogenization of the solid solution phases would involve appreciable rates (e.g., in periods of 10 to 30 min) of melting and resolidification of the solid solution phases.

2. Obviously, long-lived solid phase domains may provide severe obstacles to the lateral diffusion of fluid phase phospholipids, if an appreciable fraction of the lipids are in the solid solution phase. The magnitude of this effect may be affected by the lifetime of the solid phase domains, mentioned earlier.

3. Density or composition fluctuations, sometimes referred to as cluster formation, may also play important roles in lateral diffusion. Two studies from our laboratory will be cited that are of interest in connection with this problem.

Brûlet and McConnell[4] have observed the ^{13}C choline methyl nuclear resonance spectra of 50:50 binary mixtures of dielaidoylphosphatidylcholine and dipalmitoylphosphatidylcholine. In one set of experiments one lipid was ^{13}C enriched in the choline methyl group, and in the other

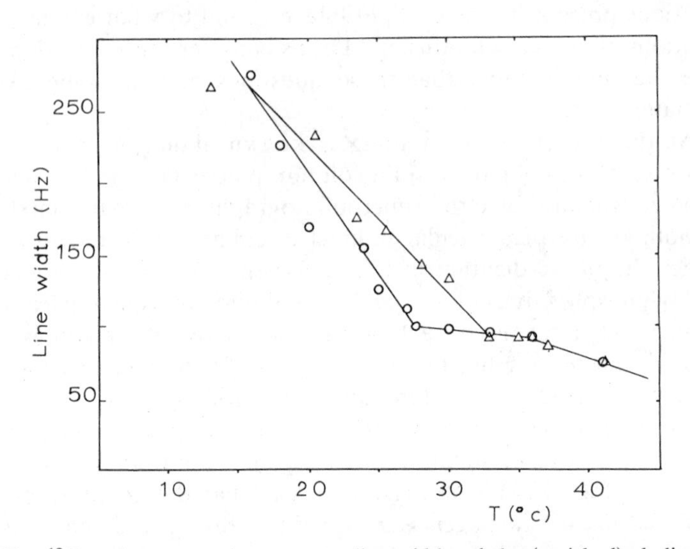

Fig. 5. The ^{13}C nuclear magnetic resonance line widths of the (enriched) choline methyl resonances in dipalmitoylphosphatidylcholine (△) and in dielaidoylphosphatidylcholine (○), as a function of temperature. Spectra taken at 90.5 MHz; similar results were also obtained at 25.2 MHz. Note that the higher-melting lipid, dipalmitoylphosphatidylcholine, shows a readily observable enhanced line broadening at temperatures $T_u \sim 32°C$, corresponding to the onset of the lateral phase separation. (Data from Ref. 4.) [Reprinted with permission from P. Brûlet and H. M. McConnell, *J. Am. Chem. Soc.*, **98,** 1314 (1977). Copyright by American Chemical Society.]

set of experiments the other lipid was ^{13}C enriched in the choline methyl group. Figure 5 shows the ^{13}C nuclear magnetic resonance line widths of the choline methyl groups of each of these lipids as a function of temperature. As expected, the ^{13}C choline methyl resonances in a 50:50 binary mixture are narrow ($\Delta\nu_{1/2} \lesssim 100$ Hz) at temperatures above 32°C where the two lipids form a homogeneous fluid phase, and the resonances are broad ($\Delta\nu_{1/2} > 250$ Hz) at temperatures where the two lipids form a homogeneous solid phase. Of particular significance is the fact that the ^{13}C choline resonance of dipalmitoylphosphatidylcholine shows a marked onset of line broadening at a temperature, 33°C, that is equal, within the experimental error, to the temperature corresponding to the onset of the lateral phase separation in Fig. 3.

Brûlet and McConnell[4] have made a theoretical calculation of the expected line broadening using the model sketched in Fig. 6. Here a planar lipid bilayer is divided into two domains, a fluid domain F and a solid domain S, with the indicated geometry. The position of the domain boundary b–b' then defines the relative proportion of the two phases, and this of course is in turn determined by the temperature and lipid composition. These investigators have made a theoretical calculation of the estimated line broadenings to be expected when the total length of the bilayer is $\sim 1\ \mu$. (The width is not important for the assumed geometry as long as it is large.) These calculations involve observed transverse nuclear relaxation rates $T_2^{-1}(S)$ $T_2^{-1}(F)$ for the two phases, and various plausible assumptions regarding diffusion in the fluid and solid phases, as well as rates of passing through the domain boundary. Typical calculated results are given in Fig. 7. It will be seen that the observed and calculated line broadenings are similar in that (*1*) with decreasing temperature the line broadenings begin at the temperature of the onset of the phase separation, and (*2*) the ^{13}C choline resonance of the higher-melting lipid broadens more rapidly than does the ^{13}C choline resonance of the lower-melting lipids. There is, however, a very significant discrepancy between observed and calculated *rates* of line width enhancement with decreasing temperature, as is obvious from comparisons of the data in the two figures. The origin of the weak line broadening at temperatures close to T_u of about 33°C in the theoretical calculation is quite clear: The time required for molecules to diffuse in the fluid state to the solid phase, when the proportion of solid phase is small, is $\sim L^2/D \sim 1$ sec, which is far too long to give a detectable line broadening. To put the problem in another form, we may ask what fraction X of the bilayer must be in the solid solution phase in order for all molecules in the solution to reach the solid phase and undergo line broadening. The time corresponding to line broadening of, say, $\Delta\nu \sim 100$ Hz, is $\tau \sim 1/\pi\Delta\nu \sim 3\times10^{-3}$ sec. In this time

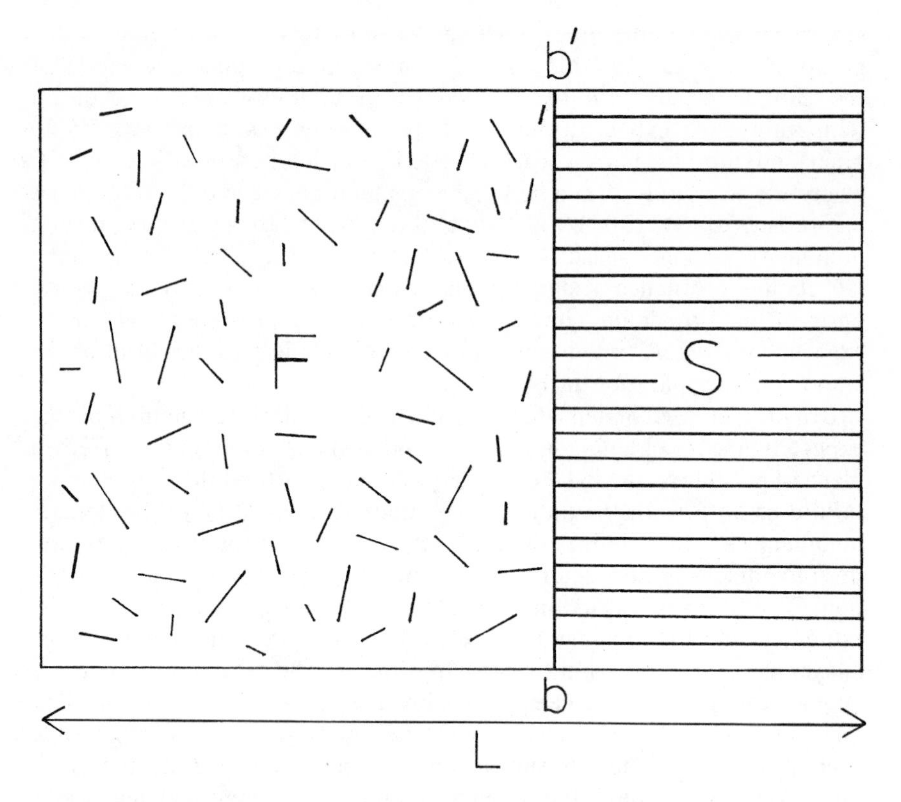

Fig. 6. Schematic model of two-phase system, with solid domain S in equilibrium with fluid domain F. The relative proportions, and compositions of F and S are determined from, for example, the phase diagram in Fig. 2. The ^{13}C nuclear resonance spectra of lipids in the solid phase S are broader than the nuclear resonance spectra of lipids in the fluid phase F; the spectra of lipids in the fluid phase F can be broadened if they diffuse to, and stick to, the solid phase. For a detailed theoretical calculation of line widths expected for this geometry, see Ref. 4. The bilayer length is L, and the domain boundary is $b-b'$. [Reprinted with permission from P. Brûlet and H. M. McConnell, *J. Am. Chem. Soc.*, **98,** 1314 (1977). Copyright by American Chemical Society.]

the molecule diffuses a distance of the order of $l \sim \sqrt{Dt} \sim 1000$ Å. In other words, if $L = 10{,}000$ Å, then only $X = l/L$, or 10% of the fluid lipids undergo significant line broadening. All the fluid lipids undergo significant broadening when 90% of the bilayer is in the solid phase. This line of reasoning accounts for the slow increase in line width shown by the theoretical calculation in Fig. 7.

The model sketched in Fig. 6 was not chosen arbitrarily. The lipid area chosen is of the magnitude appropriate to the liposomes employed in the

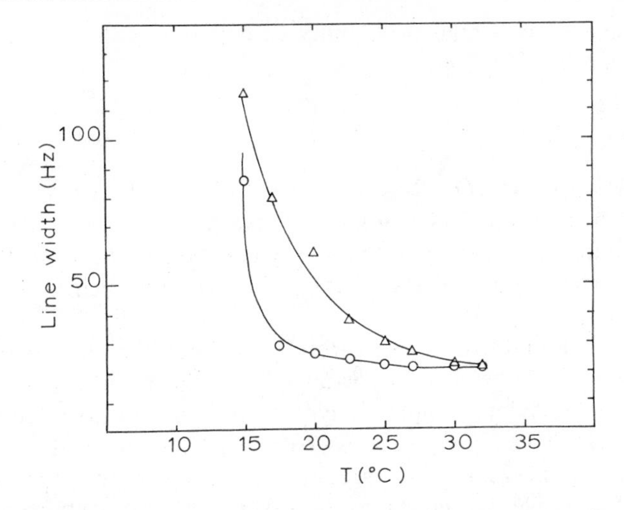

Fig. 7. Theoretically calculated line widths for the ^{13}C-choline nuclear magnetic resonance signals for dipalmitoylphosphatidylcholine (△) and dielaidoylphosphatidylcholine (○) in a 50 mol% binary mixture of the two lipids. Calculation appropriate for a ^{13}C resonance frequency of 25.2 MHz.

experimental work. Freeze-fracture microphotographs of mixtures of phosphatidylcholines consistently show large extended areas of solid and/or fluid domains, indicating that the boundary free energies are relatively high, so that the boundary lengths b-b' are minimized.[18,19,52] On the other hand, the freeze-fracture technique might very well be unable to resolve small, solidlike clusters of lipids. As shown below, small clusters of solid phase lipids do appear to be compatible with the observed enhanced line broadening.

Let us assume some small number n of lipid molecules can form a relatively stable solid phase cluster when the temperature and composition of the lipid mixture is such that, according to the phase diagram, solid phase can exist in equilibrium with the fluid phase. (For example, we later assume that $n \sim 10$.) Let us further assume that (*1*) the temperature and composition of the lipid mixture is such that X is small, $X \ll 1$, and (*2*) all the solid phase present is in the form of clusters of n molecules each. If the clusters are randomly distributed in the plane of the membrane, then each cluster will be surrounded by a number of fluid molecules of the order of magnitude of $N \sim n/X$. The area occupied by the surrounding fluid phase molecules is then NA_0 where, $A_0 \sim 60$ Å^2. Let us now calculate lower limit on X, X_{min}, such that each molecule in the fluid phase has a high probability of colliding with the solid phase, during a time $T_2^{-1}(F)$. Thus

$$X \sim \frac{nA_0}{DT_2} \tag{3}$$

If $A_0 = 60\ \text{Å}^2$, $n \sim 10$, $D \sim 2 \times 10^{-8}\ \text{cm}^2/\text{s}$, and $T_2 \sim 3 \times 10^{-3}$ s, we obtain $X \sim 10^{-3}$. Thus with a cluster model one can readily account for substantial line broadening at temperatures very close to the onset of the phase separation, where only ~0.1% of the lipids are present in solid solution phase.

The reader should distinguish carefully between the clusters described here, and those that have been postulated elsewhere[27] in connection with pretransition behavior in lipid bilayers. The clusters postulated in the present work have compositions corresponding to states of thermodynamic equilibrium. They are unstable relative to larger domains because of positive boundary free energy, but are stabilized by the entropy gain when large domains break up to form small clusters. Cluster formation in the sense considered here must surely affect the lateral mobility of lipids in bilayers, but further experimental and theoretical work is required before the order of magnitude of the effect becomes known.

All the aforementioned evidence for cluster formation is indirect. In this laboratory we have recently observed a special type of cluster formation that supports the notion that such clusters may exist. Rey and McConnell[41] have described the preparation of two spin labels, *N,N'*-dipalmitoyl-*N,N'*-bis-(1-oxyl-2,2,6,6-tetramethylpiperidin-4-yl)-1, 10-diaminodecane (VII) and *N,N'*-dimethyl-*N,N'*-dihexadecyl-*N,N'*-bis-(1-oxyl-2,2,6,6-tetramethylpiperidin-4-yl)-1,10-diammoniumdecane di-iodide (VIII).

These biradicals bind to lipid bilayers. The resonance spectra of the charged lipid (VIII) has been observed at various temperatures (15 to 65°C) and in various lipid mixtures (0 to 50% cholesterol in dimyristoyl-

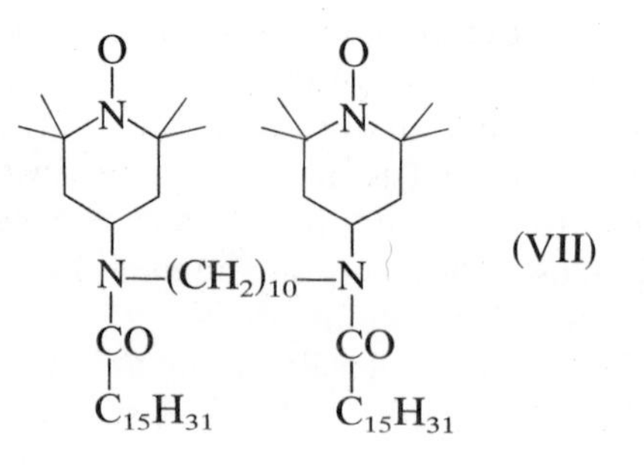

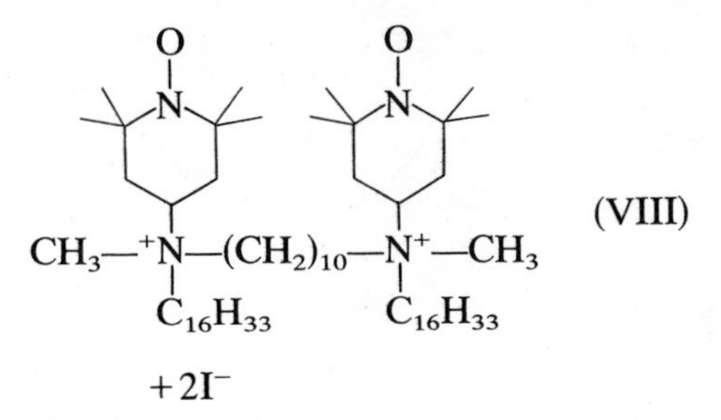

phosphatidylcholine). The distinctive feature of the observed spectra for (VIII) is that under all the various conditions there is no evidence whatever for significant spin-spin interactions arising from interactions between these radicals in the plane of the bilayer membrane. (Biradical present at a concentration of 0.9 mole % in the plane of the membrane.)

In contrast to (VIII), biradical (VII) shows a strong concentration, cholesterol- and temperature-dependent spin-spin interaction. Rey and McConnell[41] have analyzed these spectra quantitatively when the concentration of (VII) is varied between 0.025 mole % and 2 mole % in bilayer membranes (70 mole % dimyristoylphosphatidylcholine and 30 mole % cholesterol) at 30°C. The surprising result was obtained that all the spectra can be accounted for *quantitatively* as the superposition of two spectra, a monomer spectrum [one molecule of (VII)] and a hexamer spectrum [a cluster containing six molecules of (VII)]. Representative data are given in Figs. 8 and 9.

These results leave little doubt that specific clustering of amphiphilic molecules in bilayer membranes can take place. An important question

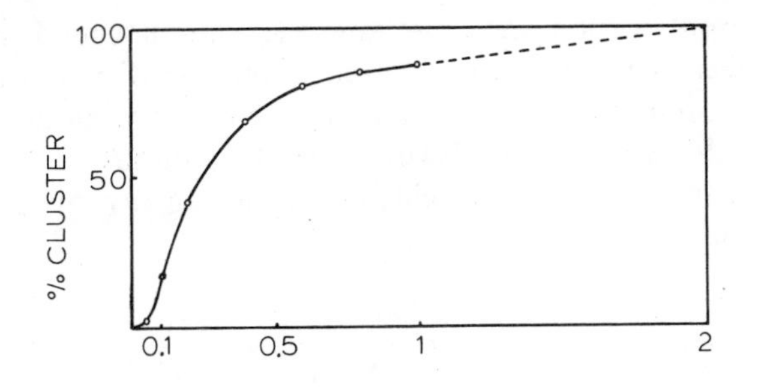

Fig. 8. The percent of membrane-bound amphiphilic spin label (VII) in a lipid bilayer host membrane that forms clusters of a specific stoichiometry (hexamers) (see text). [Reprinted with permission from P. Rey and H. M. McConnell, *J. Am. Chem. Soc.*, in press (1977). Copyright by American Chemical Society.]

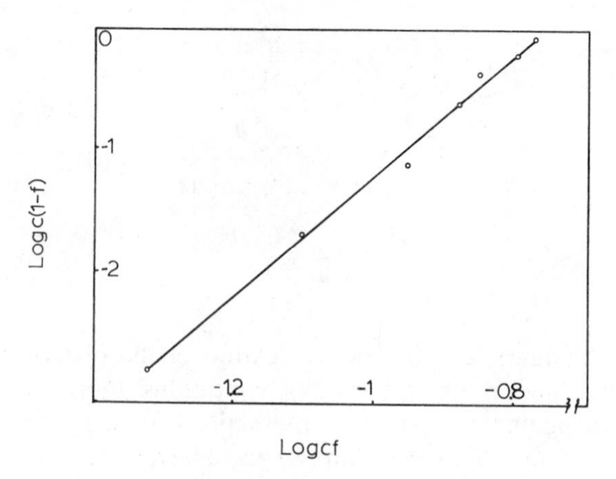

Fig. 9. Plot demonstrating that clusters of VII in bilayer membranes are hexamers. Here *c* is the moles of label VII per 100 moles of host lipid; *f* is the fraction of labels present as hexamers. The linearity of this plot is an indication that the clusters have a unique stoichiometry, and the slope (6) shows that the clusters are hexamers.

concerning these as well as the clusters mentioned earlier is their lifetimes. The line width of the hexamers of (VII) show that these clusters certainly last for times longer than the paramagnetic resonance line width (i.e., longer than 10^{-8} to 10^{-9} s). Studies of the rates of lateral diffusion of these clusters should provide more information on their lifetimes.

IV. MEMBRANE IMMUNOCHEMISTRY

In the present section we show how the immunochemistry of model membranes can depend on the lateral mobilities of membrane components. Most of our studies of the immunochemistry of model membranes have taken advantage of the discovery by Humphries and McConnell[21] that it is possible to prepare specific antibodies against the paramagnetic nitroxide group

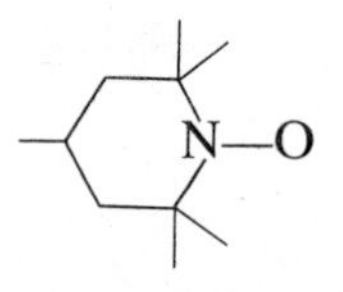

This is a fortunate result, since the paramagnetic resonance spectra of phospholipid spin labels such as (V), (VI), (IX), and (X) are sensitive to their state of aggregation, concentration, and rates of lateral motion. This

information is absolutely essential for the interpretation of the immunochemical results. The present discussion is limited to IgG antibodies, which are bivalent and have a molecular weight of the order of 150,000.

$$
\begin{array}{l}
\mathrm{C{-}O{-}CO{-}(CH_2)_{14}{-}CH_3} \\
\mathrm{CH_3{-}(CH_2)_{14}{-}CO{-}O{-}CH} \\
\mathrm{C{-}O{-}P(=O)(O^-){-}O{-}CH_2{-}CH_2{-}N(H){-}CH_2{-}C(=O){-}N(H){-}}\text{[piperidine N—O]}
\end{array}
\tag{IX}
$$

$$
\begin{array}{l}
\mathrm{CH_2{-}O{-}CO{-}(CH_2)_{14}{-}CH_3} \\
\mathrm{CH_3{-}(CH_2)_{14}{-}CO{-}O{-}CH} \\
\mathrm{CH_2{-}O{-}P(=O)(O^-){-}(CH_2)_2{-}NH{-}CO{-}(CH_2)_3{-}CO{-}NH{-}}\text{[piperidine N—O]}
\end{array}
\tag{X}
$$

Rey and McConnell[40] have shown that the affinity constants k of specific IgG antibodies for simple nitroxide haptens such as

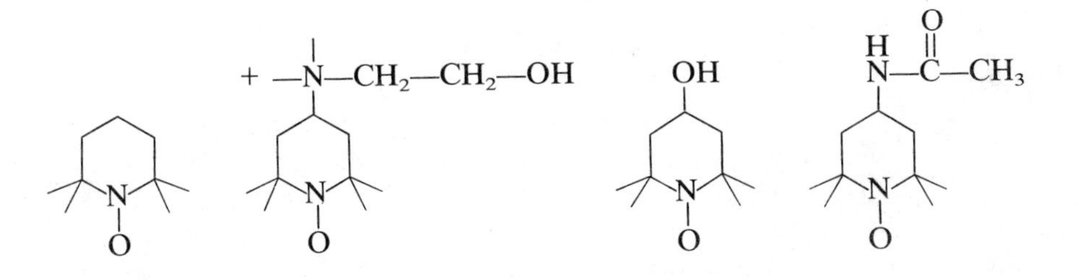

are approximately equal to one another, and of the order of approximately 3×10^6 1/mole

$$\mathrm{IgG + H \rightarrow IgGH} \tag{4}$$

$$k = \frac{[\mathrm{IgGH}]}{[\mathrm{IgG}][\mathrm{H}]} \tag{5}$$

We assume that the affinities of these specific IgG molecules for spin-labeled lipid haptens such as (V), (IX), or (X) are of the same order of magnitude. Figure 10 illustrates the effect of antibody binding on the paramagnetic resonance spectra of (X) incorporated into lipid membrane vesicles.

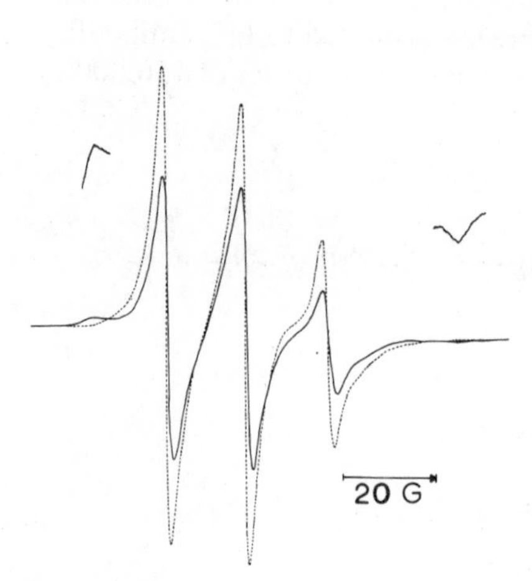

Fig. 10. Effect of antibody binding on the paramagnetic resonance spectrum of spin-label hapten (*X*) in phospholipid vesicles. The dashed line (----) spectrum is due to label (*X*) in the absence of antibodies directed against the nitroxide group, and the solid line spectrum (——) is observed after the addition of these antibodies. Antibody binding gives rise to a reduction in the sharp signal due to the membrane-bound but otherwise free lipid hapten on the outer surface of the bilayer membrane, and to the appearance of the "strongly immobilized" outer wings. These outer wings in the resonance spectrum thus correspond to nitroxide groups that are both membrane-bound and antibody-bound (see Humphries Refs. 21 and 6).

Many studies of the binding of monovalent haptens in solution to IgG antibodies have led to the conclusion that these bindings are essentially independent of one another. Thus the affinity constant for the second binding

$$\mathrm{IgGH} + \mathrm{H} \rightarrow \mathrm{IgGH_2} \tag{6}$$

is essentially equal to the equilibrium constant k for reaction (4). It is well known from many immunochemical studies that di- or multivalent haptens (e.g., H—H or H_n) bind to antibodies with much higher affinities ("avidities") than do the monovalent haptens, provided that the distance between the haptenic groups is compatible with the distance between the binding sites on the antibody molecule

$$\mathrm{IgG} + \mathrm{H{-}H} \rightarrow \mathrm{IgG(H{-}H)} \tag{7}$$

$$\frac{[\mathrm{IgG(H{-}H)}]}{[\mathrm{IgG}][\mathrm{H{-}H}]} = K \tag{8}$$

$$K \gg k \tag{9}$$

The ratio K/k can be 10^3 or greater. For a theoretical discussion, see Carothers and Metzger.[7]

Let us now consider a "fluid" membrane where the lipid haptens can move laterally, and consider the probability that if an IgG molecule is bound to one hapten, it will also be bound to a second hapten. (We assume that there are no steric constants preventing this second binding.) As a very crude estimate, we say that the second IgG binding site is located somewhere within 30 Å of the surface of the bilayer, and within this area there is a three-dimensional concentration (mole/liter) of hapten C that is

$$C \simeq \left(\frac{c}{100}\right)\left(\frac{1}{A \times 10^{-16}}\right)\left(\frac{1}{30 \times 10^{-8}}\right)\left(\frac{1}{6 \times 10^{23}}\right) \times 10^{3} \tag{10}$$

where c is the mole % hapten in the plane of the membrane, and $A \simeq 60$ Å^2 is the average area per lipid molecule. Thus $C \simeq c/100$. With these assumptions it is clear that the second site is bound to hapten as long as $c \gg 100\ k^{-1}$.

Let us now consider a second, related problem. Suppose a membrane is rigid and contains a random distribution of membrane haptens in the plane of the membrane. What is the probability ρ that if an antibody molecule is bound to the membrane at one hapten, it will also be bound at a second hapten?

If R is a typical distance between the two combining sites in an antibody molecule ($R \sim 150$ Å) and if an antibody molecule can bind to the membrane with both combining sites if the two haptens are a distance between R and $R + \Delta R$ apart, then the a priori probability of binding with both combining sites is of the order of

$$\frac{2\pi R\, \Delta R c}{100 A} \tag{11}$$

where c is again the mole % hapten in the plane of the membrane and A is the average area of a lipid molecule. If ΔR is about 20 Å (depending on the flexibility of the antibody molecule and the hapten) and $A = 60$ Å^2, we see that this a priori probability of two-site binding is of the order of one when c is of the order of 0.3 mole %, $c = c^* \sim 0.3$ mole %. In other words, when $c \ll c^*$, there is a low probability of two-site binding in a rigid membrane, whereas the degree of two-site binding can be high if the membrane is fluid, and condition $c \gg 100\ k^{-1}$.

We now consider very briefly the kinetics of antibody binding, leading to two-site attachment. Obviously, for a completely rigid membrane, there will be two-site binding for all antibodies when $c \gg c^*$. When $c \ll c^*$, the only mechanism for two-site binding is by diffusion from solution to the small proportion of sites that are paired. For a *fluid* membrane (lipid hapten diffusion constant $D \simeq 2 \times 10^{-8}$ cm^2/s) we expect paired binding whenever condition $c \gg 100\ k^{-1}$ holds. An appropriate discussion of the kinetics of this process follows.

For simplicity, suppose that a membrane-bound IgG complex is stationary and lipid haptens diffuse in its presence. If the free haptens are imagined to diffuse on a square lattice, with sides l, then the jump rate τ^{-1} is of the order of magnitude

$$\tau^{-1} = \frac{2D}{l^2} \tag{12}$$

The probability per unit time that a hapten will appear at a particular lattice site is

$$\rho = \frac{2D}{l^2}\left(\frac{c}{100}\right) \tag{13}$$

If the hapten binds to the antibody when it appears at a particular site, then $\rho = 2 \times 10^4\ \mathrm{s}^{-1}$ when $l^2 = 60 \times 10^{-16}\ \mathrm{cm}^2$, $D = 2 \times 10^{-8}\ \mathrm{cm}^2/\mathrm{s}$, and $c =$ 0.3 mole %.

Substantially the same result is obtained using the solutions of the two-dimensional diffusion equation given by Adam and Delbrück.[1] We suppose that at time $t = 0$ an IgG antibody molecule binds to a lipid hapten with one site. What is the average rate at which a second hapten diffuses up to the second combining site? From (18) of Adam and Delbrück[1] this is

$$\rho = \frac{Dy_1^2}{b^2} \tag{14}$$

where πb^2 is the average area occupied by *two* haptens; the term y_1 is in the range between 1.1 and 0.49 when a/b is in the range 10^{-1} to 10^{-4}; when $k' < 10^{-4}$,

$$\frac{1}{y_1^2} = 1.15 \log_{10}(k^{-1}) - 0.250 \tag{15}$$

to within an error of less than 2%. Here $k' = a/b$, where πa^2 is the effective area of the antibody combining site and/or hapten, taking into account segmental flexibilities. Taking $\pi b^2 = 2A(100/c)$, $A =$ $60 \times 10^{-16}\ \mathrm{cm}^2$, $c = 0.3$ mole %, $k' = 10\ \text{Å}/100\ \text{Å}$, we obtain $\rho \simeq 10^4\ \mathrm{s}^{-1}$.

If the *on* rates for small hapten binding to antibody are close to diffusion limited, say, 10^8 *l*/mole-s, then with $k \simeq 3 \times 10^6$ *l*/mole we see that the lifetime of the state in which antibody is bound to lipid hapten is approximately 10^{-2} to 10^{-3} s. Thus for nitroxide hapten concentrations of the order of 0.3 mole %, each monovalent binding is expected to be followed by divalent binding, in the absence of steric restraints. This conclusion depends strongly on the assumptions, and the kinetics of divalent binding must decrease markedly for lower values of the hapten concentration and diffusion constant.

From the preceding discussion it is clear that both the degree and rate of antibody binding to a model membrane can depend on membrane fluidity, when the hapten concentration in the plane of the membrane is low (i.e., $c^* \leq 0.3$ mole %). We have not yet made studies of the kinetics of antibody binding.

In recent work we have, however, studied a more complex immunochemical reaction, complement fixation, and this does show a dependence on lateral hapten mobility in the plane of the membrane, when the hapten concentration is low, $c \sim 0.3$ mole %.[5] Complement consists of some 11 serum proteins (designated $C_1, C_2, C_3, \ldots, C_9$). These proteins can kill (lyse) a target cell in the presence of antibodies directed against haptens or antigens on the surface of the target cell. In this process the activity of the complement is depleted and the complement is said to be fixed. Studies of the lysis of sheep red blood cells by complement in the presence of specific antibodies of the IgG type has led to the conclusion that at least two IgG molecules must bind next to one another on the red blood cell surface in order to activate the complement system, leading to complement depletion and red blood cell lysis.[39] It is likely but not proved that each of these IgG molecules needs to bind with both combining sites. Thus we can again anticipate a dependence of an immunochemical reaction on hapten lateral mobility, when the hapten concentration is low. For the complement to be activated by the "classical pathway" (the only one discussed here), C_1 must be activated to form an active serine esterase that acts on C_2. In order for C_1 to be activated, one component of this protein, C_{1q}, must bind to IgG molecules that are in turn bound to the membrane haptens. Based on the known dimensions of C_{1q} we again conclude that interhapten distances of the order of 150 to 200 Å, for groups of four haptens, are necessary for complement activation and fixation. Again a strong dependence of complement fixation on lateral mobility is expected when the hapten concentration is below about 0.3 mole %. Brûlet and McConnell[5] have shown this effect as described in Fig. 11. In this experiment the hapten concentration in the plane of the membrane is varied between 0.25 and 0.025 mole %. As expected, for the lower hapten concentrations, complement fixation is far more effective in the fluid membrane than in the solid membrane. The fluid membrane is dimyristoylphosphatidylcholine at a temperature (32°C) above its chain melting transition temperature (23°C). The solid membrane is dipalmitoylphosphatidylcholine at a temperature (32°C) below its chain melting transition temperature (42°C).

It is clearly desirable to study the activation of complement in more detail, particularly to show an explicit dependence of the rate of C1 activation on the two-dimensional diffusion constant D. This would

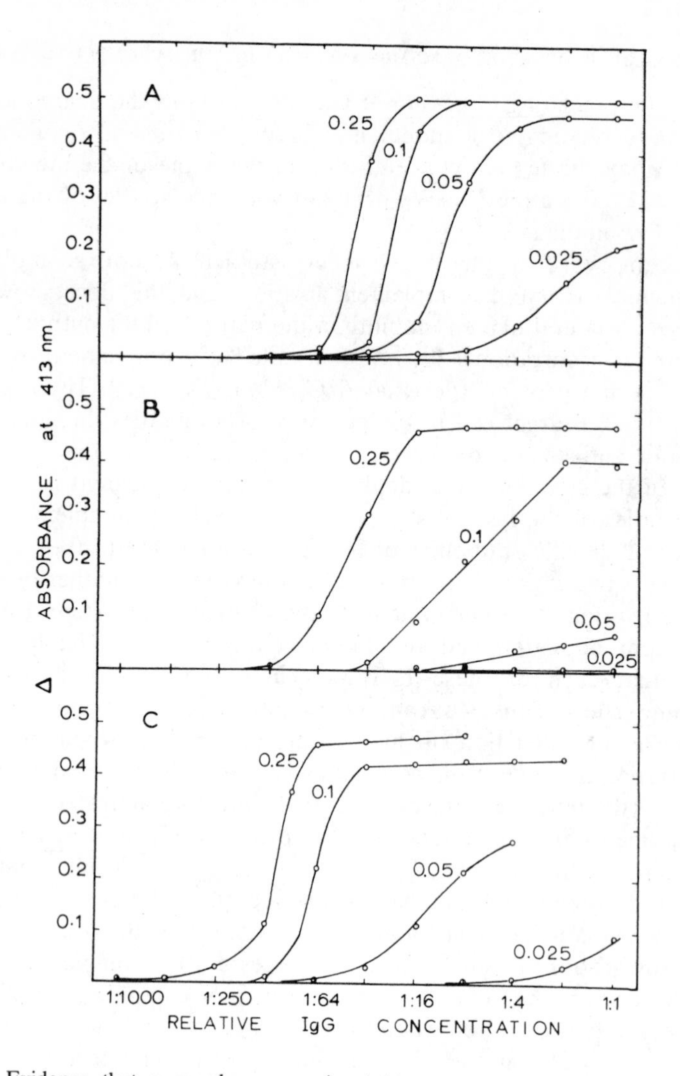

Fig. 11. Evidence that a membrane-associated immunochemical reaction (complement fixation) depends on the mobility of the target hapten (IX) in the plane of a model membrane. The extent of the immunochemical reaction, complement fixation, is measured by "Δ Absorbance at 413 nm." Temperature is always 32°C, which is above the chain-melting temperature (23°C) of dimyristoylphosphatidylcholine used for the data given in *A* and below the chain-melting transition temperature (42°C) of dipalmitoylphosphatidylcholine used for the data in *B*. Thus *A* refers to a fluid membrane and *B* refers to a solid membrane. The numbers by each curve are equal to *c*, the mole % of spin-label hapten IX in the plane of the lipid membrane. It will be seen that complement fixation, as measured by "Δ Absorbance at 413 nm" is far more effective in the fluid membrane than in the solid membrane at low hapten concentrations (i.e., $c \ll 0.3$ mole%). In *C* the lipid membrane host is a 50:50 mole ratio mixture of cholesterol and dipalmitoylphosphatidylcholine. The immunochemical data suggest that this membrane is in a state of intermediate fluidity. Specific affinity-purified IgG molecules were used in these experiments. (For further details, see Ref. 5.)

require the preparation of pure, inactivated Cl from serum and the careful removal of other serum components that can either activate Cl or inhibit activated Cl. Consider an experiment in which a membrane containing a low hapten concentration is saturated with high-affinity antibodies and enough time elapses so that at $t=0$ all membrane-bound antibodies have both binding sites occupied. If Cl is added to this membrane preparation at time $t=0$, then we can anticipate that complement fixation will proceed by a two-step reaction sequence,

$$(IgGH_2)+Cl \rightarrow (IgGH_2)Cl \tag{16}$$

$$(IgGH_2)+(IgGH_2)Cl \rightarrow (IgGH_2)_2Cl \tag{17}$$

$$(IgGH_2)_2Cl \rightarrow (IgGH_2)_2C\bar{l} \tag{18}$$

The rate constant of the second reaction is clearly proportional to the membrane diffusion constant D. For *this* dependence on the diffusion constant to be detected, it is necessary that the lifetimes of $IgGH_2$ and $(IgGH_2)Cl$ be sufficiently long and the lifetime of $(IgGH_2)_2Cl$ be sufficiently short (relative to the formation of $(IgGH_2)_2C\bar{l}$) so that the overall rate is limited by reaction (17).

A possible example of a biological membrane exhibiting a mobility-dependent susceptibility to attack by specific antibodies and complement is provided by the Rh(D) antigen in the human erythrocyte. In a detailed study of IgG (and Fab) binding to Rh(D) positive cells, Hugh-Jones[20] concluded that there are about 30,000 antigens per red blood cell, and that this antigen is monovalent (i.e., only one binding site of each of the IgG molecules bound to the membrane is occupied). These antigens are "dilute" in the plane of the membrane in that on the average each antigen occupies about 30,000 Å^2 if they are randomly distributed. All studies of the intact red blood cell indicate that the rates of lateral diffusion (or other motion) of membrane components are very low. Klum and Muschel[25] found that even though anti-Rho(D) antibodies (of the IgM type) react with red cell D antigens, this reaction does not fix complement and does not lead to immune lysis. On the other hand, trypsin-treated membrane fragments (stroma) from these cells did fix complement in the presence of anti-D sera (IgM antibodies). These results are therefore consistent with the mobility-dependent complement reaction, but of course do not prove this relationship, since trypsin treatment of cell membranes could have effects other than merely enhancing the lateral mobilities of the Rh(D) antigens.

In our opinion it is likely that a number of important membrane immunochemical reactions will be found to be dependent on lateral mobility. It will be a challenge to discover whether the lateral mobility of membrane components is under biochemical and biophysical control, in response to cell function, and whether or not changes in mobility can be involved in immune response and cell-cell recognition.

Acknowledgments

This work was supported by NIH Grant no. 1 R01 AI13587-01 BBCB.

References

1. G. Adam and M. Delbrück, in *Structural Chemistry and Molecular Biology*, Freeman, San Francisco, 1968, pp. 198–215.
2. P. Brûlet, Ph.D. thesis, Stanford University, 1975.
3. P. Brûlet and H. M. McConnell, *Proc. Nat. Acad. Sci.* (*U.S.*), **72,** 1451–1455 (1975).
4. P. Brûlet and H. M. McConnell, *J. Am. Chem. Soc.*, **98,** 1314–1318 (1976).
5. P. Brûlet and H. M. McConnell, *Proc. Nat. Acad. Sci.* (*U.S.*), **73,** 2977–2981 (1976).
6. P. Brûlet, G. M. K. Humphries, and H. M. McConnell, in *Structure of Biological Membranes*, S. Abrahamsson and I. Pascher, Eds., Plenum, New York, 1977, pp. 321–329.
7. D. M. Carothers and H. Metzger, *Immunochemistry*, **9,** 341–357 (1972).
8. R. J. Cherry, A. Bürkli, M. Busslinger, G. Schneider, and G. R. Parish, *Nature*, **263,** 389–393 (1976).
9. P. Devaux and H. M. McConnell, *J. Am. Chem. Soc.*, **94,** 4475–4481 (1972).
10. P. Devaux and H. M. McConnell, *Ann. N.Y. Acad. Sci.*, **222,** 489–496 (1973).
11. P. Devaux, C. J. Scandella, and H. M. McConnell, *J. Magn. Res.*, **9,** 474–485 (1973).
12. G. M. Edelman, *Science*, **192,** 218–226 (1976).
13. M. Ediden, *Ann. Rev. Biophys. Bioeng.*, **3,** 179–201 (1974).
14. E. L. Elson and W. W. Webb, *Ann. Rev. Biophys. Bioeng.*, **4,** 311–334 (1975).
15. R. Fettiplace, L. G. M. Gordon, J. B. H. Ladky, J. Requena, H. R. Zingsheim, and D. A. Hayden, in *Methods in Membrane Biology*, E. D. Korn, Ed., Plenum, New York, 1975, pp. 1–75.
16. L. D. Frye and M. Ediden, *J. Cell Sci.*, **7,** 319–333 (1970).
17. H. J. Galla and E. Sackmann, *Biochim. Biophys. Acta*, **339,** 103–115 (1974).
18. C. W. M. Grant and H. M. McConnell, *Proc. Nat. Acad. Sci.* (*U.S.*), **71,** 4653–4657 (1974).
19. C. W. M. Grant, S. H. W. Wu, and H. M. McConnell, *Biochim. Biophys. Acta*, **363,** 151–158 (1974).
20. N. C. Hugh-Jones, *Nature*, **227,** 174–175 (1970).
21. G. M. K. Humphries and H. M. McConnell, *Biophys. J.*, **16,** 275–277 (1976).
22. M. J. Janiak, D. M. Small, and G. G. Shipley, *Biochem.* **15,** 4575–4580 (1976).
23. W. Kleemann, C. W. M. Grant, and H. M. McConnell, *J. Supramolecular Struct.* **2,** 609–616 (1974).
24. W. Kleemann and H. M. McConnell, *Biochim. Biophys. Acta*, **419,** 206–222 (1976).
25. M. J. Klum L. H. Muschel, *Nature*, **212,** 159–161 (1966).
26. R. D. Kornberg and H. M. McConnell, *Proc. Nat. Acad. Sci.* (*U.S.*), **68,** 2564–2568 (1971).
27. A. G. Lee, N. J. M. Birdsall, J. C. Metcalfe, P. A. Toon, and G. B. Warren, *Biochemistry*, **13,** 3699–3705 (1974).
28. E. J. Luna and H. M. McConnell, *Biochim. Biophys. Acta*, **466,** 381–392 (1977).
29. E. J. Luna and H. M. McConnell, *Biochim. Biophys. Acta*, submitted for publication.
30. S. Mabrey and J. Sturtevant, *Proc. Nat. Acad. Sci.* (*U.S.*) **73,** 3862–3866 (1976)

31. H. M. McConnell, in *The Neurosciences: Second Study Program*, F. O. Schmitt, Ed., Rockefeller University Press, New York, 1970, pp. 697–706.
32. H. M. McConnell, P. Devaux, and C. Scandella, in *Membrane Research*, C. F. Fox, Ed., Academic, New York, 1972, pp. 27–37.
33. H. M. McConnell, in *Spin Labeling: Theory and Applications*, L. Berliner, Ed., Academic, New York, 1976, pp. 525–560.
34. G. Nicolson, *Biochim. Biophys. Acta*, **457,** 57–108 (1976).
35. R. Peters, J. Peters, K. H. Tews, and W. Bähr, *Biochim. Biophys. Acta*, **367,** 282–294 (1974).
36. M. C. Phillips, B. D. Ladbrooke, and D. Chapman, *Biochim. Biophys. Acta*, **196,** 35–44 (1970).
37. M. Poo and R. A. Cone, *Nat. New Biol.*, **247,** 438–441 (1974).
38. M. C. Raff, *Sci. Am.*, **234,** 30–39 (1976).
39. H. J. Rapp and T. Borsos, *Molecular Basis of Complement Action*, Appleton-Century-Crofts, New York, 1970.
40. P. Rey and H. M. McConnell, *Biochem. Biophys. Res. Commun.*, **73,** 248–254 (1976).
41. P. Rey and H. M. McConnell, *J. Am. Chem. Soc.*, **99,** 1637–1642 (1977).
42. C. J. Scandella, P. Devaux, and H. M. McConnell, *Proc. Nat. Acad. Sci.* (*U.S.*), **69,** 2056–2060 (1972).
43. J. Schlessinger, D. E. Koppel, D. Axelrod, K. Jacobson, W. W. Webb, and E. L. Elson, *Proc. Nat. Acad. Sci.* (*U.S.*), **73,** 2409–2413 (1976).
44. E. J. Shimshick and H. M. McConnell, *Biochemistry*, **12,** 2351–2360 (1973).
45. E. J. Shimshick and H. M. McConnell, *Biochem. Biophys. Res. Commun.*, **53,** 446–451 (1973).
46. E. J. Shimshick, W. Kleemann, W. L. Hubbell, and H. M. McConnell, *J. Supramolecular Struct.*, **1,** 285–294 (1973).
47. S. J. Singer, in *Advances of Immunology*, Vol. 19, F. J. Dixon and H. G. Kunkel, Eds., Academic, New York, 1974, pp. 1–62.
48. S. J. Singer and G. L. Nicolson, *Science*, **175,** 720–731 (1972).
49. L. Steck, *J. Cell Biol.*, **62,** 1–19 (1974).
50. H. Träuble and E. Sackmann, *J. Am. Chem. Soc.*, **94,** 4499–4510 (1972).
51. J. D. Verhoeven, *Fundamentals of Physical Metallurgy*, Wiley, New York, 1975, p. 246.
52. A. J. Verkeleij and P. H. J. Ververgaert, *Ann. Rev. Phys. Chem.*, **26,** 101–122 (1975).
53. G. Weismann and R. Claiborne, Eds., *Cell Membranes; Biochemistry, Cell Biology and Pathology*, HP Publishing Company, Inc. New York (1975).
54. S. H. W. Wu and H. M. McConnell, *Biochemistry*, **14,** 847–854 (1975).

LIST OF INTERVENTIONS

1. Ubbelohde
2. Ter Minassian
   2.1 Ubbelohde
   2.2 McConnell
   2.3 Ubbelohde
   2.4 McConnell
   2.5 Ubbelohde
   2.6 McConnell

1. Intervention of Ubbelohde

It may be desirable to define certain basic physical processes afresh, when we are dealing with systems essentially subject to two-dimensional conformations and hence two-dimensional constraints. This is the case for membranes, and also for a number of alkali salts of alkali *n*-alkane carboxylates. These "melt" to give mesophases, in which the anions and cations are arranged in layerlike structures. At considerably higher temperatures the mesophases pass into isotropic ionic melts, but in the intervening temperature range they exhibit marked anisotropy of optical and physical properties. In these mesophases, which are ordered fluid states, some remarkable diffusion parameters are found. For example, parallel to the layers the ionic mobilities are unusually high, and the "activation" energy for ionic diffusion is lower than that even in aqueous solutions. In fact, "cooperative" displacements can sometimes take place faster than displacements that are fully dissipative, because they are only partly randomized [cf. H. J. Michels and A. R. Ubbelohde, *Proc. Roy. Soc.*, **A338,** 447 (1974); J. J. Duruz and A. R. Ubbelohde, *Proc. Roy. Soc.*, **A347,** 301 (1976), which give earlier references to these novel "fluids"].

2. Intervention of Ter Minassian

Concerning the electrostatic interaction of charged macromolecules with lipids and segregation in mixed films. I would like to point out that (to be published in *Electrical Phenomena at Membrane Level*, p. 273 Elsevier, Amsterdam, 1977):

Mixed monolayers of PVPC6 poly-(2-methyl-5-vinyl-hexylpyridinium bromide) and L: phosphalidyl inositol monophosphate, were studied as model of complex membranes.

Surface pressure and potential were measured for various ratios of PVPC6/L.

It was found that when the negatively charged anionic lipid L is added, the repulsion between the positively charged PVPC6 polyions is reduced and the polyion resists compression to higher surface—"lateral"—pressures.

The resulting variation of the ejection or collapse pressure with the mole fraction x_L of L is shown in Fig. 1.

This curve resembles a phase diagram. It may be seen that the compound PVPC6-L is formed when $x_L = 0.3$. For $0 < x_L < 0.3$, the L

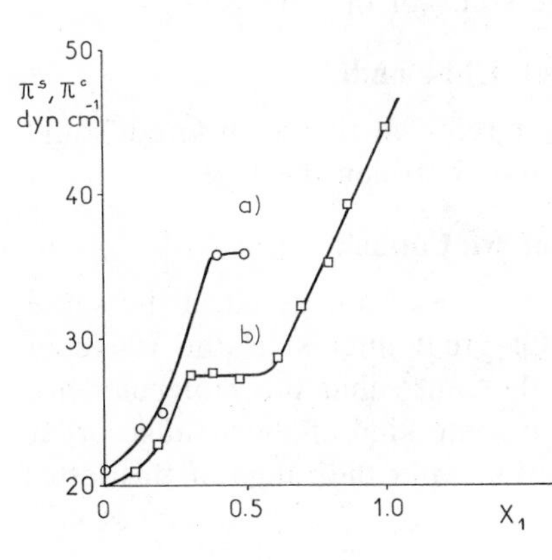

Fig. 1. Phase diagram of mixed PVPC6+L monolayers. Variation of PVPC6 "collapse" pressure with segment fraction of lipid x_L.

molecules screen the repulsion between PVPC6. For $0.3 < x_L < 0.5$, there are two phases corresponding to the nonmiscibility of the neutral PVPC6-L complex and of the mixture of charged PVPC6 and L molecules.

From the analysis of the partial areas of PVPC6 and L in the mixed films, it follows that for $0 < x_L < 0.3$, the lipid molecules are "condensed." But the excess of lipid molecules in films for which $x_L > 0.5$ is "fluid," presumably in a liquid crystalline state.

2.1 Intervention of Ubbelohde

In such mixed films will diffusion be enhanced at defect sites in the membrane, as happens (by analogy) at defect sites in three-dimensional solids?

2.2 Intervention of McConnell

As indicated in my report, we now know the rates of lateral diffusion of phospholipids in lipid bilayers in the "fluid" state, and in a few cases the rates of lateral diffusion of proteins in fluid lipids are also known. At the present time nothing is known about the rates of lateral diffusion of phospholipids in the crystalline, "solid" phases of the substances. As mentioned in my report, there are reasons to suspect that the rates of lateral diffusion of phospholipids in the *solid solution* crystalline phases of binary mixtures of phospholipids may be appreciable on the experimental time scale. Professor Ubbelohde may well be correct in pointing out the possibility of diffusion caused by defects. However, such defects, if present, apparently do not lead to significant loss of the membrane permeability barrier, except at domain boundaries.

An extremely interesting class of problems concerns the behavior of guest molecules in host lipids, particularly when the host lipid system is in a condition where a solid fluid equilibrium exists in the plane of the membrane. Spin labels are ideal for the study of this problem.

2.3 Intervention of Ubbelohde

The organic molten salt mesophases I referred to are ordered liquid crystals with strong electrostatic constraints between the layers.

2.4 Intervention of McConnell

Multilamellar bilayers in the fluid phase are also ordered in the sense that they are smectic liquid crystals. Of great interest is the range of molecular order; this is long-range in the sense that the molecules are confined to two dimensions; there is also some kind of short-range order in molecular orientations and conformations, but the range of this latter ordering is not known at present.

2.5 Intervention of Ubbelohde

There is no diffusion *constant*, I prefer to say.

2.6 Intervention of McConnell

As just mentioned, there are a large number of unsolved problems in membrane biophysics, including the questions of local anisotropic diffusion, hysteresis, protein-lipid phase separations, the role of fluctuations in membrane fusion, and the mathematical problems of diffusion in two dimensions Stokes paradox).

The ^{13}C nuclear resonance studies in my report provide some informations on lipid membrane fluctuations in binary mixtures. Totally unsolved problems include an appropriate two-dimensional Debye-Huckel theory for membranes, and theoretical treatments of boundary free energies (between proteins and lipids, and between solid and fluid phase lipids).

3. Intervention of Mathot

I should like to remark that in the case of a phase diagram as reported by Fig. 2 in Dr. McConnell's paper, one is bound to observe concentration inhomogeneities both in the liquid and in the solid phases during solidification (see Fig. 1.).

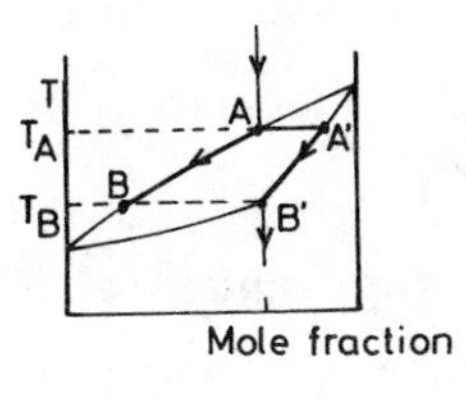

Fig. 1

Indeed for a given mixture the fluid composition changes continuously from A to B, whereas the solid composition varies correspondingly from A' to B', as the temperature decreases from T_A to T_B. So *unless diffusion is extremely efficient* (as in plastic crystals, for example), the two phases are likely to be fairly inhomogeneous.

3.1 Intervention of McConnell

I believe that Dr. Mathot has raised the question of phase equilibration in our phase diagrams. If we consider a simple solid solution, fluid solution phase diagram for a binary mixture, there are two limiting consequences of lowering the temperature from above the fluidus curve (T_1) to below the solidus curve (T_2). The solid phase may or may not have a uniform composition, depending on the rate of cooling and the rate of diffusion in the solid phase. In many of our experiments we are quite confident that the solid phase has a uniform composition because the solid phase has a texture (seen in the electron microscope) that is quite uniform (see Fig. 3 in my report as an example).

3.2 Intervention of Ubbelohde

What is the domain size in these systems?

3.3 Intervention of McConnell

As mentioned in my report, in addition to large coexisting domains of solid and fluid phase lipids, there may be fluctuations of composition and density in the fluid-lipid phase that are not seen in the electron microscope but that may affect nuclear magnetic resonance spectra.

A special type of cluster formation has been observed for certain spin-label solutes in host lipid bilayers, corresponding to the equilibrium

$$nX \rightleftharpoons Xn$$

as discussed in my report.

3.4 Intervention of Ubbelohde

Is the domain structure a macrofluctuation structure?

3.5 Intervention of Mayer

What is their order of magnitude?

3.6 Intervention of McConnell

Their order of magnitude is $n = 6 \pm 0.1$ or 0.2. The formula is given in the report (VII).

3.7 Intervention of Ubbelohde

In the domain structures found for mesophases of molten alkali carboxylates, isomerism of the fatty acid chain makes an enormous difference to domain size; for example, it is much larger for *n* than for isoalkane carboxylates.

3.8 Intervention of McConnell

The chemical compositions and isomeric structures of the fatty acid chains of phospholipids is well known to have large effects on the physical properties of lipid bilayers, such as the temperatures of endothermic chain melting phase transitions. Lipid vesicles "sensitized" with lipid haptens can be agglutinated with specific antibodies directed against the haptens (see Fig. 1).

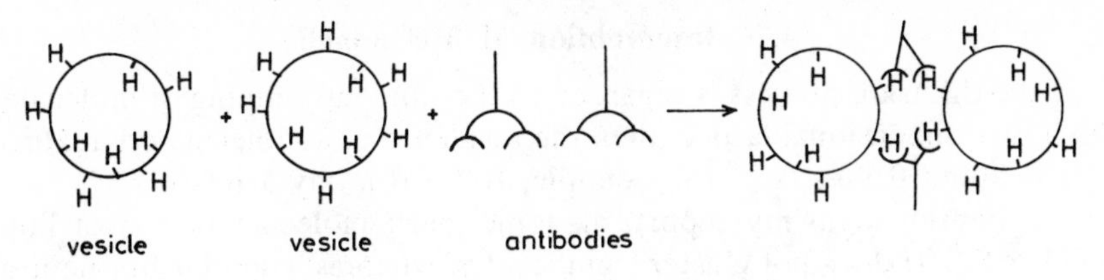

Fig. 1

3.9 Intervention of Ubbelohde

Is there any evidence for mechanical deformation?

3.10 Intervention of McConnell

The extent and speed of this agglutination depends on the hapten concentration in the plane of the membrane and on the lateral mobility of the hapten (G. K. H. Humphries, P. Brûlet, H. McConnell, unpublished). It is extremely probable that this simple agglutination reaction leads to a change in the membrane shape.

4. Intervention of Sanfeld-Steinchen

As was stressed by Professor Ubbelohde, in the process of cell recognition not only the lateral diffusion of the binding sites has to be considered, but also the mechanical effects resulting from the local change of surface tension, inducing convection at the cell surface. It is well known, in the cell-to-cell contact inhibition of motion, in tissue culture, that a cell approaches another cell by "touching" it by means of microvilli and that this process can be affected when adding surfactants to the culture. Now the point is, "What is the relative importance of both diffusion and convection?" Well, in binary surface films, it was observed that the transport process induced by two-dimensional convection is much more rapid than the two-dimensional diffusion.

4.1 Intervention of Hess

Please discuss whether there are any means to control the rate of lateral diffusion in the membrane system you have studied.

What is the effect of charge distribution and density, and of the ionic strength and the bulk temperature on the cluster formation, equivalent to those observed by Hermann Traüble and his group?*

*H. Traüble, Membrane Electrostatics, p. 509–550 in "Structure of Biological Membranes," Eds. S. Abrahamson and I. Pascher, Plenum Publ. New York (1976).

4.2 Intervention of McConnell

If the diffusion process is regarded as the random jumping of molecules on a two-dimensional lattice, then the residence of a molecule on a lattice site is about $0.1\,\mu$s (see, for example, Ref. 10 in my article).

As mentioned in my report, an ionic guest molecule in a host lipid bilayer (VIII) does not cluster significantly, whereas a similar but neutral guest (VII) clusters extensively as hexamers. I would expect this clustering to depend strongly on ionic strength only if the host lipids were charged.

Laser pulse techniques may enable us to determine the diffusion times of clusters.

4.3 Intervention of Ubbelohde

Is there "hopping" diffusion?

4.4 Intervention of McConnell

Studies by Strittmatter at Connecticut have provided examples of lateral diffusion-limited, as well as nonlimited, biochemical reactions involving two membrane-bound proteins.

5. Intervention of Bangham

I am curious to know by what mechanism simple phospholipids, such as liposomes or vesicles, are instantly recognized by within blood as "non-self"?

5.1 Intervention of McConnell

There is certainly some nonspecific binding between proteins such as immunoglobulin molecules and some lipid membranes.

5.2 Intervention of Bangham

The recognition of hapten-free phospholipid liposomes as nonself, when introduced into the bloodstream, still seems to me to be very important. I cannot believe that antitumor drugs will be successfully delivered until this problem is solved.

5.3 Intervention of McConnell

Studies by Kriss and colleagues in the Stanford Medical School have shown how lipid vesicles are "cleared" from the bloodstream by various organs.

5.4 Intervention of Bangham

I am glad to hear that Dr. McConnell agrees with me that the accretion of lipid vesicles by the liver is nonspecific.

5.5 Intervention of McConnell

There can certainly be nonspecific binding, or fusion in this case.

6. Intervention of Hauser

Is it possible to use simple bilayer vesicles (liposomes) to test the involvement of other modes of motion than lateral diffusion of lipids (e.g., motions across the bilayer that would be important in the transmission of signals from the cell interior to the external cell surface)?

6.1 Intervention of McConnell

There are certainly plausible mechanisms whereby intracellular events could provide extracellular signals involving the display and mobility of surface antigens. Transmembrane proteins such as those shown in Fig. 1 could be held in place, or allowed to move, by submembranous "regulator protein complexes," sensitive to Ca^{2+}, cAMP, and so on.

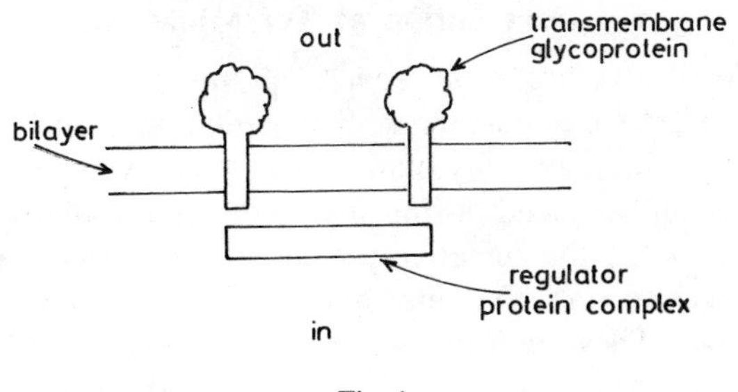

Fig. 1

6.2 Intervention of Welch

I would like to note a particular phenomenon that supports Professor McConnell's comments on the importance of lateral mobility of membrane components. This concerns the observed motion of membrane agglutination sites in viral transformed cells. Apparently, this mobility, which does not exist in normal cells, is responsible for the agglutination effect. The point I am stressing is the relationship between the mobility, say, of concanavalin A binding sites and agglutinability. The motion usually does not exist in normal cells, which presumably have a control mechanism governing membrane stability. And fixation of the surface membrane (e.g., by glutaraldehyde) in transformed cells inhibits agglutination [e.g., I. Vlodavsky, M. Inbar, and L. Sachs, *Proc. Nat. Acad. Sci.* (*U.S.*), **70,** 1780 (1973)].

6.3 Intervention of McConnell

There is much indirect evidence that transformation of cells by viruses leading to enhanced agglutination by lectins may be due to enhanced mobilities of cell surface lectin receptors. The immune response to such cell surfaces may also be affected by transformation.

7. Intervention of Hess

Can you comment on "Cap" formation?

7.1 Intervention of McConnell

It is well known, for example, from the work of Raff and colleagues, that plant lectins produce two types of motion in lymphocytes, for example. These motions lead to "patching" (not requiring ATP) and "capping" (requiring ATP). For details, see the *Scientific American* article by Raff in my report, list of references.

8. Intervention of Ter Minassian

An example of attachment on several points of the membrane is the Ca^{2+} binding. Using monolayers with negative sites carboxyls or phosphates, it is seen that Ca^{2+} may bind either by one valence on one site or by two valences on two neighboring sites. The type of binding depends on the distance between the negatively charged sites: Only when this distance is very close to Ca^{2+} diameter does the two-valence or cross-linking attachment occur. These two types of Ca^{2+} attachment to the monolayer lead to very different surface potential and viscosity of the monolayers of phospholipids or fatty acids.

8.1 Intervention of McConnell

There are a large number of factors that can potentially affect the mobility and accessibility of membranes antigens.

8.2 Intervention of Eisenman

I assume there must be some temperature dependence of hapten binding. Is there any correlation (or absence of correlation) with solid-liquid lipid phase transitions (i.e., with membrane fluidity)?

8.3 Intervention of McConnell

We have no evidence for a significant temperature dependence of antibody binding, although one might anticipate a strong dependence for very "short" haptens (close to the membrane).

8.4 Intervention of Bangham

Were the antibody versus mole % cholesterol uptakes made at constant hapten concentration?

8.5 Intervention of Ter Minassian

The mixed lipid-cholesterol monolayers are unstable. Recent studies of these systems show that for molecular fractions of cholesterol larger than 30% the cholesterol separates from the film, as if there were a limited misability in two dimensions.

8.6 Intervention of McConnell

Many studies show changes in the physical and even biochemical properties of phosphatidylcholine cholesterol membranes, when the cholesterol concentration is increased above 20 to 40%.

8.7 Intervention of Ubbelohde

It is encouraging to find a convergence in our discussions between the biochemical and the mechanical approach to these problems. As one passes to molecules of ever-increasing size, such convergence is to be sought for, but we need to appreciate effects of molecular probability statistics on the rigid determinism of mechanics.

SELECTIVE TRANSPORT PROCESSES IN ARTIFICIAL MEMBRANES

W. SIMON

Department of Organic Chemistry, Swiss Federal Institute of Technology, Zurich, Switzerland

I. INTRODUCTION

Much of the rather slow information transfer in biological systems is achieved by the release and subsequent transport of messenger molecules. For a fast information transfer over large distances, however, a combination of electrical and chemical transport processes is involved (in part cited from Ref. 1). Since the body is an aqueous organization, rather hostile to free electrons, it comes as no surprise that the carriers of charge are predominantly ions.[1] Among the inorganic ions, Na^+, K^+, Ca^{2+}, and Cl^- play an essential role in nerve transmission.

Biological membranes are usually very thin and suggest a comparison of their transport behavior with the one of correspondingly thin model membranes. Such artificial, ultrathin membranes[2–4] [bilayer lipid membranes, black lipid membranes (BLM)] can be made from a variety of well-characterized chemicals and a vast amount of data has been reported on such systems. For a recent review on the electrochemistry of artificial, ultrathin lipid membranes see Ref. 1 (and also Refs. 5 to 7). Straightforward experimental work on ion transport is cumbersome; this is all the more true if mechanisms are to be studied. This is one of the reasons why the present report deals with relatively thick solvent polymeric membranes (bulk membranes). To further simplify the discussion the system is restricted to bulk membranes, the ion selectivity of which is modified by electrically neutral ion-selective components (ion carriers, ionophores) only. The effect of classical ion exchangers (charged ligands) is not discussed here.

II. PROPERTIES OF SOLVENT POLYMERIC BULK MEMBRANES*

Solvent polymeric membranes[9] (SPM) were prepared by dissolving polyvinyl chloride (approximately 30 wt.%), plasticiser (approx. 65 wt.%),

* The structures of the different ligands discussed are given at the end of this report.

TABLE I. Properties of Solvent Polymeric and Black Lipid Membranes

Property	Solvent polymeric membrane 35 wt.% polyvinyl chloride, 65 wt.% dioctyladipate	Black lipid membrane (Lecithin)
Thickness	0.004 cm	$(50\text{–}80) \cdot 10^{-8}$ cm, Ref. 6
Dielectric constant of major component	4.2, Ref. 11	2–3, Ref. 6
Resistance of unmodified membrane	$5 \cdot 10^{7}\ \Omega\ \text{cm}^2$, Ref. 11	$10^{8}\text{–}10^{10}\ \Omega\ \text{cm}^2$, Ref. 6
Resistance of valinomycin-modified membrane	$8 \cdot 10^{5}\ \Omega\ \text{cm}^2$,[a] Ref. 11	$2.5 \cdot 10^{3}\ \Omega\ \text{cm}^2$,[b] Ref. 12
Electric field at membrane potential difference of 60 mV	$15\ \text{Vcm}^{-1}$	$10^{5}\ \text{Vcm}^{-1}$
Diffusion coefficient of valinomycin in membrane phase	$2 \cdot 10^{-8}\ \text{cm}^2\ \text{sec}^{-1}$	$3 \cdot 10^{-8}\ \text{cm}^2\ \text{sec}^{-1}$ Ref. 12 (estimated for triolein)

[a] $4 \cdot 10^{-3}$ *M* KCl (aqueous phase): 1.0 wt.% valinomycin ($8.9 \cdot 10^{-3}$ mole kg^{-1}) in dioctyl-adipate

[b] $4 \cdot 10^{-3}$ *M* KCl, 10^{-7} *M* valinomycin (aqueous phase).

and a neutral ion-selective component (approx. 1 to 5 wt.%) in tetrahydrofuran and pouring the solution into a glass ring resting on a sheet of plate glass.[10] The resulting membrane, about 0.01 to 0.1 mm thick, was then peeled off the glass and used as described later. In Table I some properties of these membranes are compared with those of black lipid membranes. Although dominating properties such as the dielectric constant of the major membrane components and the diffusion coefficients are comparable, there are larger differences in the electrical membrane resistance and in other parameters depending on membrane thickness.

For similar solvent polymeric membranes (78 wt.% dicresyl butyl phosphate in polyvinyl chloride) self-diffusion coefficients of the order of $10^{-7}\ \text{cm}^2\ \text{s}^{-1}$ have been reported.[13] These diffusion coefficients, as well as measurements of rotational mobilities,[14] indicate that the solvent polymeric membranes studied here are indeed liquid membranes. This liquid phase is so viscous, however, that convective flow is virtually absent. This contrasts with pure solvent membranes where an organic solvent is interposed between two aqueous solutions either by sandwiching it between two cellophane sheets or by fixing it in a hole of a Teflon sheet separating the aqueous solutions.[15] The extremely high convective flow is one of the reasons why the term *membrane* for extraction systems consisting of thick layers of low viscosity solvents is extremely inappropriate.[16,17]

III. SELECTIVITY IN MEMBRANE POTENTIAL ("EQUILIBRIUM DOMAIN")

The selectivity of membranes between two cations of the same charge (in the absence of other interfering species) can be conveniently characterized by the observed zero-current membrane potential E through a measurement of the EMF of a cell of the type shown in Fig. 1. The potential difference between solutions 1 and 2, the so-called membrane potential E, is ideally described by:[18–21]

$$E = \frac{RT}{zF} \ln \frac{a'_I + K^{\text{Pot}}_{IJ} a'_J}{a''_I + K^{\text{Pot}}_{IJ} a''_J} \qquad (1)$$

where $\frac{RT}{zF}$ = Nernst factor (z, charge of cations)

K^{Pot}_{IJ} = potentiometric selectivity factor of membrane

a'_I, a'_J = activities of the cations in solution 1

a''_I, a''_J = activities of the cations in solution 2

If the composition of solution 2 is assumed to remain unchanged and if K^{Pot}_{IJ} is constant, the Goldman–Hodgkin-Katz type equation (1)[20,21] (see

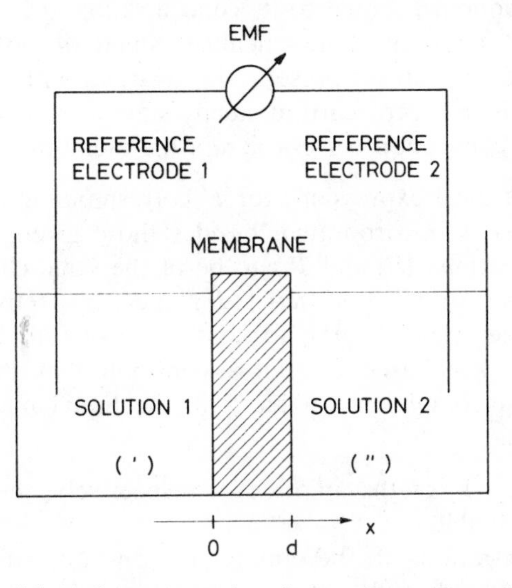

Fig. 1. Schematic representation of a membrane electrode assembly.

also Nicolsky[22]) reduces to

$$E = \text{const} + \frac{RT}{zF} \ln [a_I' + K_{IJ}^{\text{Pot}} a_J'] \tag{2}$$

An inspection of (1) and (2) shows the selectivity behavior of the membranes to be fully specified by K_{IJ}^{Pot}. Through measurements of the EMF of the cell shown in Fig. 1 and by using as aqueous solutions 1 e.g., fixed activities of the chlorides of the cations I^{z+} and J^{z+} respectively, K_{IJ}^{Pot} may be determined experimentally (for details, see Ref. 23).

The theoretical treatment of the selectivities of various types of liquid membranes is mainly due to efforts of Eisenman and his school and others.[24–32] To derive simple expressions for the membrane potential E, a number of assumptions have to be made.[19,32] Using the assumptions 1 to 7 a universal description of E becomes possible.[19,32]

1. The same solvent is used for the electrolyte solutions on either side of the membrane.
2. A thermodynamic equilibrium exists between the membrane and each of the outside solutions at the respective phase boundaries.
3. Every cell component is homogeneous with respect to a direction perpendicular to the cell axis; therefore concentration gradients and the concomitant potential differences are possible only along this cell axis (x-coordinate, Fig. 1). There is no pressure gradient.
4. The membrane phase approximates ideal behavior, that is, the activity of any component is equal to its concentration.
5. Within the membrane, the chemical standard potentials and the mobilities of all particles present are invariant with space and time.
6. The system is in a zero-current steady state.
7. There is no convection current across the membrane.

To obtain theoretical expressions for E corresponding to (1), additional restrictions must be incorporated, besides those given in the preceding list. Assuming cations I^{z+} and J^{z+} to be of the same charge $z+$ and the formation of any or all $1:n$ ($n = 1, 2, \ldots, N$, respectively, M) cation-carrier complexes possible, the expressions given in Table II are obtained.[33] These are based on the assumption that the concentration $[c_S(x)]$ of the uncomplexed carrier S in the liquid membrane remains constant and that

A. The mobility u_X of the anions X^-, respectively, their concentration $c_X(x)$ is negligible, *or*
B. The concentration of the anions is constant, furthermore, electroneutrality holds within the membrane phase, *or*

TABLE II. Model for Neutral Carrier (S) Liquid Membranes, Cations of Equal Charge (I^{z+}, J^{z+}), and Anions X^- [a]

Species in membrane	A $u_X \cdot c_X(x) = 0$	No ion pair formation, $c_S(x)$ constant, and[b] B $c_X(x)$ constant electroneutrality holds in membrane	C $\sum_{n=0}^{N} c_{IS_n}(x) + \sum_{n=0}^{M} c_{JS_n}(x)$ constant no diffusion potential difference
I $\begin{cases} I^{z+}, IS^{z+}, \ldots, IS_N^{z+}, \\ J^{z+}, JS^{z+}, \ldots, JS_M^{z+}, \\ X^-, S \end{cases}$	$K_{IJ}^{\text{Pot}} = \frac{k_J}{k_I} \cdot \frac{u_J + \sum_{n=1}^{M} u_{JS_n} c_S^n \beta_{JS_n}}{u_I + \sum_{n=1}^{N} u_{IS_n} c_S^n \beta_{IS_n}}$	$K_{IJ}^{\text{Pot}} = \frac{k_J}{k_I} \cdot \frac{zu_J + u_X + \sum_{n=1}^{M} (zu_{JS_n} + u_X) c_S^n \beta_{JS_n}}{zu_I + u_X + \sum_{n=1}^{N} (zu_{IS_n} + u_X) c_S^n \beta_{IS_n}}$	$K_{IJ}^{\text{Pot}} = \frac{k_J}{k_I} \cdot \frac{1 + \sum_{n=1}^{M} c_S^n \beta_{JS_n}}{1 + \sum_{n=1}^{N} c_S^n \beta_{IS_n}}$
II $\begin{cases} IS^{z+}, \ldots, IS_N^{z+}, \\ JS^{z+}, \ldots, JS_M^{z+}, \\ X^-, S \end{cases}$	$K_{IJ}^{\text{Pot}} = \frac{\sum_{n=1}^{M} u_{JS_n} c_S^n \beta_{JS_n} k_J}{\sum_{n=1}^{N} u_{IS_n} c_S^n \beta_{IS_n} k_I}$	$K_{IJ}^{\text{Pot}} = \frac{\sum_{n=1}^{M} (zu_{JS_n} + u_X) c_S^n \beta_{JS_n} k_J}{\sum_{n=1}^{N} (zu_{IS_n} + u_X) c_S^n \beta_{IS_n} k_I}$	$K_{IJ}^{\text{Pot}} = \frac{\sum_{n=1}^{M} c_S^n \beta_{JS_n} k_J}{\sum_{n=1}^{N} c_S^n \beta_{IS_n} k_I}$
III $\begin{cases} I^{z+}, \\ J^{z+}, \\ X^- \end{cases}$	$K_{IJ}^{\text{Pot}} = \frac{k_J}{k_I} \cdot \frac{u_J}{u_I}$	$K_{IJ}^{\text{Pot}} = \frac{k_J}{k_I} \cdot \frac{zu_J + u_X}{zu_I + u_X}$	$K_{IJ}^{\text{Pot}} = \frac{k_J}{k_I}$

[a] From Ref. 33.

[b] u = mobilities in membrane phase; $c(x)$ = concentrations in membrane phase; z = charge of ions I and J; k = coefficient of partition between outside solution and membrane; β = cumulative or gross (overall) stability constants of the carrier cation complexes in the membrane phase.

C. The sum of the positive charges is invariant of the location within the membrane. There is no diffusion potential difference.

Equations IIA to IIC in Table II are of special interest for the present discussion. Assuming a 1:1 stoichiometry and comparable radii r of the complexes formed, we may write

$$r_{IS} = r_{JS} \tag{3}$$

and therefore (Stokes law[34])

$$u_{IS} = u_{JS} \tag{4}$$

According to Born's equation,[35] the partition coefficients may then be described by

$$k_{IS} = k_{JS} \tag{5}$$

Results of X-ray analysis of complexes of alkali- and ammonium ions with valinomycin *1*[36,37] and the macrotetrolide antibiotics *2* to *6*[38–42] support this reasoning. For this idealized situation the Equations IIA to IIC (Table II) reduce to:[33]

$$K_{IJ}^{\mathrm{Pot}} = \frac{k_J \cdot \beta_{JS}}{k_I \cdot \beta_{IS}} = \frac{\beta_{JS}^{w}}{\beta_{IS}^{w}} \tag{6}$$

where

k_I, k_J = partition coefficients of the uncomplexed ions between outside solution and membrane

β_{IS}, β_{JS} = stability constants of the carrier cation complexes in the membrane phase

$\beta_{IS}^{w}, \beta_{JS}^{w}$ = stability constants of the carrier cation complexes in the outside solution (in water)

Independent of the assumptions A to C the cation selectivity of the membranes in the "equilibrium domain" is therefore controlled by the ratio of the complex formation constants (6) and should therefore be identical for different types of neutral carrier membranes.[18] Figure 2 indicates that there is indeed a close parallelism between the selectivities of solvent polymeric membranes (SPM) and bilayer lipid membranes (BLM) modified with valinomycin *1*, nonactin *2*, trinactin *5*, and tetranactin *6* (see also Ref. 18). This is in good agreement with findings from Eisenman's[45] and Lev's[15] research groups.

For carriers forming complexes of different stoichiometries with a given cation, no such simple behavior can be expected. The ligands *7* to *9* form

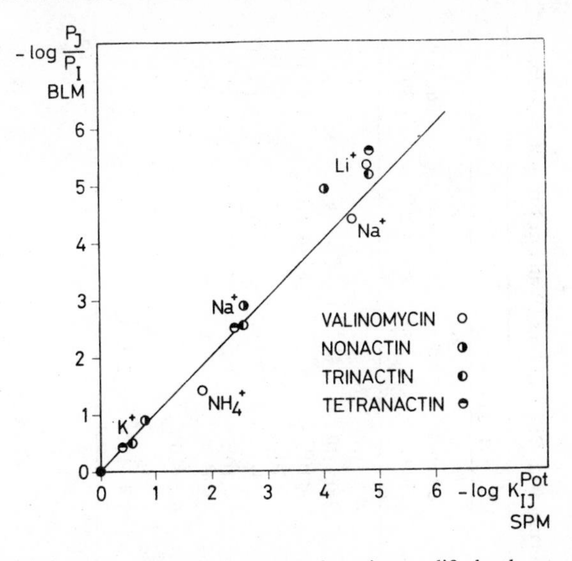

Fig. 2. Comparison of the selectivities of neutral-carrier-modified solvent polymeric- [43] and bilayer membranes. The permeability ratios P_J/P_I (at "equilibrium" (Ref. 18) as far as available) fulfilled for the glyceryl dioleate BLM's are taken from Figs. 10 and 11 in Ref.18. Values on the SPM's were obtained using 0.1 *M* solutions of the aqueous chlorides and membranes of the composition: 33.1 wt.% polyvinyl chloride, 66.2 wt.% dioctyl adipate, 0.7 wt.% carrier. For the macrotetrolides $I^{z+}:NH_4^+$; for valinomycin $I^{z+}:K^+$.

1:1 (cation to carrier) as well as 1:2 complexes with certain alkaline earth metal cations and 7 forms such complexes even with alkali metal ions.[33,44] For $M=N=2$ and comparable ionic mobilities for all cationic species in the membrane phase, Equation IIC (Table II) leads to:[33]

$$K_{IJ}^{\mathrm{Pot}} = \frac{k_J}{k_I} \cdot \frac{\beta_{JS} + \beta_{JS_2} \cdot c_S}{\beta_{IS} + \beta_{IS_2} \cdot c_S} \tag{7}$$

respectively,

$$K_{IJ}^{\mathrm{Pot}} = \frac{k_S \cdot k_{JS} \cdot \beta_{JS}^{w} + k_{JS_2} \cdot \beta_{JS_2}^{w} \cdot c_S}{k_S \cdot k_{IS} \cdot \beta_{IS}^{w} + k_{IS_2} \cdot \beta_{IS_2}^{w} \cdot c_S} \tag{8}$$

where the concentration c_S of the free ligand *S* in the membrane phase is assumed to be constant. Equation 8 clearly demonstrates that

$$\beta_{IS}^{w} > \beta_{JS}^{w}$$

and/or

$$\beta_{IS_2}^{w} > \beta_{JS_2}^{w}$$

TABLE III. Stepwise Stability Constants for the Interaction of Some Carrier Molecules with Different Ions (Water or Ethanol[a] in kg/mole)

Cations	Ligand *7* K_1 water 25°C	Ligand *7* K_1 ethanol 30°C	Ligand *7* K_2 ethanol 30°C	Ligand *8* K_1 ethanol 30°C	Ligand *8* K_2 ethanol 30°C	Ligand *9* K_1 ethanol 30°C	Ligand *9* K_2 ethanol 30°C	Ligand *1* K_1 water 25°C	Ligand *1* K_1 ethanol 30°C
Li^+	<0.18	$<10^1$	–	$6.0 \cdot 10^1$	–	$6.1 \cdot 10^1$	–	[b]	[b]
Na^+	<0.08	$2.2 \cdot 10^2$	$2.3 \cdot 10^1$	$1.3 \cdot 10^2$	–	$3.0 \cdot 10^1$	–	[b]	–
K^+	<0.12	$1.6 \cdot 10^2$	$4.1 \cdot 10^1$	$7.0 \cdot 10^1$	–	$<10^1$	–	2.3	$1.6 \cdot 10^6$
Rb^+	<0.17	$1.2 \cdot 10^2$	$2.8 \cdot 10^1$	$8.0 \cdot 10^1$	–	$<10^1$	–	5.9	$2.0 \cdot 10^6$
Cs^+	<0.17	[b]	[b]	[b]	[b]	[b]	[b]	0.12	$5.1 \cdot 10^5$
NH_4^+	<0.17	$<10^1$	–	$<10^1$	–	$<10^1$	–	[b]	[b]
Mg^{2+}	<0.05	$2.4 \cdot 10^2$	–	$9.9 \cdot 10^2$	–	[b]	[b]	[b]	[b]
Ca^{2+}	0.9	$8.9 \cdot 10^2$	–	$1.4 \cdot 10^3$	$8.0 \cdot 10^1$	$6.1 \cdot 10^2$	–	[b]	[b]
Sr^{2+}	<0.05	$K_1 \cdot K_2 > 10^8$		$5.0 \cdot 10^3$	$2.7 \cdot 10^3$	[b]	[b]	[b]	[b]
Ba^{2+}	1.39	$1.2 \cdot 10^4$	$3.0 \cdot 10^3$	$3.0 \cdot 10^3$	$7.6 \cdot 10^2$	$1.4 \cdot 10^3$	$2.3 \cdot 10^1$	[b]	[b]

[a] See Refs. 33, 44, 46, 47.
[b] Not measured.

is neither a necessary nor a sufficient criterion for a membrane to prefer the ion I relative to the ion J [$K_{IJ}^{\mathrm{Pot}} < 1$; see (2)]. High selectivity for I may be obtained if the carrier forms only 1:1 complexes with J but gives both 1:1 and 1:2 complexes with I (see (8)).[33] This might be important in the explanation of carrier *8*'s high selectivity of Ca^{2+} over Mg^{2+} and the monovalent cations[48] (see Table III). In addition, discrepancies between the complex stability constant measured in ethanol (Table III) and the values β^{w} for water as solvent [Eq. (8)] may be of importance, especially in the case of alkaline earth metal cations.

Both Table II and (8) show clearly that membrane selectivities may depend highly on the concentration c_S of the carrier within the membrane phase. Although the ligands *10*[23] and *11*[49] induce Ca^{2+}- and Na^{+}-selectivities, respectively, in bilayer membranes,[50,51] the correlation of these selectivities with those of bulk membranes is not as pronounced as indicated in Fig. 2.[50] In bulk membrane experiments Ca^{2+} and Na^{+} are found to be preferred by *10* and *11*, respectively.[23,49,52] For a more careful comparison of the selectivities, the concentration c_S of the ligand in the membrane phase should be adjusted.

IV. SELECTIVITY IN TRANSPORT PROCESSES ("KINETIC DOMAIN" AND "EQUILIBRIUM DOMAIN")

To achieve a transport of ions I^{z+} across a membrane, there must be an interfacial transfer of the type

$$I^{z+}\,(\text{free, aqueous}) \overset{K_I}{\rightleftharpoons} I^{z+}\,(\text{complexed, membrane}) \tag{9}$$

that is independent of the membrane-internal transport mechanism (see Fig. 3). In the case of pores there seems to be no direct complex formation between ligand and cation but rather a modification of the membrane by an interaction of the membrane matrix with the ligands.[8] In contrast to the simple situation described in Section III studies on different bilayers[12,18,53,54] indicate that it is possible to encounter situations ("kinetic domain") for which the rates of loading and unloading of the carriers at the interfaces become comparable to the rates of transport of the complexes across the membrane interior.[18] The permeability ratios therefore depend on the transmembrane potential (potential difference between solutions 1 and 2 in Fig. 3). Such a behavior has indeed been found for bilayer lipid membranes[18] (see also Ref. 53).

Based on the Nernst–Planck flux equation and Eyring's rate theory, a simple theoretical model was evolved for the description of the transport of ions through thick carrier membranes[55] (see also Ref. 15). The primary

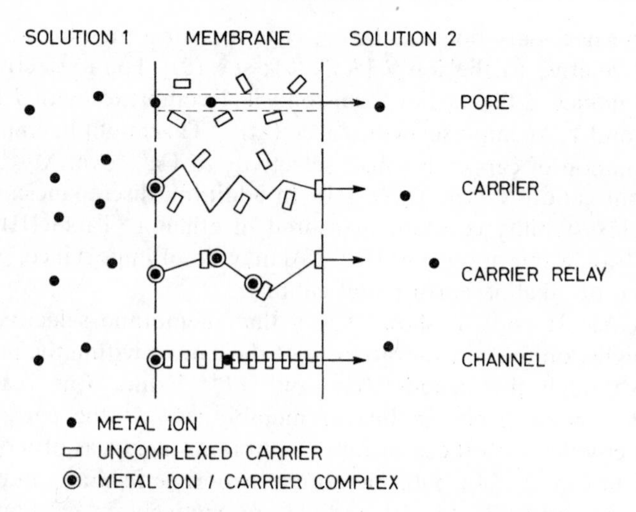

Fig. 3. Schematic representation of transport processes.

aim of this theoretical analysis of the ion transport was to describe the ionic fluxes (i.e., contributions to the electrical current) as a closed function of the ion activities in contact with the membrane and the applied voltage.[55] The assumptions made are

1. The cell is symmetrical (the membrane is interposed between two aqueous solutions of identical composition). For an extension see Section IV.C.
2. The current density with respect to the transmembrane potential (applied voltage) is kept rather low. Thus a thermodynamic equilibrium may be assumed to exist between the membrane boundaries and the adjoining solutions ("equilibrium domain"; for an extension see Section IV.D).
3. The membrane phase is in a steady state, that is, the flux of any species is constant throughout.
4. Electroneutrality holds.
5. Within carrier membranes permselective for cations, the anions of low permeability can be treated as fixed charges.[56,57]
6. The mobility is assumed to be the same for all cations.

The results may be summarized as follows.

A. Two Permeating Cations of the Same Charge, I^{z+} and J^{z+}

For permeating carrier complexes of one and the same stoichiometry, the transference number ratio is obtained as follows:[55]

$$\frac{t_I}{t_J} = \frac{\beta^w_{IS_n} \cdot k_{IS_n} \cdot a_I}{\beta^w_{JS_n} \cdot k_{JS_n} \cdot a_J} \approx \frac{\beta^w_{IS_n} \cdot a_I}{\beta^w_{JS_n} \cdot a_J} \tag{10}$$

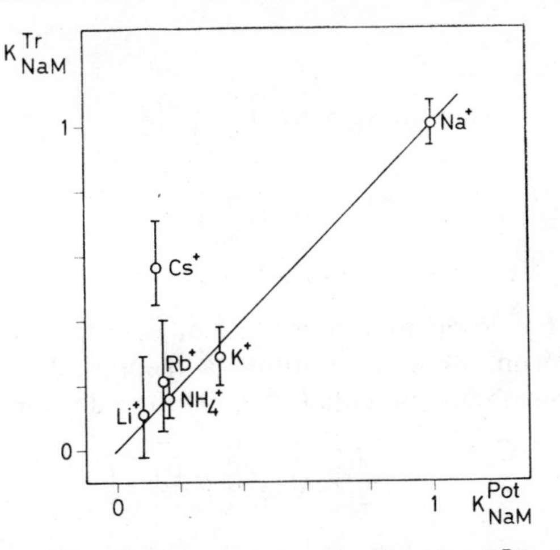

Fig. 4. Transport selectivity K_{IJ}^{Tr} and potentiometric selectivity K_{IJ}^{Pot} of a Na^+-selective neutral carrier membrane using ligand *11*. Experimental coefficients K_{NaM} obtained with (2) and (11) respectively given for different cations *M*. Membrane composition: 32 wt.% polyvinyl chloride, 65 wt.% dibutyl sebacate, 3 wt.% carrier *11*. Thickness of membrane $\approx$100 μm. Current density approx. 0.1 μA mm^{-2}.

Obviously, the ion-selective transport behavior exhibited by neutral carrier membranes is based on the complexation specificity of the carrier ligands used. A comparison with (6) shows that (10) is equivalent to

$$\frac{t_I}{t_J} = \frac{a_I}{K_{IJ}^{Tr} \cdot a_J} \tag{11}$$

$$t_I + t_J = 1 \tag{12}$$

The selectivity factors determined in potentiometric studies (K_{IJ}^{Pot}) should therefore be identical to the ones (K_{IJ}^{Tr}) determined in transport experiments. In Fig. 4 selectivities obtained potentiometrically on a membrane containing ligand *11* (3 wt.% carrier *11*, 65 wt.% dibutyl sebacate; 32 wt.% polyvinyl chloride, thickness $\approx$100 μm) are compared with those obtained in electrodialytic transport experiments.[55] Although widely different methods have been used to determine the ion selectivity, the agreement between the two sets of data is evident and corroborates the model presented. The deviation for Cs^+ may possibly be due to kinetic limitations suggesting a loss in transport selectivity (see Section IV.D).

B. Two Permeating Cations of Different Charge, I^{2+} and J^+

The transference number ratio is given by[55]

$$\frac{t_I}{t_J} = \sqrt{\frac{8cK_I a_I}{(K_J a_J)^2} + 1} - 1 \tag{13}$$

where

$$K_I = \beta_{IS_n}^w \left(\frac{c_S}{k_S}\right)^n k_{IS_n} \tag{14a}$$

$$K_J = \beta_{JS_n}^w \left(\frac{c_S}{k_S}\right)^n k_{JS_n} \tag{14b}$$

$$c = 2c_{IS_n} + c_{JS_n} \tag{15}$$

The term c refers to the total concentration of cationic charges within the membrane. Again, we can substitute a theoretical expression for the zero-current membrane potential ("equilibrium domain")[58,59] to rewrite

$$\frac{t_I}{t_J} = \sqrt{\frac{a_I}{\frac{1}{4}K_{IJ}^{\mathrm{Tr}} a_J^2} + 1} - 1 \tag{16}$$

$$t_I + t_J = 1 \tag{17}$$

where the monovalent-divalent cation selectivity K_{IJ}^{Tr} is also obtained through the EMF relationship[58,59]

$$E = \text{const} + \frac{RT}{F} \ln\left[\sqrt{a_I + \frac{1}{4}K_{IJ}a_J^2} + \sqrt{\frac{1}{4}K_{IJ}a_J^2}\right] \tag{18}$$

These theoretical results again reveal a parallel between ion-selectivity parameters obtained by potentiometric measurements and such obtained

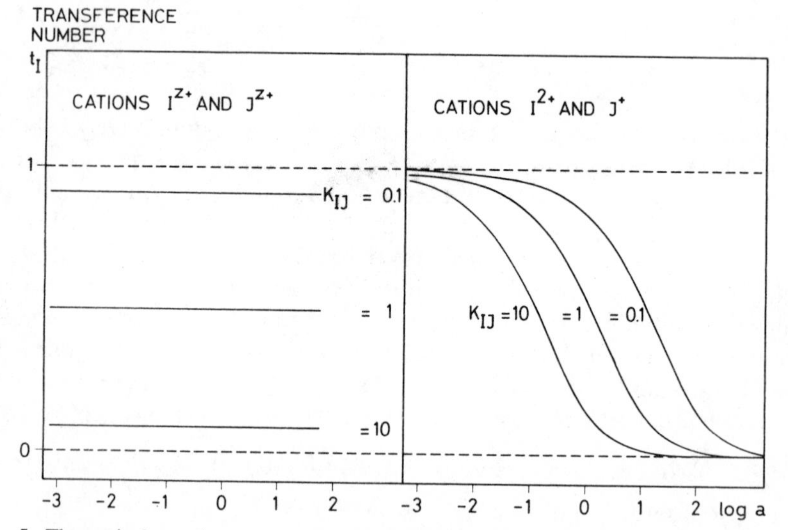

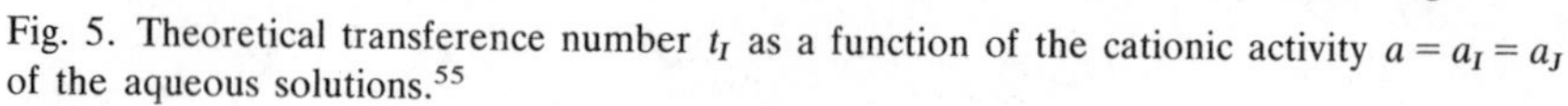

Fig. 5. Theoretical transference number t_I as a function of the cationic activity $a = a_I = a_J$ of the aqueous solutions.[55]

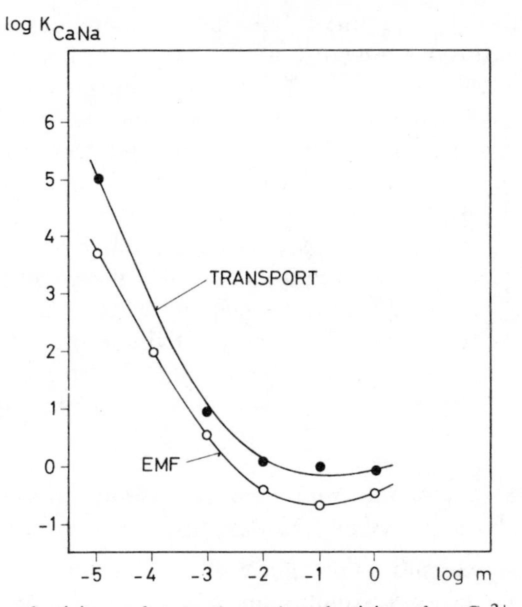

Fig. 6. Transport selectivity and potentiometric selectivity of a Ca^{2+}-selective neutral carrier membrane (3 wt.% carrier *10*, 65 wt.% *o*-nitrophenyl-octyl ether, 32 wt.% polyvinyl chloride). Experimental selectivity coefficients K_{CaNa} obtained with (16) and (18), respectively, as a function of the cationic concentration m (in moles/liter).

by electrodialysis experiments. For solutions containing different cations at the same activity a, the membrane is found, according to (16) and Fig. 5, to be permselective for the multivalent species in the range of low activities (depending on K_{IJ}) whereas it is permselective for monovalent ions at relatively high activities. In contrast, the transference number ratio for cations of the same valence is not influenced by diluting or concentrating the solution [(11) and Fig. 5]. A careful experimental investigation of the monovalent-divalent cation selectivity[55] has shown that another drastic change in the observed transport behavior occurs at very low sample activities. The same behavior is found in the potentiometric study of ion-selective liquid membrane electrodes and is related to the processes determining the lower detection limit of such systems. In fact, there is a close correlation between the experimental selectivity parameters obtained potentiometrically and those found by transport studies (see Fig. 6).

C. Extension of the Model to a Nonsymmetrical Cell Assembly[55]

For a description of membranes in contact with two aqueous solutions of different composition, the simple model is not appropriate and, thus, an extended theory is required. The most comprehensive treatment of electrodiffusion through aqueous diffusion layers or bulk membranes was offered by Schlögl[60] in a pioneering but, unfortunately, scarcely cited work. Based on the steady-state assumption, the electroneutrality condition, and the assumption of interfacial equilibria, this model permits an implicit description of the ionic fluxes. For general applications, however, the numerical evaluation becomes rather grueling.

In the limiting case of two classes of permeating ions and high voltages (values in the order of 1 V), Schlögl's extended theory leads to a description of the ion transference that is analogous to the relationships derived above for symmetrical cells. Thus the selectivity in the cation transport is generally given by (11) and (16), respectively, where the ion activities refer to the aqueous solution on the side of the cation uptake into the membrane.[55]

D. Extension of the Model for Deviations from Interfacial Equilibrium (Kinetic Limitations)

For *thick membranes* and sufficiently fast reaction kinetics at the membrane boundaries ("equilibrium domain"), diffusion through the membrane interior is rate limiting and we obtain for the mass flux of ions:

$$J_I = -z_I u_{IS_n} c_{IS_n} F \frac{E}{d} \tag{19}$$

where d is the membrane thickness. Equation 19 suggests ohmic behavior of the membrane (see below).

For *thin membranes* the ionic fluxes may become large at rather low values of E. Thus large displacements from the equilibrium state must be considered to occur at the membrane-solution interfaces. The same effect has to be expected for slow interfacial reactions. A more detailed discussion of this "kinetic domain" is given in Ref. 55.

It is obvious that large displacements from the interfacial equilibria should lead to serious changes in the selectivity behavior of a membrane, as is observable when two or more cations are simultaneously involved in the partitioning reactions. Thus, in the "kinetic domain" of membrane properties (see also Ref. 53), which is generally recognizable by a strong nonlinearity of the current-voltage characteristic, the ion transfer from the aqueous phase into the membrane becomes rate determining and the ionic fluxes are no longer given by (19). For cations of the same charge we may then characterize the ion transport selectivity by

$$K_{IJ}^{\mathrm{Tr}} = \frac{\mathbf{k}_J}{\mathbf{k}_I} \tag{20}$$

where $\mathbf{k}_I$ and $\mathbf{k}_J$ are the rate constants of the process by which I^{z+} and J^{z+} cross the interface from the aqueous phase into the membrane. The same limiting value was derived by Eisenman's group[53] for the permeability ratio of carrier membranes; the parameters $\mathbf{k}$ were found to be determined by the rate constants of the complexation reaction. The pronounced ion specificity in the complex formation by carriers, however, is to a large degree given by the rate of the decomplexation reaction.[61,62] Therefore a certain loss of ion selectivity has to be tolerated in the "kinetic domain."

In contrast, a drastic change in the ion selectivity due to kinetic limitations must be expected if cations of different charge are present. The reason is that the Eyring-type description for the interfacial ion fluxes involves an exponential potential dependency.[55] This term gives a strong preference for the species of the highest valence when the applied voltage is increased within the "kinetic domain."

E. Extension of the Model to Limitations from Closed-Circuit Flux of the Carrier within the Membrane

For thick carrier membranes we may usually apply the assumption of a thermodynamic equilibrium at the phase boundaries as a good approximation (see above). On the other hand, it should be considered that a positive flux of cationic complexes within the membrane (as induced by an applied voltage $E<0$) leads to a certain accumulation of free carriers at $x=d$, respectively to a depletion at $x=0$ (see Fig. 1). For practical purposes the carriers are confined to the membrane phase in the case of ideal bulk membranes (in contrast, a supply of carriers from the outside solutions is stipulated for transport studies on bilayers[45,53]).

A more detailed discussion (given elsewhere[55]) shows that a saturation of the ionic flux, respectively the electrical current, is predicted at high voltages. The existence of such an asymptotic behavior is due to the fact that the back diffusion of free carriers within the membrane cannot be accelerated beyond a certain limit. When the charge concentration $c = z_I \cdot c_{IS_n}$ within the membrane is increased, the saturation flux, respectively the saturation current, decreases to the same extent (see Fig. 7). For membranes with *predominantly free carriers* an ohmic behavior of the membrane is found in the lower voltage range.[55] The membrane conductance is controlled by the concentration and mobility of the cationic complexes and increases with the same factor as the charge concentration (A and B in Fig. 7). Contrasting behavior is found for membranes with *predominantly complexed carriers.* Saturation is attained at rather low voltages. The electrical properties are directly related to the mean concentration and the mobility of the free carriers. Thus, in contrast to the

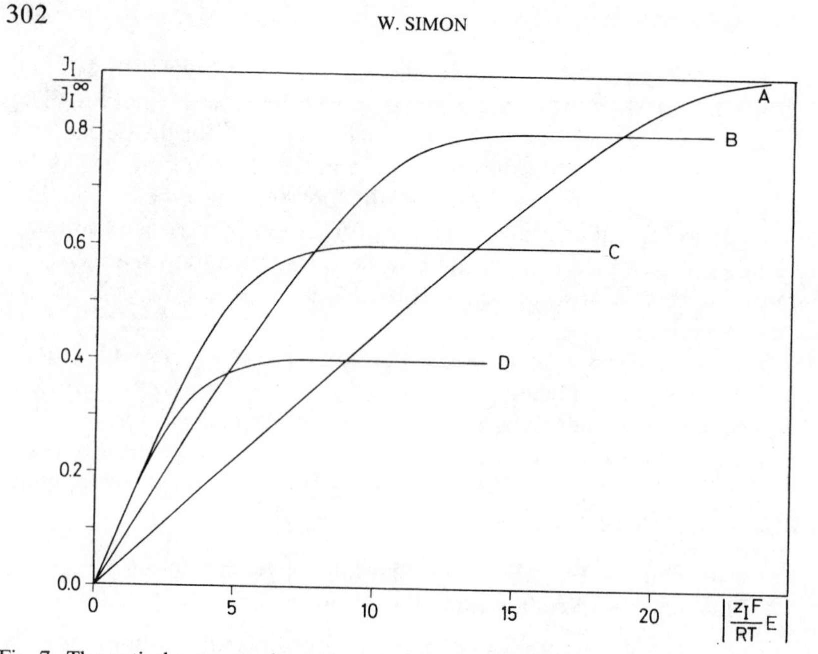

Fig. 7. Theoretical current-voltage characteristics of thick carrier membranes with different charge concentrations $c = z_I c_{IS_n}$. The values for c are arbitrary and increase from curve A to D (for details see [55]).

behavior predicted earlier, the zero-current membrane conductance is here found to increase with decreasing complex concentration (i.e., with increasing carrier concentration). This implies that a maximum must exist for the zero-current conductance of the membrane at intermediate charge concentrations c (see curves C and D in Fig. 7). The current-voltage characteristic of a thick Ca^{2+}-selective carrier membrane is shown in Fig. 8 (see also Ref. 63). A comparison with Fig. 7 reveals a striking parallelism between theoretical and experimental curves. It is evident that, in practice, the membrane parameter c is not a real constant but increases to a certain degree with increasing activity of the external solutions. A more detailed experimental investigation of the electrical properties of thick carrier membranes was performed by Lev et al.;[15] the authors also offered an interesting theoretical approach, which, however, is limited to 1:1 electrolytes. For polar membranes with the carrier valinomycin the internal membrane conductance in the ohmic limit was found to vary moderately (in analogy to Fig. 8), whereas for nonpolar carrier membranes the electrical properties were nearly independent of the composition of the aqueous solutions contacting the membrane.[15] This is in striking contrast to the findings for bilayer membranes[45,53,64] where the zero-current conductance is usually directly proportional to the

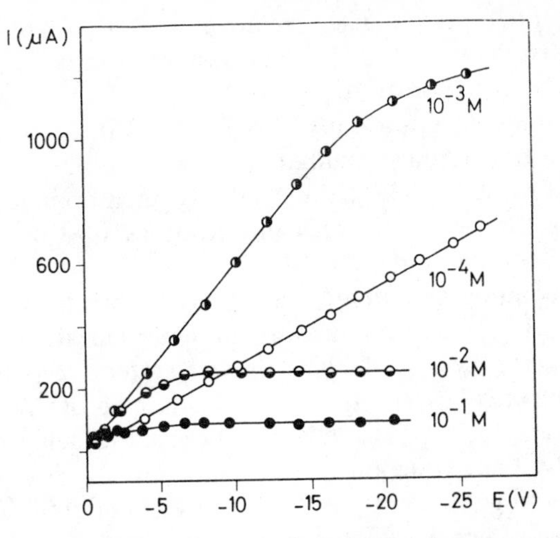

Fig. 8. Experimental current-voltage curves of a solvent polymeric membrane containing the neutral carrier *10* (3 wt.% carrier *10*, 65 wt.% *o*-nitrophenyl-octyl ether, 32 wt.% polyvinyl chloride, thickness of membrane approximately 100 μm, 25°C). The membrane was in contact with aqueous $CaCl_2$-solutions of the molarity indicated.[55]

outside activity of the permeating ion as well as to its overall partition coefficient K_I, which mimics the ion selectivity of the neutral carrier. So far, there is no cogent explanation for the "unusual extraction behavior"[15] observed for thick carrier membranes. However, a qualitative understanding of many phenomena may be obtained by postulating that the membrane behaves as a phase with fixed negative charges of roughly constant concentration *c*.

F. Selective Transport Systems

A series of ion-selective membrane electrodes based on neutral carrier solvent polymeric membranes has been designed for the potentiometric determination of ion activities (for reviews see Refs. 52, 65). Systems with analytically relevant selectivities for Li^+, Na^+, K^+, NH_4^+, Ca^{2+}, and Ba^{2+}, are available. In agreement with the treatment given in Sections III and IV, the ions preferred in potentiometric studies may be transported preferentially through the same membranes in electrodialytic experiments. So far, selective carrier transports have been realized for Li^+, Na^+, K^+, and Ca^{2+}.

Since the transfer reaction (9) largely depends on the dielectric constant of the membrane phase and since (other parameters kept constant) the transfer is favored with increasing dielectric constant ϵ, some predictions in respect to the influence of ϵ on K_{IJ} are possible.[58,65,66] One and the same carrier (*10*) prefers Ca^{2+} relative to Na^+ when dissolved in a membrane solvent of high dielectric constant (*o*-nitrophenyl-octyl ether (*o*-NPOE), $\epsilon \approx 25$) whereas it prefers Na^+ relative to Ca^{2+} when a low dielectric constant solvent (dibutyl sebacate (DBS), $\epsilon \approx 5$) is used. Correspondingly, the transference numbers for Ca^{2+} are 0.98 ± 0.04 and 0.63 ± 0.04 for an SPM with *o*-NPOE and DBS as membrane solvents, respectively (carrier *10;* equal molar concentrations of $CaCl_2$ and NaCl in contact with the membrane).[66]

The overwhelming enantiomer selectivity of biochemical processes as well as the role of neuroregulators in neuronal functions stimulated efforts toward the design of chiral model systems capable of inducing enantiomer-selective cation transport. Both electrically charged[67] and electrically neutral[17,68,69,70] ligands may be used to achieve separation of enantiomers by the extraction from an aqueous to an organic phase. In view of an electrogenic ion transport electrically neutral ligands acting as ion carriers are of special interest.[71] Measurements of ^{13}C-nmr chemical shifts as well as spin-lattice relaxation times T_1 clearly indicate an interaction of the ligand *12* with α-phenylethylammonium ions.[72] Since the chemical shifts change only up to a molar ratio of 1:1 when adding the ammonium chloride to the ligand (in $CDCl_3$), evidence for the existence of a 1:1 carrier-cation complex is corroborated (see Ref. 73). The introduction of chirality into ligands of the type *12* should therefore lead to diastereomeric interactions with enantiomeric cations and thus to corresponding chiral recognition. Extensive studies of this phenomenon in complexation, using conformationally chiral ligands (e.g., *13*), have been reported by D. J. Cram et al.[17,74] Although remarkable enantiomer differentiation has been achieved in conventional liquid-liquid extraction devices,[74] chiral recognition in electrochemical membrane systems has been realized only recently with configurationally chiral ionophores.[71] In the meantime ligands of the latter type of chirality have been reported to be capable of chiral recognition in extraction experiments.[75]

In view of models for stereospecific transport processes in biological systems the behavior of solvent polymeric membranes incorporating the compounds *1* and *14* to *18* has been studied. Compounds *17* and *18* are representatives of a series of molecules designed and prepared in the laboratories of V. Prelog.[76,77] Figure 9 displays selectivity sequences obtained with a liquid membrane free of carrier (DOA: dioctyladipate) as well as in the presence of ligands *15* and *1* (valinomycin), respectively. Through the incorporation of ligands the selectivity range increases substantially. As expected,[7] the valinomycin-containing membrane considerably prefers K^+ over Na^+ (see also Fig. 2). In the presence of *15*, however, (*RS*)-α-phenylethylammonium is singled out. Such EMF

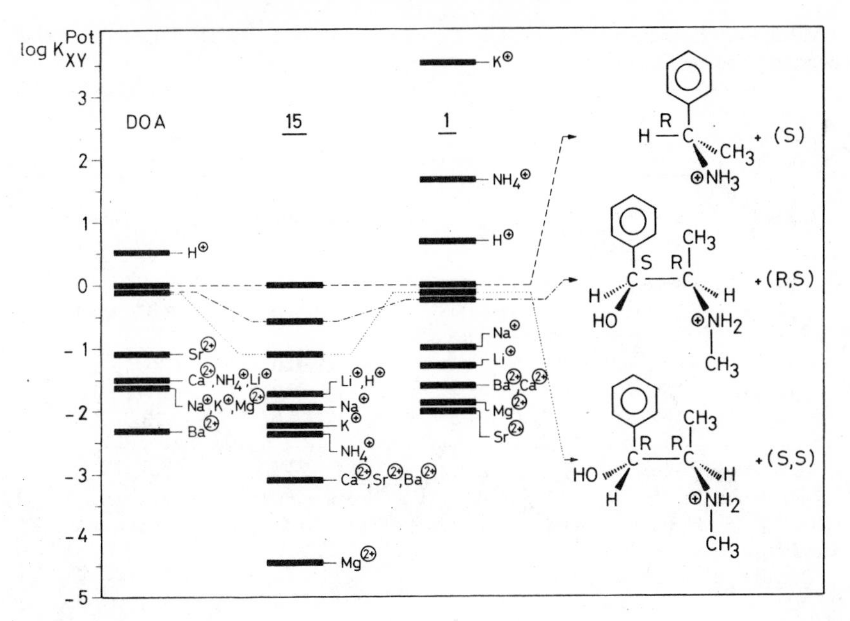

Fig. 9. Results of the potentiomeric determination of selectivity coefficients K_{XY}^{Pot} using 0.1 *M* chloride solutions. Membranes consist of 66 wt.% membrane solvent (here DOA, i.e., dioctyladipate was used), 33 wt.% polyvinyl chloride and 1 wt.% of ligand; if no ligand is incorporated in the membrane, the weight ratio of membrane to solvent polyvinyl chloride is 67 to 33 (see the column on the left in this figure).

studies indicate interactions of ligands *1* and *14* to *18* with the biogenic ammonium ions shown in Fig. 9. Since all these molecules (*1*, *14* to *18*) are chiral, they are theoretically capable of inducing enantiomer selectivity in membranes. In Table IV results of emf-measurements on the cell shown in Fig. 1 in the presence of 0.1 *M* solutions of (*R*)- and (*S*)-α-phenylethylammonium chloride, respectively, are given.

The values given in column 3 of Table IV were obtained from the data in column 2 ((2) and (6)]. A comparison of the results for *14* and *15* indicates that the introduction of methyl groups at sites 4 and 5 (see *12*), leading to central chirality *R* at both carbon atoms, is the main cause of enantiomer selectivity. This is in agreement with the only slightly different enantiomer selectivity of *16* relative to *15*. As expected, the effect of *17* is reversed by *18*. Although valinomycin *1* is chiral, no enantiomer selectivity was detectable (see Table IV). The potentiometrically determined enantiomer selectivity ΔEMF is correlated to the transport selectivity [(2), (6), (10), and (11)]

$$\frac{t_R}{t_S} = \frac{\beta_{RL}^{w}}{\beta_{SL}^{w}} \tag{21}$$

TABLE IV. Enantiomer Selectivity of Ligands *1* and *14* to *18* Determined by Potentiometry[a] and by Electrodialysis[b]

	Potentiometry		Transport $\frac{t_R}{t_S}$	
Ligand	ΔEMF	$\frac{\beta_{RL}^{w}}{\beta_{SL}^{w}}$	racemate X	racemate Y
14	−0.3±0.4	0.99±0.02		
15	2.3±0.1	1.09±0.003	1.067±0.015	1.070±0.026
16	1.2±0.5	1.05±0.02		
17	−3.5±0.7	$(1.15\pm0.03)^{-1}$	$(1.093\pm0.027)^{-1}$	$(1.094\pm0.012)^{-1}$
18	2.9±0.8	1.12±0.04	1.124±0.017	1.121±0.012
1[c]	0.0±0.4	1.00±0.02	1.00±0.02	1.00±0.02

[a] $\Delta emf = emf_R - emf_S$ in mV; $T = 25\pm0.1$°C.

[b] Racemate X: ^{14}C-(R)-, ^{3}H-(S)-α-phenylethylammonium chloride; racemate Y: ^{3}H-(R)-, ^{14}C-(S)-α-phenylethylammonium chloride; $T = 21$°C; confidence limits are 95%. Membrane solvent is o-nitrophenyl-octyl ether.

[c] Membrane solvent: dioctyladipate.

with

t_R, t_S = transport numbers of the enantiomeric cations R and S through the liquid membrane

β_{RL}^{w}, β_{SL}^{w} = stability constants of the interaction of the cations R and S with the chiral ligand L

By applying a potential difference across neutral carrier membranes (see legend to Fig. 9) in contact with similar solutions of racemic α-phenylethylammonium chloride, a carrier-mediated cation transport through the membrane is expected. According to (21) a ratio $t_R/t_S \neq 1$ has to be expected for ionophores with $\Delta EMF \neq 0$. Using the double labeling of racemates shown in Fig. 10 and described in detail in Ref. 71, transport numbers may easily be determined for the enantiomers and may be confirmed by reversing the labeling. Results of these electrodialytic transport experiments are shown in the last two columns of Table IV and are in good agreement with the data obtained potentiometrically (see column 3).

V. MECHANISMS OF THE TRANSPORT PROCESS

Transport mechanisms of the pore type (Fig. 3) seem to be induced (e.g., by amphotericin B) in bilayer membranes[8] and there is very good

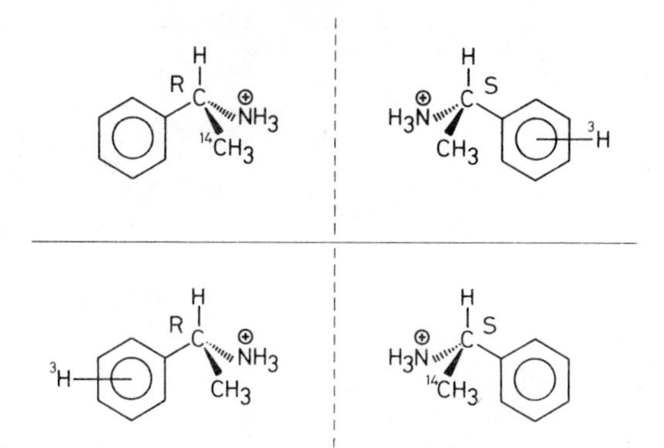

Fig. 10. Labeling of the enantiomeric cations for transport studies.

evidence available for accepting channel mechanisms induced by gramicidin A-type antibiotics in similarly thin membranes.[1,8,78,79] For thick liquid membranes the existence of carrier-relay mechanisms has been substantiated (valinomycin *1* in octane-2-ol[80] and nonactin *2*–monactin *3*–dinactin *4* in the same solvent.[81,82] Under these conditions valinomycin clearly transports K^+ as a 1:1 cation-carrier complex.[80,81]

A more detailed study of transport processes in solvent polymeric membranes was initiated recently.[72] One aim was to get information on the distribution within the membrane of the carrier and the cation transported after a steady state has built up during an electrodialysis experiment. A further objective was the demonstration of a relaxation of the concentration gradients of both carrier and cation. To this end the transport properties of solvent polymeric membranes containing the carrier ^{14}C-valinomycin (66 wt.% dioctyladipate, 33 wt.% polyvinyl chloride, 1 wt.% ^{14}C-valinomycin) in contact with aqueous solutions of ^{3}H-α-phenylethylammonium chloride were studied.

Five membranes (thickness, 40 μm) were stacked and the concentration of ligand and cation in each membrane was measured before and immediately after the transport experiment as well as 5 days after restacking the membranes. Since a concentration gradient of valinomycin developed (Fig. 11, $t = 3$ hr), which decayed almost completely after a relaxation period (Fig. 11, $t = 5$ days), the α-phenylethylammonium cation had obviously been transported by a carrier mechanism. During the transport process a cation profile built up (Fig. 12, $t = 3$ hr) that had the same trend as the ligand profile. This cation gradient disappeared after some time (Fig. 12, $t = 5$ days).

The cross-section of the membrane involved in the transport process

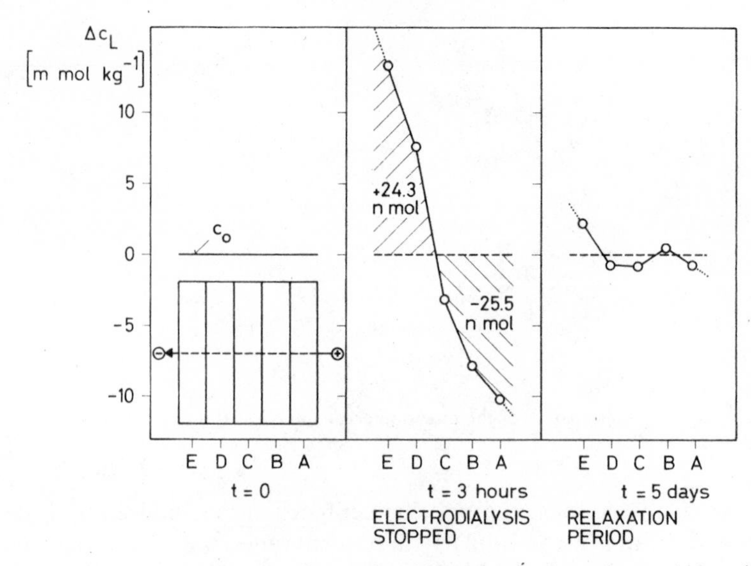

Fig. 11. Study of the mobility of valinomycin within a stack of membranes before ($t = 0$), immediately after ($t = 3$ hr) electrodialysis as well as after a five-day period of restacking the membranes ($t = 5$ days). c_0: initial ligand concentration; Δc_L: change of total ligand concentration. The size of the circles denotes 95% confidence limits.[11,72]

contained approximately 50 nanomoles of valinomycin. Therefore, according to Fig. 12, less than 12% of the carrier molecules were in the complexed form. Since about 25 nanomoles of carrier left the entry sections (anode) of the membrane stack to move toward the exit sections (cathode), the entrance phase boundary must have become nearly depleted of carrier. This is in perfect agreement with the situation discussed (Section IV.E; see Ref. 55) where the back diffusion of the carrier is the controlling parameter ("kinetic domain"). Because the concentration of the uncomplexed carrier is at least 10 times that of the complexed ligand in section *E*, this proves that the decomplexation reaction at the exit phase boundary is not rate limiting for the transport process.

These results are at variance with Läuger's interpretation of the saturation current-voltage characteristics for valinomycin-K^+ in phosphatidylserine bilayer membranes, which he purports to be due to limiting decomplexation rates.[83]

In the experiment described earlier 145 nanomoles of cations were transported from the anode to the cathode compartment. This corresponds to a transference number of 0.78.

In a similar electrodialysis experiment only the carrier in section *A* was labeled (Fig. 11). In the course of the transport experiment only about

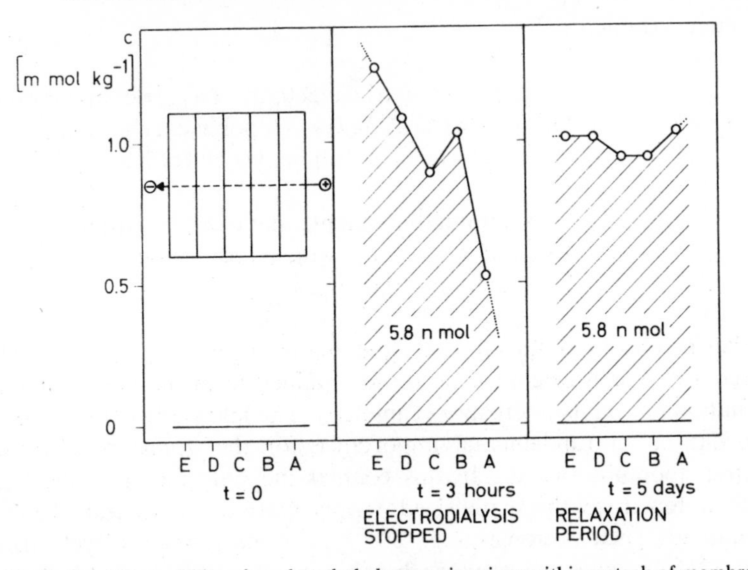

Fig. 12. Study of the mobility of α-phenylethylammonium ions within a stack of membranes before ($t = 0$), immediately after ($t = 3$ hr) electrodialysis as well as after a five day period of restacking the membranes ($t = 5$ days). c: cation concentration in the membrane sections. The size of the circles denotes 95% confidence limits.[11,72]

20% of the labeled ligand passing the A/B boundary reached section E. This suggests nothing less than an exchange of ligands during the transport, i.e., a carrier relay mechanism, which is in agreement with earlier findings.[80,81]

VI. SELECTIVITY OF EXTRACTION SYSTEMS

The discussion in Sections III and IV indicates that the constants of complex formation of carrier ligands with cations are important but are not the only parameters that describe the selectivity of carrier membranes in the "equilibrium domain." In the general expressions given for K_{IJ}^{Pot}, partition coefficients of cationic species are also involved (Table II). The equilibrium (9) relevant for the ion selectivity, which includes both complex formation and partition reactions, is accessible through extraction studies. The importance of such techniques was recognized early.[28,84,85] Even for carriers that clearly form both 1:1 and 1:2 complexes there is a parallel between their ability to extract cations from an aqueous phase into an organic solvent and the selectivity of the corresponding carrier membranes. Accordingly ligand 7 shows the following selectivity sequences[33,86]

$$\mathrm{Ba} > \mathrm{Sr} = \mathrm{Ca} > \mathrm{Mg}\begin{cases}\text{potentiometric selectivity of a carrier membrane}\\ \text{(0.7 wt.\% 7, 34.0 wt.\% polyvinyl chloride,}\\ \text{65.3 wt.\% o-nitrophenyl-octyl ether)}\end{cases}$$

$$\mathrm{Ba} > \mathrm{Sr} > \mathrm{Ca} > \mathrm{Mg}\begin{cases}\text{amount of metal picrates extracted from}\\ \text{water into o-nitrophenyl-octyl ether}\\ \text{containing 0.07 wt.\% 7}\end{cases}$$

Similar results are obtained when studying the ligands *8*, *9*, and *19*. These selectivity sequences contradict those obtained from the measurement of formation constants in ethanol (Table III). The knowledge of such extraction data is of fundamental importance for the design of electrically neutral lipophilic ligands. In this context the outstanding properties of quite a few recently synthesized representatives of several classes of compounds, such as crowns (e.g., *20*[17,85]) and macroheterobicyclic ligands (e.g., *21*[87,88]) must be seen as being partially or largely due to their extraction behavior.

VII. CONCLUSIONS

The transport selectivity of cation permselective neutral carrier membranes is to a large extent given by the inherent capability of the carrier molecules to complex selectively the corresponding cations. Structural parameters responsible for such a selective complexation have been discussed in detail elsewhere.[19,89,90,91] One and the same neutral carrier showing a given selectivity sequence in the complexation of cations in homogeneous systems may, however, lead to widely different selectivity sequences in the transport behavior of the corresponding neutral carrier membranes. The following parameters may influence this transport selectivity:

1. Activity of the cations in contact with the membrane (IV.B).
2. Concentration of the carrier within the membrane phase (III, IV)
3. Dielectric constant of the membrane phase (IV.F)
4. Rate constant of the transfer of the ion from the aqueous phase to the membrane phase (IV.D)

In the present report theoretical and/or experimental evidence is given for the effect of these parameters in neutral carrier liquid membranes.

STRUCTURES OF THE LIGANDS DISCUSSED

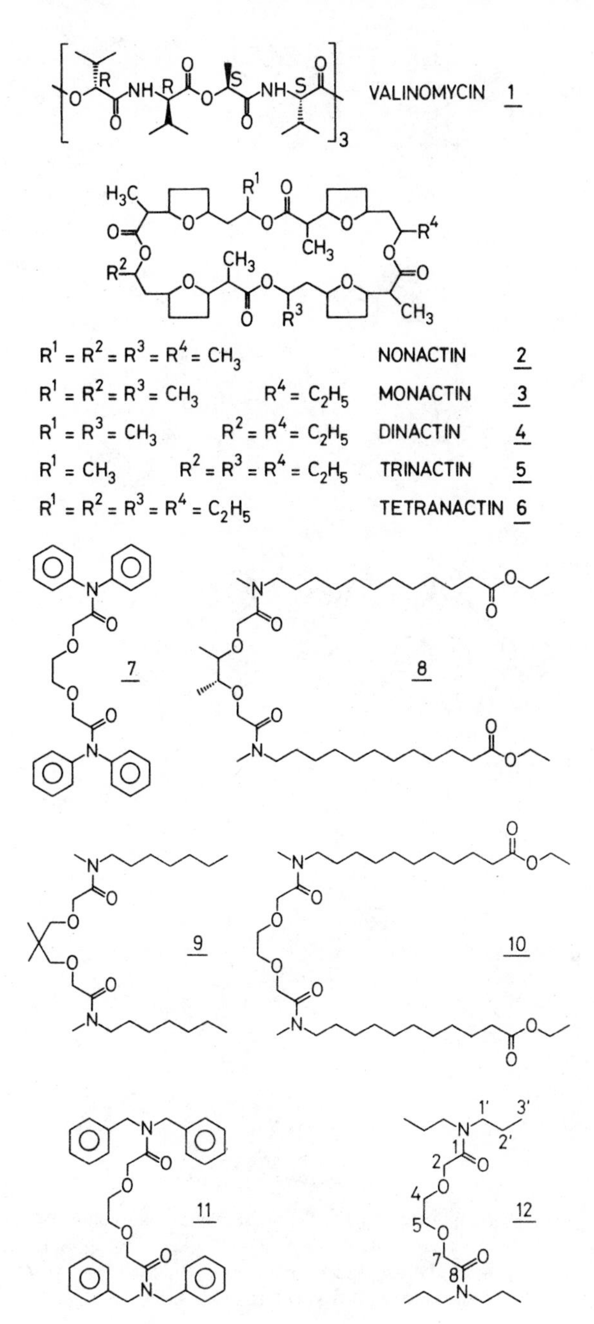

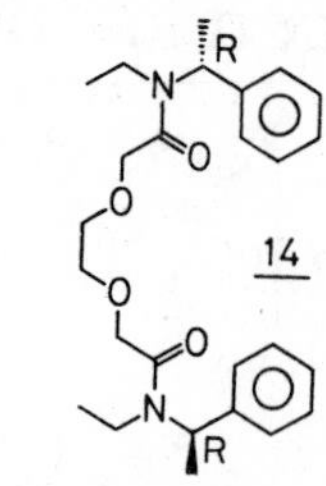

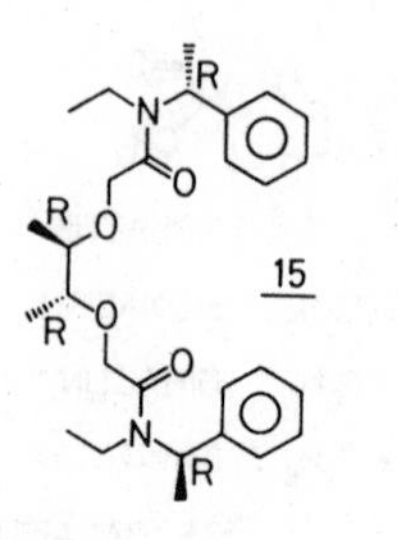

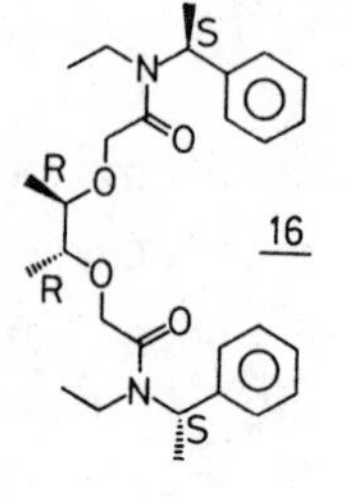

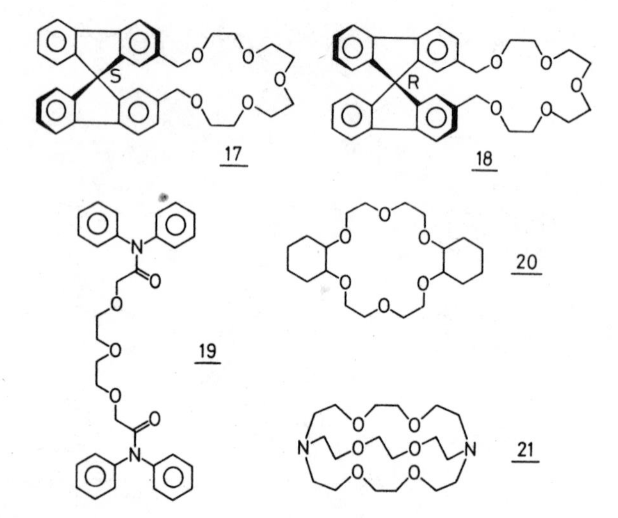

Acknowledgments

This work was partly supported by the Schweizerischer Nationalfonds zur Förderung der wissenschaftlichen Forschung. I thank Prof. Dr. Yu. A. Ovchinnikov for a supply of ^{14}C-labeled valinomycin. I should like to thank Drs. W. Morf and P. C. Meier for their critical reading of the script and the invigorating discussions we had.

References

1. Robert De Levie, *J. Electroanal. Chem.*, **69,** 265 (1976).
2. P. Mueller, D. O. Rudin, H. Ti Tien, and W. C. Wescott, *Nature*, **194,** 979 (1962).
3. P. Mueller, D. O. Rudin, H. Ti Tien, and W. C. Wescott, *Circulation*, **26,** 1167 (1962).
4. P. Mueller, D. O. Rudin, H. Ti Tien, and W. C. Wescott, *J. Phys. Chem.*, **67,** 534 (1963).
5. M. K. Jain, *The Bimolecular Lipid Membrane: A System*, Van Nostrand Reinhold, New York, 1972.
6. H. Ti Tien, *Bilayer Lipid Membranes* (*BLM*), *Theory and Practice*, Marcel Dekker, New York, 1974.
7. Yu. A. Ovchinnikov, V. T. Ivanov, and A. M. Shkrob, BBA Library 12, Membrane-Active Complexones, Elsevier, Amsterdam, 1974.
8. P. B. Chock and E. O. Titus, in *Current Research Topics in Bioinorganic Chemistry*, S. J. Lippard, Ed., Wiley, 1973.
9. R. Bloch, A. Shatkay, and H. A. Saroff, *Biophys. J.*, **7,** 865 (1967).
10. G. J. Moody, R. B. Oke, and J. D. R. Thomas, *Analyst*, **95,** 910 (1970).
11. A. P. Thoma and W. Simon, in preparation.
12. P. Läuger and G. Stark, *Biochim. Biophys. Acta*, **211,** 458 (1970).
13. J. Jagur-Grodzinski, S. Marian, and D. Vofsi, *Sep. Sci.*, **8,** 33 (1973).
14. R. Büchi, E. Pretsch, and W. Simon, *Helv. Chim. Acta*, **59,** 2327 (1976).
15. A. A. Lev, V. V. Malev, and V. V. Osipov, in *Membranes*, Vol. 2, G. Eisenman, Ed., Marcel Dekker, New York, 1973, p. 479.
16. J.-P. Behr, J.-M. Lehn, *J. Am. Chem. Soc.*, **95,** 6108 (1973).
17. D. J. Cram, in *Applications of Biochemical Systems in Organic Chemistry*, J. B. Jones, D. Perlman, and C. J. Sih Eds., in *Techniques of Chemistry*, A. Weissberger, Ed., Wiley-Interscience, New York, 1976.
18. G. Eisenman, S. Krasne, and S. Ciani, *Ann. N.Y. Acad. Sci.*, **264,** 34 (1975).
19. W. Simon, W. E. Morf, and P. Ch. Meier, *Struct. Bonding*, **16,** 113 (1973).
20. D. E. Goldman, *J. Gen. Physiol.*, **27,** 37 (1943).
21. A. L. Hodgkin and B. Katz, *J. Physiol.* (*London*), **108,** 37 (1949).
22. B. P. Nicolsky, *Z. Fiz. Khim.*, **10,** 495 (1937).
23. D. Ammann, E. Pretsch, and W. Simon, *Anal. Lett.*, **5,** 843 (1972).
24. J. Sandblom, G. Eisenman, J. L. Walker, Jr., *J. Phys. Chem.*, **71,** 3862 (1967).
25. G. Eisenman, *Anal. Chem.*, **40,** 310 (1968).
26. J. Sandblom, *J. Phys. Chem.*, **73,** 249 (1969).
27. S. Ciani, G. Eisenman, G. Szabo, *J. Membrane Biol.*, **1,** 1 (1969).
28. G. Eisenman, S. Ciani, G. Szabo, *J. Membrane Biol.*, **1,** 294 (1969).
29. G. Szabo, G. Eisenman, S. Ciani, *J. Membrane Biol.*, **1,** 346 (1969).
30. G. Eisenman, in *Ion-Selective Electrodes*, R. A. Durst, Ed., Nat. Bur. Stand. Spec. Publ. 314, Washington, 1969.
31. J. Sandblom and F. Orme, in *Membranes*, Vol. 1, G. Eisenman, Ed., Marcel Dekker, New York, 1972, p. 125.
32. H.-R. Wuhrmann, W. E. Morf, and W. Simon, *Helv. Chim. Acta*, **56,** 1011 (1973).
33. N. N. L. Kirsch, Thesis Swiss Federal Institute of Technology, Zurich (ETHZ). Nr. 5842 (1976).
34. J. J. Hermans, *Z. Phys.*, **97,** 681 (1935).
35. M. Born, *Z. Phys.*, **1,** 45 (1920).
36. M. Pinkerton, L. K. Steinrauf, and P. Dawkins, *Biochem. Biophys. Res. Commun.*, **35,** 512 (1969).

37. K. Neupert-Laves and M. Dobler, *Helv. Chim. Acta*, **58,** 432 (1975).
38. B. T. Kilbourn, J. D. Dunitz, L. A. R. Pioda, and W. Simon, *J. Mol. Biol.*, **30,** 559 (1967).
39. M. Dobler and R. P. Phizackerley, *Helv. Chim. Acta*, **57,** 664 (1974).
40. Y. Iitaka, T. Sakamaki, and Y. Nawata, *Chem. Lett.* (*Japan*), 1225 (1972).
41. K. Neupert-Laves and M. Dobler, *Helv. Chim. Acta*, **59,** 614 (1976).
42. Y. Nawata, T. Sakamaki, and Y. Iitaka, *Chem. Lett.* (*Japan*), 151 (1975).
43. G. Horvai, A. P. Thoma, and W. Simon, unpublished results.
44. N. N. L. Kirsch and W. Simon, *Helv. Chim. Acta*, **59,** 357 (1976).
45. G. Szabo, G. Eisenman, R. Laprade, S. M. Ciani, and S. Krasne, in *Membranes*, Vol. 2, G. Eisenman, Ed., Marcel Dekker, New York, 1973, p. 179.
46. M. B. Feinstein and H. Felsenfeld, *Proc. Nat. Acad. Sci.* (*U.S.*), **68,** 2037 (1971).
47. M. M. Shemyakin, Yu. A. Ovchinnikov, V. T. Ivanov, V. K. Antanov, E. I. Vinogradova, A. M. Shkrob, G. G. Malenkov, A. V. Evstratov, I. A. Laine, E. I. Melnik, and I. D. Ryabova, *J. Membrane Biol.*, **1,** 402 (1969).
48. D. Ammann, M. Güggi, E. Pretsch, and W. Simon, *Anal. Lett.*, **8,** 709 (1975).
49. D. Ammann, E. Pretsch, and W. Simon, *Anal. Lett.*, **7,** 23 (2974).
50. G. Amblard and C. Gavach, *Biochim. Biophys. Acta*, in press.
51. G. Eisenman and K.-H. Kuo, private communication.
52. D. Ammann, R. Bissig, Z. Cimerman, U. Fiedler, M. Güggi, W. E. Morf, M. Oehme, H. Osswald, E. Pretsch, and W. Simon, in *Ion and Enzyme Electrodes in Biology and Medicine*, M. Kessler, L. C. Clark, Jr., D. W. Lübbers, I. A. Silver, and W. Simon, Eds., Proceedings of an International Workshop at Schloss Reisensburg, Germany, September 15–18, 1974, Urban & Schwarzenberg, Munich, 1976, p. 136.
53. S. M. Ciani, G. Eisenman, R. Laprade, and G. Szabo, in *Membranes*, Vol. 2, G. Eisenman, Ed., Marcel Dekker, New York, 1973, p. 61.
54. R. Laprade, S. M. Ciani, G. Eisenman, and G. Szabo, in *Membranes*, Vol. 3, G. Eisenman, Ed., Marcel Dekker, New York, 1975, p. 127.
55. W. E. Morf, P. Wuhrmann, and W. Simon, *Anal. Chem.*, **48,** 1031 (1976).
56. O. Kedem, M. Perry, and R. Bloch, IUPAC International Symposium on Selective Ion-Sensitive Electrodes, paper 44, Cardiff, 1973.
57. J. H. Boles and R. P. Buck, *Anal. Chem.*, **45,** 2057 (1973).
58. W. E. Morf, D. Ammann, E. Pretsch, and W. Simon, *Pure Appl. Chem.*, **36,** 421 (1973).
59. W. E. Morf and W. Simon, in preparation.
60. R. Schlögl, *Z. Phys. Chem.* (*Frankfurt am Main*), **1,** 305 (1954).
61. M. Eigen and R. Winkler, in *The Neurosciences: Second Study Program*, F. O. Schmitt, Ed., Rockefeller University Press, New York, 1970.
62. E. Grell, Th. Funck, and F. Eggers, in *Molecular Mechanisms of Antibiotic Action on Protein Biosynthesis and Membranes*, E. Munoz, F. Garcia-Ferrandiz, and D. Vazquez, Eds., Proceedings of a Symposium held at the University of Granada (Spain), June 1–4, 1971, Elsevier, Amsterdam, 1972.
63. P. Wuhrmann, A. P. Thoma, and W. Simon, *Chimia* (*Switzerland*) **27,** 637 (1973).
64. P. Läuger and B. Neumcke, in *Membranes*, Vol. 2, G. Eisenman, Ed., Marcel Dekker, New York, 1973, p. 1.
65. W. Simon, E. Pretsch, D. Ammann, W. E. Morf, M. Güggi, R. Bissig, and M. Kessler, *Pure Appl. Chem.*, **44,** 613 (1975).
66. W. Simon, W. E. Morf, E. Pretsch, and P. Wuhrmann, in *Calcium Transport in Contraction and Secretion*, E. Carafoli, F. Clementi, W. Drabikowski, and A. Margreth, Eds., Proceedings of the International Symposium on Calcium Transport in Contrac-

tion and Secretion held at Bressanone, Italy, May 12–16, 1975, North-Holland Publishing Co., New York, 1975, p. 15.

67. J. M. Lehn, A. Moradpour, and J. P. Behr, *J. Am. Chem. Soc.* **97,** 2532 (1975).
68. D. J. Cram and J. M. Cram, *Science*, **183,** 803 (1974).
69. R. C. Helgeson, J. M. Timko, and D. J. Cram, *J. Am. Chem. Soc.*, **95,** 3023 (1973).
70. W. D. Curtis, D. A. Laidler, J. F. Stoddart, and G. H. Jones, *J. Chem. Soc. Chem. Commun.*, 833 (1975).
71. A. P. Thoma, Z. Cimerman, U. Fiedler, D. Bedeković, M. Güggi, P. Jordan, K. May, E. Pretsch, V. Prelog, and W. Simon, *Chimia*, **29,** 344 (1975).
72. A. P. Thoma, E. Pretsch, G. Horvai, and W. Simon, Proceedings of the FEBS Symposium on the Biochemistry of Membrane Transport held in Zurich, July 18–23, 1976, Springer-Verlag, Berlin, 1977.
73. R. Büchi, E. Pretsch, and W. Simon, *Tetrahedron Lett.*, **20,** 1709 (1976).
74. M. Newcomb, R. C. Helgeson, and D. J. Cram, *J. Am. Chem. Soc.* **96,** 7367 (1974).
75. W. D. Curtis, D. A. Laidler, J. F. Stoddart, and G. H. Jones, *J. Chem. Soc. Chem. Commun.*, 835 (1975).
76. D. Bedeković, Thesis. Swiss Federal Institute of Technology, Zurich (ETHZ) *Nr.*, 5777 (1976).
77. D. Bedeković and V. Prelog, *Helv. Chim. Acta*, in preparation.
78. E. Bamberg, H. Alpes, H.-J. Apell, R. Benz, K. Janko, H.-A. Kolb, P. Läuger, and E. Gross, Proceedings of the FEBS Symposium on the Biochemistry of Membrane Transport, held in Zurich, July 18–23, 1976, Springer-Verlag, Berlin, 1977.
79. S. Krasne, G. Eisenman, and G. Szabo, *Science*, **174,** 412 (1971).
80. H.-K. Wipf, A. Olivier, and W. Simon, *Helv. Chim. Acta*, **53,** 1605 (1970).
81. H.-K. Wipf, W. Pache, P. Jordan, H. Zähner, W. Keller-Schierlein, and W. Simon, *Biochem. Biophys. Res. Commun.*, **36,** 387 (1969).
82. H.-K. Wipf and W. Simon, *Biochem. Biophys. Res. Commun.*, **34,** 707 (1969).
83. P. Läuger, *Science*, **178,** 24 (1972).
84. B. C. Pressman, *Fed. Proc.*, **27,** 1283 (1968).
85. C. J. Pedersen, *Fed. Proc.*, **27,** 1305 (1968).
86. P. Oggenfuss, unpublished results.
87. J. M. Lehn, J. P. Sauvage, *Chem. Commun.*, 440 (1971).
88. B. Dietrich, J. M. Lehn, and J. P. Sauvage, *Chem. in unserer Zeit*, **4,** 120 (1973).
89. J. M. Lehn, *Struct. Bonding*, **16,** 1 (1973).
90. R. M. Izatt, D. J. Eatough and J. J. Christensen, *Struct. Bonding*, **16,** 161 (1973).
91. W. E. Morf and W. Simon, *Helv. Chim. Acta*, **54,** 2683 (1971).

LIST OF INTERVENTIONS

1. Eisenman
2. Caplan
    2.1 Simon
    2.2 Ubbelohde
    2.3 Simon
    2.4 Ubbelohde
    2.5 Simon
3. Prelog
4. Eisenman

5. Stucki
   5.1 Simon
   5.2 Gabellieri
   5.3 Simon
6. Baranowski
   6.1 Simon

1. Intervention of Eisenman

I would like to extend Prof. Simon's characterizations of these beautiful new molecules to include a description of the effects on lipid bilayers of his Na^+ selective compound number 11, which my post-doctoral student, Kun-Hung Kuo, and I have found to induce an Na^+ selective permeation across lipid bilayer membranes [K.-H. Kuo and G. Eisenman, "Na^+ Selective Permeation of Lipid Bilayers, mediated by a Neutral Ionophore," Abstracts 21st Nat. Biophysical Society meeting (*Biophys. J.*, **17,** 212a (1977))]. This is the first example, to my knowledge, of the successful reconstitution of an Na^+ selective permeation in an artificial bilayer system. (Presumably the previous failure of such well known lipophilic, Na^+ complexing molecules as antamanide, perhydroantamanide, or Lehn's cryptates to render bilayers selectively permeable to Na^+ is due to kinetic limitations on their rate of complexation and decomplexation).

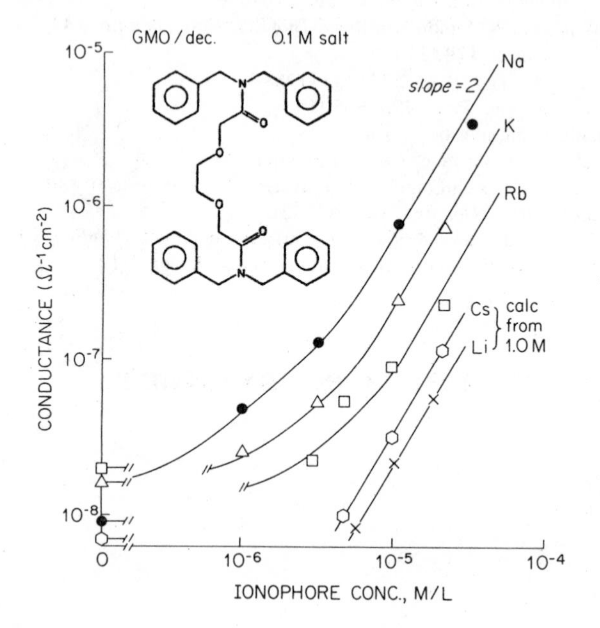

Fig. 1

In glycerol monooleate/decane bilayers we find the steady-state conductance at zero current to be proportional to the first power of the ion concentration and to the second power of the ionophore concentration, as illustrated in Fig. 1. (The current-voltage characteristic is "hyperbolic" for all ionic species indicating that this molecule is in the "equilibrium domain" for the interfacial reactions, with the rate-limiting step being the ion translocation across the membrane interior.) The conductance selectivity sequence is seen to be Na>K>Rb>Cs, Li.

A Nernst response is observed for Na^+ in Cl^- or NO_3^- solutions; and the zero-current membrane potential obeys the Goldman-Hodgkin-Katz equation where i and j are cations, x is an anion, and a' and a'' signify the activities in the different solutions on the two sides of the membrane. The P's are permeabilities, and the ratio P_i/P_j between cations is found to be concentration and voltage independent (as is the ratio P_x/P_y between anions); although the ratio P_x/P_i between anions and cations depends upon concentration (but not voltage) as summarized in Fig. 2. The finding of an ionophore-induced anion permeability to NO_3^- which can be greater even than the permeability for the less permanent cations is unprecedented and has not heretofore been seen with conventional carriers such as valinomycin or the nactin homologues. The anion permeability leads to an initially puzzling concentration dependence of the apparent selectivity among cations, which however is resolved in terms of the analysis of Fig. 2. The concentration-independent selectivities among

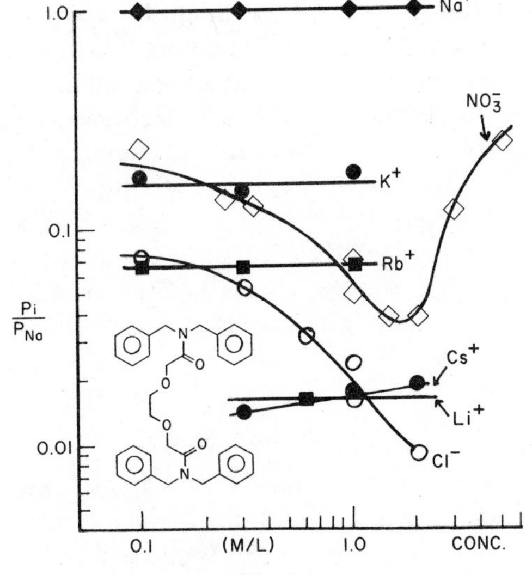

Fig. 2

PRINCIPAL EFFECT IS ON BLACK BILAYERS

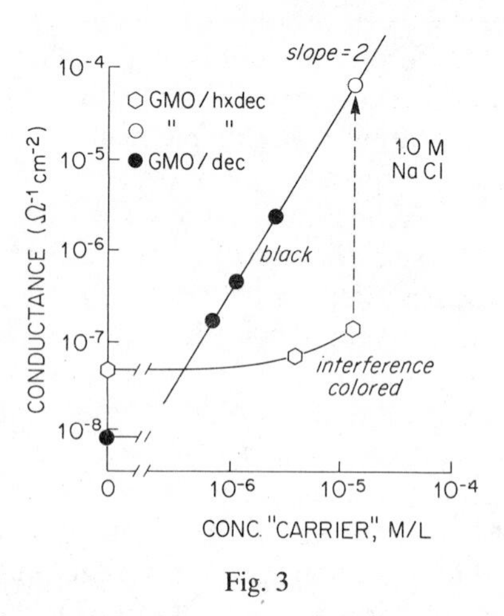

Fig. 3

cations of Fig. 2 are in good agreement with the cationic contributions to the conductances and also with the selectivity reported by Ammann, Pretsch, and Simon (*Anal. Lett.*, **7**, 23 (1974)) for bulk solvent polymer membranes.

All of the above behavior is exactly as expected if this neutral molecule acts as an ideal 2:1 carrier of cations and of anions, in the "equilibrium domain."

Finally, one additional observation should be of interest to this group, namely the use of stable interference-colored membranes (ca. 2000 Å thick) to enable thicknesses to be studied intermediate between that of the above bilayer (60 Å) and Simon's bulk electrodes. This is illustrated in Fig. 3 where the steady-state conductance of stable interference-colored membranes made from GMO/hexadecane (bottom) could be compared with that in black bilayers of GMO/decane (filled circles). At the highest carrier concentration the interference-colored membrane was caused to go black with an applied voltage, giving the 500x increase in conductance indicated by the open circle, which is nicely in agreement with that of the GMO/decane bilayers.

2. Intervention of Caplan

I realize from your earlier remarks that it may not be possible to answer this question at present, but I am anxious to know what information you have related to the presence of anions in your membranes. As you pointed out yourself, electroneutrality requires that the concentration of anions or anionic groups in the membranes match that of the cations. If I remember the data correctly, you obtained very high transport numbers for cations. This seems to indicate that the anions within the membrane are in some way fixed and immobilized.

2.1 Intervention of Simon

The question hits the big unknown in explaining the cation permselectivity of the neutral carrier systems mentioned in my report. In the solvent polymeric membranes we studied, the contribution of anions to the electrical current (anion transport number) is usually negligible if hydrophilic anions (e.g., Cl^-) are involved. In the presence of lipophilic anions (e.g., SCN^-) there exists some contribution of anions to the total ion flux across the membrane [see *Anal. Chem.*, **48,** 1031 (1976)]. The reasons for such a behavior may be:

1. The membrane contains permanent anions that are chemically bound to the supporting material, as was suggested by Kedem, Perry, and Bloch.
2. The membrane contains permanent anions that are immobilized because of their poor water solubility.
3. The membrane extracts sample anions but the integral mobility of these species across the membrane is low as compared to the cationic complexes.

We have some preliminary results indicating that reason 3 is the likely explanation for conventional carrier membranes.

2.2 Intervention of Ubbelohde

Can you test whether selectivity is exhibited between Na^+ and K^+ in nonaqueous systems?

2.3 Intervention of Simon

We have to expect such selectivity changes and it would be very interesting to measure them. Unfortunately, the solvent polymeric membranes mentioned can usually not be contacted with nonaqueous solvent systems because of the expected increased solubility of the membrane components involved.

2.4 Intervention of Ubbelohde

Can you insolubilize by radiation?

2.5 Intervention of Simon

We have only very modest experience with the influence of radiation on the membrane systems discussed. We know, however, that the sterilization of such membranes for their clinical application in ion-selective electrodes may be achieved by using X-rays. The electric properties seem to remain unchanged.

3. Intervention of Prelog

In reply to Professor Ubbelohde, research on ionophores reported in Professor Simon's paper emerged from studies of the properties of natural compounds such as valinomycin and macrotetrolides. That confirms once more the importance of such studies as initial points for the investigations of the multidimensional world of biology. I would like to add also that the natural ionophores resemble in several respects the much larger enzymes: (*1*) they are specific and stereospecific, (*2*) their active sites are located in cavities, (*3*) their stable secondary structure, which is essential for activity, is determined by the primary structure, and (*4*) they transport hydrophilic substrates through lipophilic membranes.

4. Intervention of Eisenman

I would like to emphasize the simplicity of Simon's molecules more than I did in my previous comment. These are by far the simplest class of ion-selective molecules that act in bilayers; and they open up very promising prospects for the study and understanding of structure-selectivity relationships.

I would also like to reemphasize the noncyclic character of these molecules, which I believe provides the way to circumvent the kinetic limitations on the rates of complexing and decomplexing such strongly hydrated ions as Na^+, Li^+, Ca^{2+}, which I suspect has stood in the way previously of translating the complexing selectivity of macrocyclic complexones into effective mediators for the selective permeation of these ions across the bilayer membrane. This may already offer a clue to the architecture of the Na^+ selective sites of the cell membrane.

5. Intervention of Stucki

Is it possible to synthesize ionophores specific for anions? These molecules would be of great importance in practical applications, especially in medicine. If they could be made stereospecific they could, in addition, offer new insights into enzyme and carrier mechanisms since they can be regarded as primitive models of such molecules.

5.1 Intervention of Simon

Although one might think of neutral anion carriers for use in membrane electrodes, I cannot offer good suggestions. It is possible, however, to use charged ligands (classical ion exchangers) as membrane components. This makes accessible the measurement of, for example, Cl^- and HCO_3^- in blood serum, without serious mutual interference.

5.2 Intervention of Gabellieri

Comparison may be interesting between Simon's membranes and homogeneous polymeric ionophores as commercially available (Nafion from duPont or G.E.).

5.3 Intervention of Simon

I assume, the membranes you are referring to are classical ion exchange membranes, the behavior of which is dictated to a large degree by charged groupings of the membrane material. The ion transport mechanism in such membranes is clearly not a carrier mechanism.

6. Intervention of Baranowski

1. What can be said about the diffusion potential inside the membrane?
2. Is the electrochemical method not suitable for the determination of the effective diffusion coefficient of the mobile component inside the membrane?

6.1 Intervention of Simon

It is, of course, not easy to make statements about the relative contributions of phase boundary and diffusion potentials. Since the electrochemical behavior of membranes is generally reflected by the total membrane potential, we did not try to differentiate in this respect. The models described in my report may, however, approximate the selectivity of certain membrane systems in the equilibrium domain even when assuming the absence of diffusion potentials.

GENERAL PANEL DISCUSSION

1. Intervention of Carafoli

It strikes me that in biological membranes, at least in eucaryotic cells, the transport mode almost universally chosen is the channel, or pore, mode, and not the mobile carrier mode. Surely there must be reasons for this, and it would seem appropriate to me if either Professor Simon or Professor Eisenman could start this discussion with a description of the respective merits of the two transport modes, with respect to selectivity, efficiency, and other parameters.

1.1 Intervention of Eisenman

Regarding the questions of channels versus carriers, a channel mechanism is advantageous when one desires all permeant ions to be "electrically active" (i.e., the transportation of all ions is seen electrically). A carrier mechanism where the carrier has a charge opposite to that of the carried ion has the permeant species in a neutral form and thus can be "electrically silent." In contrast, all the permeant ions in a channel are electrically active regardless of the charge intrinsic to the channel. A second advantage of a channel is that it enables more ions to move per unit time per molecule of "permease"; so channels are useful when one needs larger currents than can be transported by a carrier mechanism. On the other hand, carriers may be capable of exhibiting higher selectivities than channels, at least this appears to be the case for the model carriers and channels known at present. However, even the gramidicin A channel, previously regarded as not being very selective, is now being found to exhibit some substantial selectivity and binding properties (G. Eisenman,

J. Sandblom, and E. Neher, in *Metal-Ligand Interactions in Organic Chemistry and Biochemistry*, B. Pullman and N. Goldblum, Eds., Part 2, 1–36 D. Reidel, Dordrecht, Holland, 1977).

1.2 Intervention of Simon

Is there clear proof for the existence of channels or clear proof for the existence of carrier transport mechanisms in biological systems?

1.3 Intervention of Thomas

The answer is in the question itself. There is no proof at all for demonstrating whether or not carrier or channel mechanisms exist in permeation phenomena among biological systems.

1.4 Intervention of Clementi

When one talks about carriers, perhaps one should not neglect the possibility that the hydration water around an ion can be a "carrier" to some extent. In other words, one should not disregard the possibility that during the transport process the ion does not necessarily cross the membrane as an isolated ion, but as a hydrated ion. In addition, the hydration number (of molecules of water) does not have to be the same as the hydration number in solution, but likely is smaller, "selectively" smaller.

1.5 Intervention of Eisenman

I disagree with Dr. Thomas that there are no known biological channels or carriers. I know of at least one example of a clearly demonstrated channel. This is the acetylcholine activated channel in denervated muscle demonstrated so elegantly by Neher and Sackmann (*Nature*, *260*, 779, 1976), who resolved unit conductance jumps that are far too large to be accounted for by a carrier mechanism. A less unambiguously demonstrated example of channels are the Na^+ and K^+ channels of nerve, which both by noise analysis and pharmacological evidence imply the movement of about 1000 times as many ions in a unit of time as is reasonable for any diffusive carrier mechanism across the entire membrane.

1.6 Intervention of Thomas

These experimental results can be explained by other mechanisms, such as coupling between transfer and metabolism.

1.7 Intervention of Carafoli

Of course it is difficult to decide at what point the evidence on this particular matter can be accepted as conclusive. It is a fact, however, that in many other biological systems, in addition to the one quoted by

Professor Eisenman, there is a large body of results indicating the existence of channel systems. One could mention the Ca^{2+} ATPase of sarcoplasmic reticulum, the H^+ transporting ATPase of the inner mitochondrial membrane, the purple protein system of halobacteria, the Na^+ and K^+ channels of the axonal membranes. Apart from the classical type of evidence provided, for example, by the noise fluctuation technique, we now even begin to see direct electron microscopic evidence for the existence of transport-related openings in biological membranes. On the other hand, solid evidence for the existence of mobile carriers in eucaryotic cell membranes is very scarce, if not outright absent.

1.8 Intervention of Simon

The properties of channels mentioned by Professor G. Eisenman could be rationalized just as well by assuming the existence of pores.

1.9 Intervention of Caplan

In a discussion of permeability it is important to recognize that we deal with operational definitions, since the act of measurement influences the state of the system. In your case, applying an electrical potential gradient and performing electrodialysis alter the distribution of ionophore within the membrane. I wonder whether you have attempted to measure permeability by isotopic tracer techniques? In this method the distribution of ionophore would not be influenced. Furthermore, information can be obtained on the question of carriers versus channels or pores. It should not be difficult to determine the extent of possible isotope interaction between tracer species and abundant species in the membrane as discussed by Kedem and Essig [*J. Gen. Physiol.*, **48,** 1047 (1965)]. Positive isotope interaction would tend to suggest the presence of channels or pores, negative isotope interaction the presence of carriers.

1.10 Intervention of Simon

Although attractive isotope effects may be expected especially in the Ca^{2+} selective solvent polymeric membranes mentioned [see B. E. Jepson and R. DeWitt, *J. Inorg. Nucl. Chem.*, **38,** 1175 (1976)], we have not so far studied such effects. As compared to the flux of ions in the electrodialytic transport experiments we carried out on solvent polymeric membranes, the flux of ions is negligible in the absence of an electric field (other parameters kept constant).

2. Intervention of Mayer

The only paper I have ever published that had any relation to biophenomena was thirty years ago, with James Franck [*Arch. Biochem.*

14, 297 (1947)], who was concerned with the question of how water rises in a tree. One now knows that in dry, hot weather the main transport, which can be enormous, is accomplished by having a negative pressure in the sap column near the top, and since the humidity is low the chemical potential difference between the moist ground and dry air is greater than the gravitational potential difference. However, the preceding mechanism is fragile and I would guess that there must be a chemically activated repair mechanism. There are many examples of very extensive transport against the chemical gradient—for instance, in the kidneys. At that time Franck asked me if I could see a mechanism by which the free energy of a chemical reaction could pump water up to the top of a tree without using wheels. We came almost immediately to a diagram essentially like that which Professor Simon had on the board:

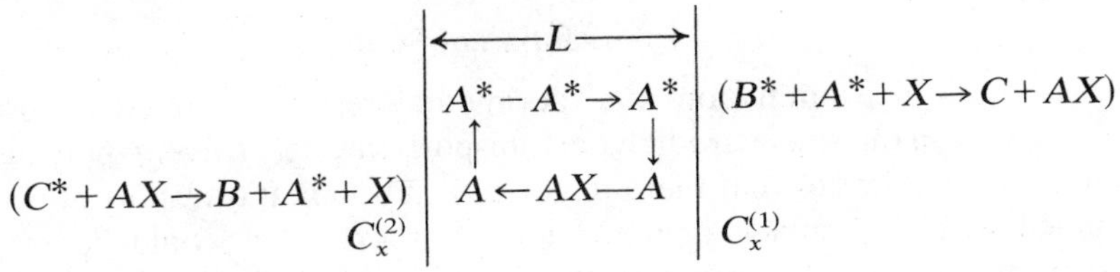

$$\begin{array}{r|c|l} & \longleftarrow L \longrightarrow & \\ & A^* - A^* \rightarrow A^* & (B^* + A^* + X \rightarrow C + AX) \\ & \uparrow \qquad\qquad \downarrow & \\ (C^* + AX \rightarrow B + A^* + X) & A \leftarrow AX — A & \\ C_x^{(2)} & & C_x^{(1)} \end{array}$$

but with a net reaction

$$C^* + B^* \rightarrow C + B$$

which releases free energy ΔG_R greater than

$$\Delta G_x \cong RT \ln \frac{C_x^{(2)}}{C_x^{(1)}} \text{ per mole}$$

For a neutral X and A and A^* the efficiency

$$q = (\Delta G_R)^{-1} \ln \frac{C_x^{(2)}}{C_x^{(1)}}$$

was very low (assuming reasonable diffusion constants) unless both

1. L is very small.
2. $\ln C_x^{(2)}/C_x^{(1)}$ is very small.

2.1 Intervention of Eisenman

This may be why cells use electric signals because these are in principle highly efficient. Electrochemical machines (i.e., storage batteries) are just the type that enable large concentration gradients to be balanced by electrical potential jumps so as to preserve the continuity of the electrochemical potential, which is the requirement for reversibility. The concentration gradients of K^+ and Na^+ across cell membranes that provide the immediate energy source for the electrical signals of nerve are large (e.g., factors of 10) yet are converted reversibly to electrical signals by utilizing the selective permeability of the membrane to K^+ at rest to hold the K^+ ions near electrochemical equilibrium in this state, while switching during the action potential to a nearly ideally selective permeability to Na^+, thus using the energy of the Na^+ concentration gradient efficiently. Indeed, this is one of the ways selectivity is utilized in biological systems.

2.2 Intervention of Mayer

We did not consider the possibility that X was charged or is carried by a carrier A and returning A^*, which have opposite charges. I think that then condition 2 is no longer required.

2.3 Intervention of Simon

A superficial examination of experimental results obtained by using labeling techniques (electrodialytic transport through solvent polymeric membranes) indicates that there might be a substantial transport of water coupled with the carrier-mediated ion transport. This would be rather surprising because the cation in the carrier-cation complex is not hydrated.

2.4 Intervention of Mayer

There must be a chemical reaction at some stage supplying the necessary free energy ΔG_R.

2.5 Intervention of Simon

The transport I mentioned is induced by an electric potential gradient across the membrane.

2.6 Intervention of Mayer

What is the free energy source for this process?

2.7 Intervention of Simon

The source creating the electric potential gradient.

3. Intervention of Caplan

The question of the efficiency of biological transport systems was examined extensively in the 1960s on the basis of linear nonequilibrium thermodynamics. I think it would be appropriate to give a brief account of the treatment here, especially since Professor Prigogine's early work was the source of most of our ideas at the time. The formal approach of

Prigogine was greatly extended and developed under the influence of the late Aharon Katzir-Katchalsky, and in studies in his laboratory it became clear that nonequilibrium thermodynamics gives rise to a natural description of efficiency and the parameters on which it depends. This is hardly surprising, since biological and other energy-converting systems are obviously not at equilibrium. Consider the transport of a single cation, say, a sodium ion, through a selective membrane as a consequence of the existence of an electrochemical potential gradient for that ion. This situation is shown in Fig. 1, where J_+ represents the flux of the cation and

$$\tilde{\mu}_+^1 \text{ and } \tilde{\mu}_+^2$$

represent the electrochemical potentials of the cation on sides 1 and 2, respectively. Supposing the system to be at constant temperature, the dissipation function or rate of dissipation of free energy Φ is given by

$$\Phi = J_+ \Delta \tilde{\mu}_+$$

According to the second law, the dissipation function must be positive if not zero, which of course is to be expected here, since we are dealing with a spontaneously occurring passive process. The thermodynamic force $\Delta \tilde{\mu}_+$, which contains both a concentration-dependent component and an electrical component, is the sole cause of the flow J_+. In a system in which more than one process occurs, each process gives rise to a term in the dissipation function consisting of the product of an appropriate force and its conjugate flow. In the case of active transport of the cation, as found, for example, in certain epithelial tissues, the cation flux is coupled to a metabolic reaction. If we represent the flow or velocity of the reaction per unit area of membrane by J_r, the appropriate force driving the reaction is

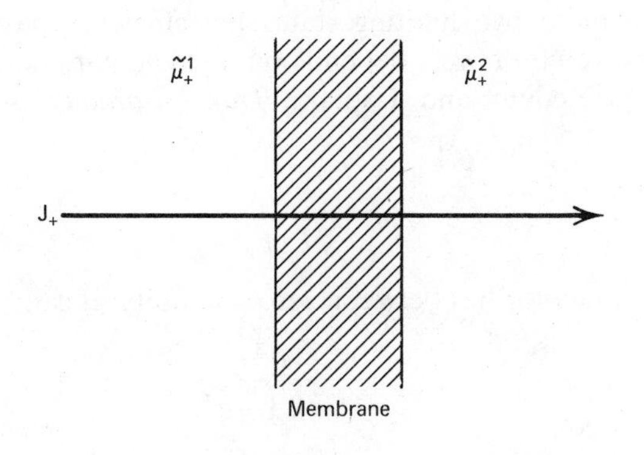

Fig. 1

essentially its negative Gibbs free energy under the circumstances, conventionally termed the affinity A. The dissipation function then becomes

$$\Phi = J_+ \Delta\tilde{\mu}_+ + J_r A$$

In this case, although the dissipation function is still required to be positive if not zero, individual terms may be negative as long as the sum is not. The reaction term now represents the spontaneous process and is positive. However, the transport term is negative, since the flow occurs in the nonspontaneous direction—against its gradient. Free energy expended by the reaction brings this about. This system is typical of an energy converter, metabolic energy being converted into osmotic energy. In the linear regime, which is remarkably wide in many biological systems, the phenomenological relations are

$$J_+ = L_+ \Delta\tilde{\mu}_+ + L_{+r} A$$

$$J_r = L_{r+} \Delta\tilde{\mu}_+ + L_r A$$

The kinetic constants of the system enter into the phenomenological L-coefficients, which are parameters of state. According to the reciprocity theorem of Onsager, the cross-coefficients L_{+r} and L_{r+} are identical. Now the definition of the efficiency η emerges directly from the dissipation function

$$\eta = \frac{\text{Output}}{\text{Input}} = \frac{-J_+ \Delta\tilde{\mu}_+}{J_r A}$$

However, the efficiency is clearly not a constant; it depends on how the system is operated (i.e., on the ratio of the forces $\Delta\tilde{\mu}_+/A$). Thus when $\Delta\tilde{\mu}_+$ is zero ("level flow"), the efficiency is zero. Similarly, when $\Delta\tilde{\mu}_+$ assumes such a value that J_+ is brought to a halt ("static head," also known as "state 4" in oxidative phosphorylation), the efficiency is also zero. Between these two limiting states the efficiency passes through a maximum. The value of $\eta_{\max}$ depends on a single parameter, the degree of coupling q [Kedem and Caplan, *Trans. Faraday Soc.*, **61,** 1897 (1965)]:

$$q = \frac{L_{+r}}{\sqrt{L_+ L_r}}$$

The absolute value of q lies between zero and unity. It can be shown that

$$\eta_{\max} = \frac{q^2}{(1 + \sqrt{1 - q^2})^2}$$

This equation indicates that the value of the maximal efficiency decreases rapidly as the degree of coupling falls below unity. A maximal efficiency of one is in principle only possible in systems that are completely coupled ($q^2 = 1$), and even then it can only be approached when the relationship between the forces is such that the system operates infinitely slowly. This is what we mean by reversible equilibrium.

3.1 Intervention of Mayer

I wonder if the mechanism does not *always* involve a charge even if the net material X that is carried is neutral.

3.2 Intervention of Caplan

Certainly there are examples of the transport of neutral molecules against their chemical potential gradients. In specific cases (e.g., amino acids) this seems to occur as a consequence of coupling between the flow of the neutral molecule and that of an ion that is actively transported. However, I do not think any generalizations can be made.

3.3 Intervention of Mandel

I may be wrong, but I seem to remember that according to the Curie principle, there can be no direct coupling between a vectorial flux and a scalar chemical reaction. That would mean that there is no coupling coefficient L_{+r}, which is essential for the calculation of the efficiency. I wonder what conditions have to be fulfilled for this coupling between a chemical reaction and the vectorial flux to occur.

3.4 Intervention of Caplan

This specter should once and for all be laid to rest. The Curie principle as it applies to the system at hand (I will not state it in its most general form) forbids, *in the linear regime*, coupling between a vectorial process such as a flow and a scalar process such as a chemical reaction in an isotropic space. However, active transport does not occur in an isotropic or symmetrical system. Clearly, the protein constituting the pump is uniquely oriented within the membrane.

3.5 Intervention of Mandel

Does that not imply that this orientation of the molecules has to be built into the flux-forces equations?

3.6 Intervention of Caplan

We do not need an oriented driving force for the chemical reaction. The quantity L_{+r} is a vectorial coefficient that reflects the structure of the system.

3.7 Intervention of Baranowski

I have two questions for Dr. Caplan:

1. What are the additional assumptions concerning the presented coupling of chemical reactions and diffusion?

2. How can this coupling be understood in a more elementary way (forgetting phenomenology)?

3.8 Intervention of Caplan

This question is not entirely fair. Thermodynamics does not elucidate mechanism, although it may well place important restrictions on the types of mechanism that may be proposed. It is essentially a "black box" approach. However, possible molecular models readily come to mind. For example, a cation may be bound to a transport enzyme at a selective site accessible from one side of the membrane. As a consequence of the driving reaction (which we may suppose to be the hydrolysis of ATP), a conformational change then occurs, which results in a movement of the site to a position such that the ion is severely hindered from returning to the side from which it originated. Perhaps there is a simultaneous change in the binding constant, so that the probability of the enzyme reverting to its original conformation with the site filled is small. Alternatively, a second ion may be preferred for the return step of the cycle. Before modeling the kinetics of such a system, it is essential to know what the degree of coupling is, that is, whether or not there is an integral and fixed stoichiometry between transport and reaction.

3.9 Intervention of Mayer

The chemical reaction *does* have a vector character along the axis normal to the membrane plane in which the transport takes place. There is the reducing *half* reaction on one side of the membrane and the oxidizing half on the other side.

3.10 Intervention of Williams

The idea of disposed redox reactions inside or across membranes for the purposes of coupling transport to metabolism has been discussed for a long time, especially in the last 15 years. The origin of these ideas goes back many more years, (e.g., Ogston and Davies, about 1950) but was revitalized by P. Mitchell and myself in 1961. A review is given by P. Mitchell, FEBS Lett. **78,** 1–20 (1977).

3.11 Intervention of Stucki

There is experimental evidence for the dependence of efficiency of transport or chemical processes in biological systems. Many studies have shown that the P/O ratio in mitochondria decreases to zero when mitochondria reach static head (i.e., state 4 respiration) [L. Ernster and K. Nordenbrand in *Dynamics of Energy Transducing Membranes*, L. Ernster, R. W. Estabrook, and E. C. Slater, Eds., BBA Library, Vol. 13, 283–288, Elsevier Scientific Publishing Company, Amsterdam, 1974; M. Klingenberg in *Mitochondria*: *Biogenesis and Bioenergetics*, S. G. van den Bergh, P. Borst, L. L. M. van Deenen, J. C. Riemersma, E. C. Slater, and J. M. Tager, Eds., 147–162, North Holland, Amsterdam, 1972; J. W. Stucki, *Eur. J. Biochem.*, **68**, 551–562 (1976)].

The same principle was found to apply to mitochondrial calcium transport [J. W. Stucki and E. A. Ineichen, *Eur. J. Biochem*, **48,** 365–375 (1974)]. This behavior is very important for bioenergetics, since the study of the flows *and* the forces might give new insights into the mechanism of oxidative phosphorylation and energy coupling in mitochondria (H. Rottenberg, S. R. Caplan, and A. Essig in *Membranes and Ion Transport*, E. E. Bittar, Ed., Vol. 1, 165–191, Wiley-Interscience, London, 1970).

3.12 Intervention of Thomas

I have presented in my report a system in which an active transport effect is observed. The effect is due not only to the reactions but also to the asymmetrical repartition of enzyme activity.

3.13 Intervention of Caplan

The model active transport system described by Dr. Thomas is based on an asymmetric arrangement of two enzymes. A model active transport system was also described by Blumenthal et al. several years ago based on a single enzyme immobilized between asymmetric boundaries [Blumenthal, Caplan, and Kedem, *Biophys. J.*, **7,** 735 (1967)]. In the latter case the phenomenological coefficients were measured, and it was possible to demonstrate Onsager symmetry and the correlation between the thermodynamic coefficients and the kinetic constants.

4. Intervention of Baranowski

I have two comments on Professor Prigogine's lecture:

1. Let me give a simple example of the creation of a collective behavior by chemical reactions. Putting an aqueous solution of copper sulfate between two horizontal plane copper electrodes, one can create a concentration gradient inside the previously homogeneous system by applying an external electrostatic potential difference. Making the upper electrode the anode, one can pass from the system at rest to a system with convection. In potentiostatic conditions this manifests itself in an increase

of the current, in galvanostatic conditions one observes a decrease of the potential difference (in an abrupt way) when convection starts. For systems at rest galvanostatic conditions are leading to a maximum entropy production when reaching stationarity. On the contrary potentiostatic conditions are equivalent with reaching a minimum entropy production at stationary states. So far only at potentiostatic conditions was some oscillating behavior observed. Some of the results mentioned were published previously [B. Baranowski, A. L. Kawczyns, *Electrochim. Acta* **17,** 695 (1972)].

2. The classical form of the thermodynamic force of diffusion ($X_i = -\text{grad } \tilde{\mu}_i$, $\tilde{\mu}_i$:electrochemical potential) can be extended if the influence of diffusion on the momentum transfer is carefully taken into account. This results if the balance equation of momentum for many component systems is formulated as the sum of contributions of all components. The final result is as follows:

$$X_i = -\text{grad } \tilde{\mu}_i - \frac{\nabla \cdot \Pi_i}{\rho_i} - \frac{D_i w_i}{Dt} - \frac{r_i \frac{1}{2} w_i}{\rho_i} - w_i \cdot \nabla V$$

when Π_i is the partial viscosity tensor, ρ_i is the partial mass density, V is the velocity of the center of mass, t is the time, w_i is the diffusion velocity, v_i is the mean velocity of component i, r_i is the partial mass source due to chemical reactions,

$$\frac{D_i}{Dt} = \frac{\partial}{\partial t} + v_i . \text{ grad}$$

where $\partial/\partial t$ denotes the local time derivative. Besides the classical term contributions are appearing due to viscosity and chemical reactions.

In applying the extended force in the balance equation of mass, one gets—in the first approximation—the following equation:

$$\frac{\partial c_i}{\partial t} = -\bar{D}_i \text{ div grad } c_i\left(1 + \frac{\bar{D}_i}{\rho_i} r_i\right)$$

$$- \frac{\bar{D}_i^2}{\rho_i} \text{grad } c_i . \text{ grad } r_i - r_i(1 + \bar{D}_i^2 \text{grad } c_i . \text{ grad } \bar{v}_i)$$

where $\bar{v}_i = 1/\rho_i$, c_i is the concentration, $\bar{D}_i$ is the diffusion coefficient. One easily sees that besides the classical terms [$\bar{D}_i$ div (grad c_i) and r_i] new contributions are appearing that may be of some importance for biological systems when the chemical reaction rate or its gradient are large enough to play some role as compared with the classical terms.

In two cases this extension was found useful: a) One can derive Prigogine's principle of independence of the entropy production caused by diffusion if the velocity of the frame of reference is changed in the presence of viscous flow. b) One can explain, without going over to nonlinear region, the possible separation effects in previously homogeneous systems when viscous flow is present [B. Baranowski, P. Romotowski, *Roczn. Chemii,* **44,** 1795 (1970); H. Kehlen, B. Baranowski, *Roczn. Chemii,* **50,** 573 (1976)].

5. Intervention of Williams

I wish to turn from general principles to specificity of uptake and binding with special reference to Professor Simon's lecture. I give two examples first:

1. The vesicles of the adrenal gland concentrate adrenaline by a factor of at least 10^6 and adenosine triphosphate by a factor of 10^2, whereas they do not concentrate K^+, Na^+, Mg^{2+}, Cl^-, and other phosphates. How is the selectivity achieved?
2. Insulin in the two zinc form binds zinc in a roughly octahedral hole. However, thiocyanate can displace the zinc to a new binding site that is created through the simultaneous anion binding in a most unusual way. Apparently, the site is a hydrophobic pocket with few dipoles anywhere in the vicinity. In fact, the hole is lined with methyl groups.

I think that we do not understand specificity very well. We should remember the importance of anion polarizability and we should not lose sight of peculiar cases such as the precipitation of potassium tetraphenyl borate, $K^+(B\phi_4)^-$.

5.1 Intervention of Simon

How is electroneutrality maintained within the vesicles mentioned by Professor Williams?

I have a statement on the use of polarizable groups in anion selective systems (earlier remark by Professor Williams). In some of the carriers mentioned in my report we replaced oxygen by sulfur atoms and unexpectedly got an anion response in EMF studies on solvent polymeric membranes modified with these ligands. Although we have a different explanation for this anion response, it might be due to the rather high polarizability of the sulfur atoms and the ligands might be carriers for anions.

5.2 Intervention of Williams

The electroneutrality is balanced roughly by the charge difference between adrenaline (1+) and ATP (4−), for they are present in concentrations of 4:1.

5.3 Intervention of Mandel

A very striking example of binding of anions through polarizability may be found in some old work on the behavior of a synthetic polyelectrolyte. I wish to refer to electrophoretic experiments of Strauss [U. P. Strauss, N. L. Gershfeld, and H. Spiera, *J. Am. Chem. Soc.*, **76,** 5909 (1954)] with poly- (vinyl pyridinium bromide) in KBr solutions. He found that the electrophoretic mobility of the polyion (to the negative electrode) was reduced by increasing concentration of KBr. For a certain concentration the electrophoretic mobility is even reversed in sign (i.e., the positively charged polyion moves toward the positive electrode). This result can only be explained if it is assumed that an excess of Br^- is bound to the positively charged macromolecule so that its net charge becomes negative. The binding of the excess of Br^- (with respect to the positive charge of the polyion) has to occur through the interaction of the highly polarizable Br^- and the highly polarizable pyridine ring.

5.4 Intervention of Klotz

I can also endorse Dr. Williams' remarks on polarizability effects by citing the early work on binding of anions by proteins (e.g., serum albumin). Some 25 years ago the extent of binding of a series of anions, Cl^-, Br^-, SCN^-, by albumin was found to parallel their polarizabilities.

I. M. Klotz, *J. Am. Chem. Soc.*, **68,** 2299–2305 (1946).
I. M. Klotz and J. M. Urquhart, *J. Am. Chem. Soc.* **71,** 847–851 (1949).
G. Scatchard, I. H. Scheinberg and S. H. Armstrong, Jr., *J. Am. Chem. Soc.* **72,** 535–540 (1950).

5.5 Intervention of Simon

Does Professor Williams have good suggestions for highly polarizable groups that might be incorporated in ligands to obtain anion carriers?

5.6 Intervention of Williams

Polarisability follows roughly the order of the size of atoms, which is also consistent with lower ionization potentials of larger atoms and ions (e.g., Se > S > O, $Se^{2-} > S^{2-} > O^{2-}$).

5.7 Intervention of Eisenman

Could Professor Williams be more specific as to the ligands for the Cl^- ion in insulin?

5.8 Intervention of Williams

The anions in insulin are bound in a hydrophobic hole with no immediate anion neighbor.

5.9 Intervention of Eisenman

I would like to clarify my previous question. What are the nearest *atoms* to the Cl^- ion?

5.10 Intervention of Williams

Some neighbors are aliphatic groups. When describing selectivity of ionic reactions we must remember that potassium tetraphenyl borate is insoluble. There are no large oriented dipoles near to the potassium.

5.11 Intervention of Prelog

It is astonishing that the potassium ion plays such an important role in biochemistry, although it was present in only low concentration in terrestrial surface waters when life originated around 10^{17} s ago.

Professor J. D. Dunitz has suggested to me that potassium ion may have replaced ammonium ion, which was ubiquitous in prebiotic times. When oxygen concentration in the atmosphere rose and ammonium ion disappeared it could have been replaced in emerging living systems by potassium ion, which in the meantime had become available and has nearly the same ionic radius as ammonium.

5.12 Intervention of Eisenman

In one regard NH_4^+ can differ from K^+. NH_4^+ likes a tetrahedral environment of the ligands that bind it. For example, using a type of individual ligand that would intrinsically not distinguish between K^+ and NH_4^+, a tetrahedral arrangement of four ligands in a site would introduce a factor favoring NH_4^+.

5.13 Intervention of Simon

It certainly would be extremely attractive to have highly selective Mg^{2+} carriers. Does someone (e.g., Professor Williams) have good suggestions for appropriate functional groups to be incorporated into ligands in view of obtaining such carriers?

5.14 Intervention of Williams

Selectivity of Mg^{2+} against Ca^{2+} is achieved by using RO^- ligands such as phenoles, enolates, or hydroxide, and using nitrogen donors (e.g., 8-hydroxy-quinolinate or diazo-phenol dyes). In biology this may not be necessary as biology can control Ca^{2+} concentration very well.

6. Intervention of Claesson

Can Professor Williams make some more comments on the motion of the protein residues in the structures he has shown us?

6.1 Intervention of Williams

Lysozyme provides one example. The motion of flipping by tyrosine or phenylalanine rings requires an oscillation of the groups of the protein around the aromatic ring. These movements if concerted could be quite small (e.g., ~1.0 Å). The lysines on the surface must sweep out large conical volumes (e.g., of a base some 5 to 10 Å across). The tryptophan motions are small and could be oscillations of 10 to 30°. Valine rotations require little space and methyl rotations almost none.

Other proteins show much bigger changes [e.g. the tyrosine of carboxypeptidase[1] and the ends of the insulin chains (Hodgkin[2])]. Still others seem to show gross changes [e.g., phospholipases (our own unpublished work)]. Finally, there are proteins that are almost random—chromogranin A and phosphovitin.[3]

Motion can be cooperative within the protein, as we have seen in cytochrome *c* (see my paper in this volume).

1. J. T. Johansen and B. L. Vallee, *Biochemistry,* **14,** 649 (1975).
2. G. Bentley, E. Dodson, D. Hodgkin and D. Mercola, *Nature,* **261,** 166 (1976).
3. A. Daniels, A. Korda, P. Tanswell, A. Williams and R. J. P. Williams, *Proc. Royal Soc. (London)* **B187,** 353 (1974).

6.2 Intervention of Welch

I have a question for Professor Williams regarding the concept of *entasis* [B. L. Vallee and R. J. P. Williams, *Proc. Nat. Acad. Sci. (U.S.)* **59,** 498 (1968)]. Is it possible that many enzymes in structured regimes in vivo exist in an "entatic state," and that such states "collapse" (e.g., the protein folds up) on extraction from the cell?

Do you consider the concept of *entasis* strictly limited to metalloenzymes?

6.3 Intervention of Williams

The entatic state hypothesis states that the protein fold generates reactivity in some region through the imposition of strain. The strain arises as the fold provides satisfactory interactions between some but not all amino acid side chains (Vallee and Williams[1]). Thus the strained groups (usually active site groove groups) are linked energetically to the whole protein. There is evidence in lysozyme that it is not just metal atoms that are activated, for in this enzyme (where there are no metals) there is a peculiar interaction between glutamate 35 and tryptophan 108. A better example of an organic entatic group is the tyrosine free radical in the ribonucleotide reductase.[2] Metal-activated groups could include the iron of cytochrome *c*[3].

A typical example of a special state is as follows. Electron transfer reactions at an atom are aided by vibrations that equilibrate the interatomic distances that differ for the two oxidation states. Thus a low-energy, high-amplitude vibration is desirable. The vibration could have the further function that it provided a time-dependent fluctuation of the redox potential. As I and Goldanskii in this volume have pointed out, this allows a precise matching of the redox potential of one redox couple with another leading to tunneling of electrons.

1. B. L. Vallee and R. J. P. Williams, *Proc. Natl. Acad. Sci. US*, **59,** 498 (1968).
2. A. Ehrenberg and P. Reichard, *J. Biol. Chem.* **247,** 3485 (1972).
3. G. R. Moore and R. J. P. Williams, *FEBS Lett.*, **79,** 229 (1977).

6.4 Intervention of Somorjai

It is known that hydrophobic surroundings can stabilize free radicals; lifetimes of up to an hour were measured for otherwise very reactive radicals. This may be of importance for the enzyme mechanism.

6.5 Intervention of Williams

There are some extraordinary examples of designed inhibition among neurotoxins and protease inhibitors. The work of Hüber and coworkers [W. Bode, P. Schwager, and R. Hüber, Proc. 10th FEBS Meeting, p. 3 (1975)] has shown that the trypsin inhibitor contains a peptide bond that is distorted into a form close to the transition state of the trypsin reaction and this makes for a very strong interaction between protein and inhibitor. Thus locally the inhibitor protein has a high-energy (*entatic*) state of an amide group generated by the tight, cross-linked protein as a whole. Neurotoxins and protease inhibitors have some close similarities in structure, being relatively rigid (see tables in my article).

6.6 Intervention of Somorjai

Does the active site have to be a rigidly maintained structure? Isn't it possible that the groups comprising the active site execute correlated, cooperative motion of some kind, perhaps a nonlinear oscillation that is sustained by energy pumping?

6.7 Intervention of Williams

The need for oscillations can be seen in another way if we look at electron transfer processes in the mitochondrial or chloroplast reaction systems. In both there are series of electron transfer centers. By studying the spin state of metal ions in these centers it has been shown that the metal atoms can flip rapidly between different spin states. As these states must have some difference in conformation and as the metals are bound to the proteins, the proteins must oscillate with the same frequency as the spin states. This could be very important, as it is to be expected that the redox potential will vary during the course of such an oscillation. If we now consider two redox centers in different adjacent proteins and remember that electron transfer by tunneling requires an almost exact matching of potentials between donor and acceptor, we see that the oscillations allow a searching for this matched situation. Any effect such as energization of the membrane carrying the proteins that altered either the redox potentials or the possibility of searching for equality by oscillating would control the possibility of electron transfer by tunneling.

Of course, we also see oscillations in other types of enzyme (e.g., lysozyme has an oscillating tryptophan).

6.8 Intervention of Somorjai

Perhaps enzyme-substrate recognition and interaction are facilitated by an oscillating active site: Its correlated motion could position more readily and reliably the catalytically essential groups of atoms. Recognition of the substrate could be visualized as a nonlinear resonance phenomenon, perhaps providing the mechanism of energy transfer from the entatic active site region to the substrate. An off-resonance condition could characterize an enzyme-inhibitor interaction.

6.9 Intervention of Goldanskii

In connection with the problem of oscillations discussed by previous speakers and other types of dynamical behavior of membranes, it would probably be timely to mention here in some more detail the experiments with vision rhodopsin that were performed in our institute by using the Mössbauer spectroscopy method [G. R. Kalamkarov et al., *Doklady Biophys.*, **219,** 126 (1974)]. These experiments manifested the existence of reversible photo-induced conformational changes in the photoreceptor membrane even at such low temperature as 77°K. We have labeled various samples of solubilized rhodopsin and of photoreceptor membranes by iron ascorbate enriched with Fe^{57} and looked for the change of Mössbauer spectra caused by the illumination of our samples.

In such a way we were able to conclude that the illumination of suspensions of photoreceptor outer segments by 450 nm light at 77°K, which was known to result in the rhodopsin→prelumirhodopsin transition (corresponding to 11-cis-retinal→transretinal photoisomerization of chromophore), leads also to the appearance of some reduction centers and to the conformational change of membrane.

The formation of reduction centers was shown by the $Fe^{3+} \rightarrow Fe^{2+}$ reduction during the subsequent defreezing of illuminated samples.

The reversible conformational change was seen by the disappearance of unresolved magnetic HFS of Mössbauer spectra after the illumination (decrease of spin-lattice relaxation time) and its restoration after the illumination of bleached samples at 77°K by light of 575 nm wavelength.

6.10 Intervention of Prigogine

Are there cooperative processes in the protein motions described by Professor Williams?

6.11 Intervention of Williams

It is not out of the question that there will be some highly cooperative events during the onset of motion in proteins. We know today of one minor example. The onset of flipping of a tyrosine in cytochrome *c* is cooperative with the onset of flipping of an adjacent phenylalanine. We could imagine this onset crossing the boundary between proteins so that many of the effects familiar in solid crystalline lattices are possible in protein aggregates (e.g., traveling waves).

6.12 Intervention of Somorjai

It might be profitable to regard a single enzyme molecule as a non-equilibrium system, maintained off equilibrium by the random energy input of colliding solvent molecules. The crucial mechanism that would prevent the rapid dissipation of this energy I postulate to be the creation of *solitons*. Solitons are nonlinear traveling waves with remarkable properties. For our purposes the most important of these are that solitons are extremely stable against finite perturbations and are capable of carrying energy over long distances and times without dissipation. Their presence in enzymes would provide the means of transforming random energy into coherent one. The hydrogen bond network of the enzyme would be the nonlinearly coupled lattice supporting these solitons and the "energy funnel" to the active site. It should be mentioned that solitons seem to be ubiquitous, and in particular, the self-trapping and nonlinear propagation of heat pulses in solids were interpreted in terms of envelope solitons. We are working on a simple model of energy transfer in enzymes, based on nonlinear wave propagation methods. The detailed results will be published elsewhere.

6.13 Intervention of Williams

The situation in the calcium-binding protein aequorin is that there is a store of energy built into a special organic molecule and into the fold of the protein. On binding calcium the protein emits lights at around 500 nm and there is a conformational change of the protein. It is not easy to write down a mechanism for this emission. However, in a more general way it is obvious that on absorption of energy a number of organized systems in biology are raised to a nonequilibrium state (e.g., in the mitochondria and chloroplasts). The properties of these states and how they store energy is not known, but they do not dissipate energy rapidly to their surroundings but utilize it to drive coupled reactions. I have proposed a possible explanation that involves the formation of a local store of high-energy protons bound in the membrane system [R. J. P. Williams, *J. Theoret. Biol.*, **1,** 1 (1961)].

INDEX